LONDON MATHEMATICAL SOCIETY LECTURE NOTE SERIES

Managing Editor: Professor J.W.S. Cassels, Department of Pure Mathematics and Mathematical Statistics, University of Cambridge, 16 Mill Lane, Cambridge CB2 1SB, England

The titles below are available from booksellers, or, in case of difficulty, from Cambridge University Press.

London Mathematical Society Lecture Note Series. 235

Number Theory

Séminaire de Théorie des Nombres de Paris
1993–4

Edited by

Sinnou David
Université Pierre et Marie Curie, Paris

CAMBRIDGE
UNIVERSITY PRESS

CAMBRIDGE UNIVERSITY PRESS
Cambridge, New York, Melbourne, Madrid, Cape Town, Singapore, São Paulo

Cambridge University Press
The Edinburgh Building, Cambridge CB2 2RU, UK

Published in the United States of America by Cambridge University Press, New York

www.cambridge.org
Information on this title: www.cambridge.org/9780521585491

First published 1996

A catalogue record for this publication is available from the British Library

ISBN-13 978-0-521-58549-1 paperback
ISBN-10 0-521-58549-X paperback

Transferred to digital printing 2006

Number Theory
Paris 1993–94

Table des Matières

Les textes qui suivent sont pour la plupart des versions écrites de conférences données pendant l'année 1993–94 au Séminaire de Théorie des Nombres de Paris. Ce séminaire est financièrement soutenu par le C.N.R.S. et regroupe des arithméticiens de plusieurs universités et est dotée d'un conseil scientifique et éditorial. Ont été aussi adjoints certains textes dont la mise à la disposition d'un large public nous a paru intéressante. Les articles présentés ici exposent soit des résultats nouveaux, soit des synthèses originales de questions récentes; ils ont en particulier tous fait l'objet d'un rapport.

Ce recueil doit bien sûr beaucoup à tous les participants du séminaire et à ceux qui ont accepté d'en réviser les textes. Il doit surtout à Monique Le Bronnec qui s'est chargée du secrétariat et de la mise au point définitive du manuscrit; son efficacité et sa très agréable collaboration ont été cruciales dans l'élaboration de ce livre.

Pour le Conseil éditorial et scientifique

S. DAVID

Liste des conférenciers

4 octobre : B. PERRIN–RIOU et D. BERNARDI. — *Fonctions L p–adiques des courbes elliptiques*

11 octobre : S. LOUBOUTIN. — *Problèmes de nombres de classes pour les corps à multiplication complexe*

18 octobre : L. MÉREL. — *Opérateurs de Hecke, symboles modulaires et courbes de Weil*

25 octobre : A. CORTELLA. — *Le principe de Hasse pour les similarités de formes bilinéaires*

8 novembre : Sir P. SWINNERTON-DYER. — *Rational points on certain intersections of two quadrics*

15 novembre : U. ZANNIER. — *Fields containing values of algebraic functions*

22 novembre : J. COUGNARD. — *Anneaux d'entiers stablement libres et non libres*

29 novembre : M. HARRISON. — *On Bloch-Kato conjecture for Hecke characters over* $\mathbb{Q}(i)$

6 décembre : R. GREENBERG. — *Two–variable Iwasawa theory*

13 décembre : P.G. BECKER. — *Transcendental values of the Douady– Hubbard function*

3 janvier : A. NITAJ. — *Conséquences et aspects expérimentaux de la conjecture* abc

10 janvier : C. BACHOC. — *Classification et construction de réseaux unimodulaires*

17 janvier : C.-G. SCHMIDT. — *Generalized Kummer congruences for Siegel modular forms*

24 janvier : J. MARTINET. — *Classification des réseaux eutactiques*

31 janvier : A. ABBÈS. — *Théorème de Hilbert–Samuel arithmétique*

7 février : M. KANEKO. — *Atkin's polynomials on supersingular $j-$invariants and hypergeometric series*

28 février : W. McCALLUM. — *Dualité en Théorie d'Iwasawa à plusieurs variables*

7 mars : B. LEMAIRE. — *Conjecture de R. Howe pour GL_N sur un corps local de caractéristique positive*

14 mars : E. URBAN. — *Congruences de formes modulaires et Théorie d'Iwasawa*

21 mars : D. HARARI. — *Obstructions de Manin transcendantes*

25 avril : F. BEUKERS. — *Units in quaternion algebras as monodromy groups*

25 avril : L. MEREL. — *Bornes pour la torsion des courbes elliptiques sur les corps de nombres*

2 mai : J. ASSIM. — *Sur les corps de nombres (p, i)–réguliers*

9 mai : P. COLMEZ. — *Fonctions zêta p–adiques en $s = 0$*

16 mai : T. NGUYEN QUANG DO. — *Sur les conjectures de Lichtenbaum*

30 mai : R. SCHULZE–PILLOT. — *Theta series and L–functions*

13 juin : J. NEKOVAR. — *p–adic regulators*

20 juin : L. FAINSILBER. — *Formes hermitiennes sur les algèbres p–adiques*

27 juin : S. FERMIGIER. — *Annulation de la cohomologie cuspidale de sous–groupes de congruence de $GL_n(\mathbb{Z})$*

Number Theory
Paris 1993–94

On the Central Critical Value of the Triple Product

L–Function

S. Böcherer and R. Schulze–Pillot[*]

Introduction

Starting from the work of Garrett and of Piatetskii-Shapiro and Rallis on integral representations of the triple product L-function associated to three elliptic cusp forms the critical values of these L-functions have been studied in recent years from different points of view. From the classical point of view there are the works of Garrett [9], Satoh [22], Orloff [21], from an adelic point of view the problem has been treated by Garrett and Harris [10], Harris and Kudla [12] and Gross and Kudla [11]. Of course the central critical value is of particular interest. Harris and Kudla used the Siegel-Weil theorem to show that the central critical value is a square up to certain factors (Petersson norms and factors arising at the bad and the archimedean primes); the delicate question of the computation of the factors for the bad primes was left open. In the special situation that all three cusp forms are newforms of weight 2 and for the group $\Gamma_0(N)$ with square free level $N > 1$, Gross and Kudla gave for the first time a completely explicit treatment of this L-function including Euler factors for the bad places; they proved the functional equation and showed that the central critical value is a square up to elementary factors (that are explicitly given).

We reconsider the central critical value from a classical point of view, dealing with the situation of three cusp forms f_1, f_2, f_3 of weights k_i ($i = 1, \ldots, 3$) that are newforms for groups $\Gamma_0(N_i)$ with $N = \mathrm{lcm}(N_i)$ a square-free integer $\neq 1$. The weights k_i are subject to the restriction $k_1 < k_2 + k_3$ where $k_1 \geq \max(k_2, k_3)$; the distinction whether this inequality holds or

[*] Both authors were supported by MSRI, Berkeley (NSF-grant DMS-9022140). R. Schulze-Pillot was also supported by the Deutsche Forschungsgemeinschaft and by the Max-Planck-Institut für Mathematik, Bonn.

not played an important role in [11] and [12] too. We start from the simplest possible Eisenstein series $\mathbb{E}$ of weight 2 for $\Gamma_0^{(3)}(N)$ on the Siegel space $\mathbb{H}_3$ ("summation over $C \equiv 0 \bmod N$"). After applying a suitable differential operator (depending on the weights k_i) to $\mathbb{E}$ we proceed in a way similar to Garrett's original approach : we restrict the differentiated Eisenstein series in a first step to $\mathbb{H}_1 \times \mathbb{H}_2$ and integrate against f_1, the resulting function on $\mathbb{H}_2$ is then restricted to the diagonal and integrated against f_2, f_3. The necessary modifications to Garrett's coset decompositions (that were for level 1) are not difficult (for the first step they have already been carried out in [2]). The actual computation of the integral is elementary and needs only standard results from the theory of newforms. It yields a Dirichlet series (2.41) whose Euler product decomposition is then computed in Section 3. The cases that p divides one, two or all three of the levels N_i or is coprime to N must all be treated separately, which makes the discussion somewhat lengthy. However, the actual computation in each of these cases is again fairly straightforward. In Section 4 we show that the Euler factors defined in Section 3 are the "right ones" by proving the functional equation. In order to exhibit the central critical value as a square (up to elementary factors) we follow a similar strategy as [12] : the Eisenstein series $\mathbb{E}$ at $s = 0$ is expressed as a linear combination of genus theta series of quaternary positive definite integral quadratic forms. At most one of these genera (depending on the levels N_i and the eigenvalues of the f_i under the Atkin-Lehner involutions) contributes to the integral. Eichler's correspondence between cusp forms for $\Gamma_0(N)$ and automorphic forms on definite quaternion algebras allows then to express this contribution as an (explicitly computable) square of an element of the coefficient field of the f_i; this element arises as a value of a trilinear form on a space of automorphic forms on the quaternion algebra and may be interpreted as the value of a height pairing similar to [11].

It may be of interest to compare the advantages of the different methods applied to this problem. Although the adelic method makes it easier to obtain general results, the explicit computations needed here appear to become somewhat simpler in the classical context. In particular, by making use of the theory of newforms and of orthogonality relations for the theta series involved from [2] we can use the same Eisenstein series $\mathbb{E}$ independent of the f_i. This is of advantage since the pullback formalism is especially simple for this type of Eisenstein series and leads to the remarkably simple computations in Sections 2 and 3.

Most of this article was written while both authors were guests of the MSRI in Berkeley during its special year on automorphic forms. We wish to thank the MSRI for its hospitality and financial support. R. Schulze-Pillot was also supported by Deutsche Forschungsgemeinschaft during a visit of one month at MSRI and was a guest of the Max-Planck-Institut für

Mathematik in Bonn in the final stage of the preparation of this manuscript.

Notations We use some standard notations from the theory of modular forms, in particular, we denote by $\mathbb{H}_n$ Siegel's upper half space of degree n (for $n = 1$, the subscript will be omitted); for functions f on $\mathbb{H}$ and $g = \begin{pmatrix} a & b \\ c & d \end{pmatrix}$ we use $(f \mid_k g)(z) = \det(g)^{\frac{k}{2}} (cz + d)^{-k} f(g < z >)$ and similarly for the action of double cosets (Hecke operators). The operators $T(p)$ and $U(p)$ however will be used in their standard normalisation. The space of cusp forms of weight k for $\Gamma_0(N) = \left\{ \begin{pmatrix} a & b \\ c & d \end{pmatrix} \in Sl_2(\mathbb{Z}) \mid c \equiv 0 \bmod N \right\}$ will be denoted by $[\Gamma_0(N), k]_0$.

1. — Differential Operators

We have to deal with two types of embeddings of products of upper half spaces into $\mathbb{H}_3$ namely

$$\iota_{12} : \begin{cases} \mathbb{H} \times \mathbb{H}_2 & \longrightarrow & \mathbb{H}_3 \\ (z, Z) & \longmapsto & \begin{pmatrix} z & 0 \\ 0 & Z \end{pmatrix} \end{cases},$$

and

$$\iota_{111} : \begin{cases} \mathbb{H}^3 & \longrightarrow & \mathbb{H}_3 \\ (z_1, z_2, z_3) & \longmapsto & \begin{pmatrix} z_1 & & \\ & z_2 & \\ & & z_3 \end{pmatrix} \end{cases}.$$

Without any danger of confusion we may denote by the same symbols the corresponding "diagonal" embeddings of groups :

$$\iota_{12} : Sl_2 \times Sp(2) \to Sp(3) \quad \text{and} \quad \iota_{111} : Sl_2^3 \to Sp(3).$$

One might try to apply Ibukiyama-type differential operators [16] in the integral representation of the triple L–functions (equivariant for $Sl_2 \times Sl_2 \times Sl_2 \hookrightarrow Sp(3)$). However in the actual computation of the integral, it is more convenient to have equivariance for $Sl_2 \times Sp(2) \hookrightarrow Sp(3)$. Therefore we use Maaß-type operators (see [20]) and the holomorphic differential operators introduced in [6]; we describe these operators here only for $Sp(3)$, but of course they also make sense for $Sp(n)$.

We start from a natural number r and three (even) weights k_1, k_2, k_3 with $k_1 = \max\{k_i\}$ and satisfying the condition

(1.1) $$k_2 + k_3 - k_1 \geq r.$$

Then we define nonnegative integers a, b, ν_2, ν_3 by

(1.2)
$$
\begin{aligned}
r + a &= k_2 + k_3 - k_1 \\
k_1 &= r + a + b \\
k_2 &= r + a + \nu_2 \\
k_3 &= r + a + \nu_3 .
\end{aligned}
$$

Then we have

(1.3)
$$ b = \nu_2 + \nu_3 . $$

We use two types of differential operators on $\mathbb{H}_3$. The first one is the Maaß operator

(1.4)
$$
\begin{aligned}
\mathcal{M}_\alpha &= \det(Z - \bar{Z})^{2-\alpha} \det(\partial_{ij}) \det(Z - \bar{Z})^{\alpha-1} \\
&= \sum_{\mu=0}^{3} \frac{\varepsilon_3(\alpha)}{\varepsilon_\mu(\alpha)} \cdot \mathrm{tr}\left((Z - \bar{Z})^{[\mu]} \cdot (\partial_{ij})^{[\mu]} \right) \\
&= \varepsilon_3(\alpha) + \cdots + \det(Z - \bar{Z}) \cdot \det(\partial_{ij})
\end{aligned}
$$

where (following [20])

$$
\varepsilon_\mu(\alpha) = \begin{cases} 1 & \mu = 0 \\ \alpha \cdot (\alpha - \frac{1}{2}) \cdots (\alpha - \frac{\mu-1}{2}) & \mu > 0 \end{cases}
$$

and for a matrix A of size n we denote by $A^{[\mu]}$ the matrix of $\mu \times \mu$-minors. We put

$$ \mathcal{M}_\alpha^{[\nu]} = \mathcal{M}_{\alpha+\nu-1} \circ \ldots \circ \mathcal{M}_{\alpha+1} \circ \mathcal{M}_\alpha . $$

We recall from [20] that

(1.5)
$$ \mathcal{M}_\alpha^{[\mu]} (f \mid_{\alpha,\beta} g) = \left(\mathcal{M}_\alpha^{[\mu]} f \right) \mid_{\alpha+\mu, \beta-\mu} g $$

for all $g = \begin{pmatrix} a & b \\ c & d \end{pmatrix} \in Sp(3, \mathbb{R})$ with $(f \mid_{\alpha,\beta} g)(Z) = \det(cZ + d)^{-\alpha} \det(c\bar{Z} + d)^{-\beta} f(g < Z >)$. Here α and β are arbitrary complex numbers, but it would be sufficient for us to take $\alpha = r + s$, $\beta = s$ with $s \in \mathbb{C}$.

The second type of differential operators was introduced in [6] : it maps scalar-valued functions on $\mathbb{H}_3$ to vector-valued functions , more precisely to $\mathbb{C}[X_2, X_3]_b$ -valued functions on $\mathbb{H} \times \mathbb{H}_2 \hookrightarrow \mathbb{H}_3$ where $\mathbb{C}[X_2, X_3]_b$ denotes the space of homogeneous polynomials of degree b; we realize the symmetric tensor representation σ_b of $Gl(2,\mathbb{C})$ on this space in the usual way. The operator $\mathbb{L}_\alpha^{(b)}$ as defined in [6] satisfies

(1.6)
$$\left(\mathbb{L}_\alpha^{(b)} f\right) |_{\alpha+b,\beta}^w g_1 = \mathbb{L}_\alpha^{(b)} \left(f \mid_{\alpha,\beta} \iota_{1,2}(g_1, 1_4)\right) (\iota_{1,2}(w, Z))$$
$$\left(\mathbb{L}_\alpha^{(b)} f\right) |_{\alpha,\beta,\sigma_b}^Z g_2 = \mathbb{L}_\alpha^{(b)} \left(f \mid_{\alpha,\beta} \iota_{1,2}(1_2, g_2)\right) (\iota_{1,2}(w, Z))$$

for all $g_1 \in Sl_2(\mathbb{R})$ and all $g_2 \in Sp(2,\mathbb{R})$, where the upper indices Z and w indicate which variable is relevant at the moment and

$$\left(\left(\mathbb{L}_\alpha^{(b)} f\right) |_{\alpha,\beta,\sigma_b}^Z g_2\right)(\iota_{1,2}(w, Z)) =$$
$$\det(cZ + d)^{-\alpha} \det(c\overline{Z} + d)^{-\beta} \sigma_b(cZ + d)^{-1} \left(\mathbb{L}_\alpha^{(b)}\right) (\iota_{12}(w, g_2 < Z >))$$

This differential operator can be described explicitly as follows :
(1.7)
$$\mathbb{L}_\alpha^{(b)} = \frac{1}{\alpha^{[b]}} \iota^\star \sum_{0 \le 2\nu \le b} \frac{1}{\nu!(b-2\nu)!(2-\alpha-b)^{[\nu]}} \cdot (D_\uparrow D_\downarrow)^\nu \cdot (D - D_\uparrow - D_\downarrow)^{b-2\nu}$$

with $\iota^\star$ denoting the restriction to $\mathbb{H} \times \mathbb{H}_2 \hookrightarrow \mathbb{H}_3$,

$$D_\uparrow = \partial_{11}$$
$$D_\downarrow = \sum_{2 \le i,j \le 3} \partial_{ij} X_i X_j$$
$$D - D_\uparrow - D_\downarrow = 2\left(\partial_{12} X_1 + \partial_{13} X_3\right)$$

and
$$\alpha^{[\nu]} = \frac{\Gamma(\alpha+\nu)}{\Gamma(\alpha)} = \begin{cases} 1 & \nu = 0 \\ \alpha(\alpha+1)\ldots(\alpha+\nu-1) & \nu > 0. \end{cases}$$

We should remark here that $\mathbb{L}_\alpha^{(b)}$ has coefficients, which are rational functions of α with no poles for $\Re(s) > 0$.

We shall use the operators
$$\mathcal{D}_\alpha^{(a,b)} := \mathbb{L}_{\alpha+a'}^{(b)} \circ \mathcal{M}_\alpha^{[a']}$$

with $2a' = a$ and $\mathcal{D}_\alpha^{\star(a,b)}$ defined by

$$\mathcal{D}_\alpha^{\star(a,b)} f = \left(\mathcal{D}_\alpha^{(a,b)} f\right) \iota_{111}(z_1, z_2, z_3)$$

Denoting by $\mathcal{D}_\alpha^{\star(a,\nu_2.\nu_3)}$ the operator which picks out of $\mathcal{D}_\alpha^{\star(a,b)}$ its $X_2^{\nu_2} X_3^{\nu_3}$-component, we get a decomposition

(1.8)
$$\mathcal{D}_\alpha^{\star(a,b)} f = \sum_{\nu_2+\nu_3=b} \left(\mathcal{D}_\alpha^{\star(a,\nu_2.\nu_3)} f \right) X_2^{\nu_2} X_3^{\nu_3}$$

with

$$\left(\mathcal{D}_\alpha^{\star(a,\nu_2.\nu_3)} f \mid_{\alpha,\beta} \iota_{111}(g_1,g_2,g_3) \right)$$
$$= \left(\mathcal{D}_\alpha^{\star(a,\nu_2.\nu_3)} f \right) \mid_{\alpha+a'+b,\beta-a'}^{z_1} g_1 \mid_{\alpha+a'+\nu_2,\beta-a'}^{z_2} g_2 \mid_{\alpha+a'+\nu_3,\beta-a'}^{z_3} g_3$$

for all $(g_1,g_2,g_3) \in Sl_2(\mathbb{R})^3$.

If f is a holomorphic function on $\mathbb{H}_3$, then

$$(y_1 y_2 y_3)^{-a'} \cdot \mathcal{D}_\alpha^{\star(a,\nu_2,\nu_3)}(f)$$

is a nearly holomorphic function (in the sense of Shimura) of all three variables z_1, z_2, $z_3 \in \mathbb{H}$.

To apply Shimura's results on nearly holomorphic functions, it is more convenient to use his differential operators δ_α^μ, which differ (in the one-dimensional case i.e. on $\mathbb{H}$) from the Maaß operators only by a factor constant $\times y^\mu$:

$$\delta_\alpha = \frac{1}{2\pi i} \left(\frac{\alpha}{2iy} + \frac{\partial}{\partial z} \right)$$

$$\delta_\alpha^\mu = \delta_{\alpha+2\mu-2} \circ \cdots \circ \delta_\alpha .$$

By elementary considerations about the degree of nearly holomorphic functions (as polynomials in y^{-1}) Shimura observed that nearly holomorphic functions on $\mathbb{H}$ are linear combinations of functions obtained from holomorphic functions by applying the operators δ_α^μ (at least if α is not in a certain finite set, for details see [24, lemma 7]. By the same kind of reasoning we get an identity

$$(y_1 y_2 y_3)^{-a'} \cdot \mathcal{D}_\alpha^{\star(a,\nu_2.\nu_3)}$$

(1.9)
$$= \sum_{0 \leq \mu_1,\mu_2,\mu_3 \leq a'} \delta_{\alpha+a+b-2\mu_1}^{\mu_1} \delta_{\alpha+a+\nu_2-\mu_2}^{\mu_2} \delta_{\alpha+a+\nu_3-2\mu_3}^{\mu_3}$$
$$\mathbb{D}_\alpha(a,\nu_2,\nu_3,\mu_1,\mu_2,\mu_3) .$$

We understand that in (1.9) the operator δ^{μ_i} acts with respect to z_i, $i = 1,2,3$; moreover $\mathbb{D}_\alpha(\ldots)$ is a holomorphic differential operator mapping functions on $\mathbb{H}_3$ to functions on $\mathbb{H} \times \mathbb{H} \times \mathbb{H}$. Following again the same line of reasoning as in lemma 7 [loc.cit], adapted to our situation, we easily get that the $\mathbb{D}_\alpha(\dots)$ satisfy

(1.10)
$$\mathbb{D}_\alpha(a, \nu_2, \nu_3, \mu_1, \mu_2, \mu_3) \left(f \mid_{\alpha,\beta} \iota_{111}(g_1, g_2, g_3) \right)$$
$$= \mathbb{D}_\alpha(a, \nu_2, \nu_3, \mu_1, \mu_2, \mu_3)(f) \mid_{\alpha+a+b-2\mu_1,\beta}^{z_1} g_1$$
$$\mid_{\alpha+a+\nu_2-2\mu_2,\beta}^{z_2} g_2 \mid_{\alpha+a+\nu_3-2\mu_3,\beta}^{z_3} g_3$$

for all $(g_1, g_2, g_3) \in Sl(2,\mathbb{R})^3$ and all holomorphic functions on $\mathbb{H}_3$ (and hence also for all C^∞-functions). The upper indices z_i on the right hand side of (1.10) indicate, on which variable g_i operates.

We have to remark here that Shimura's condition "$k > 2r$" in Lemma 7 [loc.cit] is satisfied in our situation as long as α is non-real or

$$\alpha + a + b > a$$
$$\alpha + a + \nu_2 > a$$
$$\alpha + a + \nu_3 > a.$$

However the coefficients of the ∂_{ij} on both sides of (1.9) are easily seen to be *rational* functions of α, therefore (1.9) (and subsequent equations) are true for all $\alpha \in \mathbb{C}$ as rational functions of α.

It is crucial for us to see that in the identity (1.10) the "holomorphic part", i.e.

$$\mathbb{D}_\alpha(a, \nu_2, \nu_3, 0, 0, 0)$$

is different from zero. For this purpose we consider $(y_1 y_2 y_3)^{-a'} \cdot \mathcal{D}_\alpha^{\star(a,b)}$ as a polynomial in $\partial_{12}, \partial_{13}, \partial_{23}$. It is easy to see that this is a polynomial of total degree $3a + b$, the component of degree $3a + b$ being given by

$$(\partial_{12}\partial_{13}\partial_{23})^a \times \frac{1}{\alpha^{[b]}} \cdot \frac{1}{b!} \cdot 2^b \cdot (y_1 y_2 y_3)^{-a'} \cdot \mathcal{D}_\alpha^{\star(a,b)}$$

as a polynomial in $\partial_{12}, \partial_{13}, \partial_{23}$. It is easy to see that this is a polynomial of total degree $3a + b$, the component of degree $3a + b$ being given by

(1.11)
$$(\partial_{12}\partial_{13}\partial_{23})^a \times \frac{1}{\alpha^{[b]}} \cdot \frac{1}{b!} \cdot 2^b \cdot (\partial_{12}X_2 + \partial_{13}X_3)^b$$

In particular, this component is free of y_i^{-1} and ∂_{ii}, so it can only come from the $\mathbb{D}_\alpha(a, \nu_2, \nu_3, 0, 0, 0)$ with $\nu_2 + \nu_3 = b$.

Now we define a polynomial Q_α of the matrix variable $S = S^t = (s_{i,j})_{1 \le i,j \le 3}$ by

(1.12) $$\mathbb{D}_\alpha(a, b, \nu_2, \nu_3, 0, 0, 0)e^{\text{trace}(SZ)} = Q_\alpha(S)e^{s_{11}Z_{11}+s_{22}Z_{22}+s_{33}Z_{33}}$$

By the same kind of reasoning as in [1, Satz 15] we see that by

(1.13) $$(x_1, x_2, x_3) \longmapsto P_r(x_1, x_2, x_3) := Q_r \begin{pmatrix} x_1^t x_1 & x_1^t x_2 & x_1^t x_3 \\ x_2^t x_1 & x_2^t x_2 & x_2^t x_3 \\ x_3^t x_1 & x_3^t x_2 & x_3^t x_3 \end{pmatrix}$$

we get a polynomial function of $(x_1, x_2, x_3) \in (\mathbb{C}^{2r})^3$ which in each variable is a harmonic form of degree $a+b$, $a+\nu_2$, $a+\nu_3$ respectively; more precisely, P_r defines a non-zero element of

(1.14) $$(\mathcal{H}_{a+b}(2r) \otimes \mathcal{H}_{a+\nu_2}(2r) \otimes \mathcal{H}_{a+\nu_3}(2r))^{\mathbb{O}(2r)}$$

For our investigation of the functional equation of triple L–functions we have to modify $\mathbb{D}_\alpha(a, b, \nu_2, \nu_3, 0, 0, 0)$ still further (we switch notation now from α to $r + s$). We consider the operator $\Delta = \Delta_{r,s}(a, \nu_2, \nu_3)$ given by

(1.15) $$F \longmapsto (y_1 y_2 y_3)^s \mathbb{D}_{r+s}(a, \nu_2, \nu_3, 0, 0, 0) \left(F \times \det(Y)^{-s} \right).$$

This operator (acting on functions on $\mathbb{H}_3$) is easily seen to satisfy

(1.16) $$\Delta \left(F \mid_r \iota_{111}(g_1, g_2, g_3) \right) = \Delta(F) \mid_{r+a+b}^{z_1} g_1 \mid_{r+a+\nu_2}^{z_2} g_2 \mid_{r+a+\nu_3}^{z_3} g_3$$

for all $g_1, g_2, g_3 \in Sl_2(\mathbb{R})$. By the same kind of argument about nearly holomorphic functions as above we get

(1.17)
$$\Delta_{r,s}(a, \nu_2, \nu_3) = \sum_{\substack{0 \le \mu_1 \le [\frac{a+b}{2}] \\ 0 \le \mu_2 \le [\frac{a+\nu_2}{2}] \\ 0 \le \mu_3 \le [\frac{a+\nu_3}{2}]}} \delta_{r+a+b-2\mu_1}^{\mu_1} \delta_{r+a+\nu_2-2\mu_2}^{\mu_2}$$

$$\times \; \delta_{r+a+\nu_3-2\mu_3}^{\mu_3} \Delta_{r,s}(a, \nu_2, \nu_3, \mu_1, \mu_2, \mu_3)$$

with holomorphic differential operators $\Delta_{r,s}(a, \nu_2, \nu_3, \mu_1, \mu_2, \mu_3)$ mapping functions on $\mathbb{H}_3$ to functions on $\mathbb{H} \times \mathbb{H} \times \mathbb{H}$. We should mention here that the differential operators coming up in (1.17) do not have poles as long as r is positive and s is non-real or $Re(s) \geq 0$. Again the "holomorphic part" $\Delta_{r,s}(a, \nu_2, \nu_3, 0, 0, 0)$ defines (as in (1.12),(1.13) an element of

$$(\mathcal{H}_{a+b}(2r) \otimes \mathcal{H}_{a+\nu_2}(2r) \otimes \mathcal{H}_{a+\nu_3}(2r))^{\mathbb{O}(2r)}$$

This space is known to be one-dimensional : by a result of Littelmann ([19], p. 145) the decomposition of $\mathcal{H}_{a+\nu_2}(2r) \otimes \mathcal{H}_{a+\nu_3}(2r)$ is multiplicity free and contains $\mathcal{H}_{a+b}(2r))$, hence there is a unique invariant line in the threefold tensor product. There exists therefore a function $c = c_r(s)$ such that

(1.18) $\Delta_{r,s}(a, \nu_2, \nu_3, 0, 0, 0) = c_r(s)\mathbb{D}_r(a, \nu_2, \nu_3, 0, 0, 0).$

By comparing coefficients of $(\partial_{12}\partial_{13}\partial_{23})^a \, \partial_{12}^{\nu_2}\partial_{13}^{\nu_3}$ on both sides of (1.17) we get

(1.19) $c_r(s) = \dfrac{(r+a')^{[b]}}{(r+s+a')^{[b]}} = \dfrac{\Gamma(r+a'+b)}{\Gamma(r+a')} \cdot \dfrac{\Gamma(r+a'+s)}{\Gamma(r+a'+s+b)} \cdot$

2. — Unfolding the integral

For a squarefree number $N > 0$ and three cuspforms

$$f = \sum a_f(n)e^{2\pi i n z} \in [\Gamma_0(N), k_1]_0$$

$$\phi = \sum a_\phi(n)e^{2\pi i n z} \in [\Gamma_0(N), k_2]_0$$

$$\psi = \sum a_\psi(n)e^{2\pi i n z} \in [\Gamma_0(N), k_3]_0$$

with k_1, k_2, k_3 as in section 1 we want to compute the threefold integral $\mathcal{A}(f, \phi, \psi, s)$, defined by

(2.1)

$$\int\int_{(\Gamma_0(N)\backslash\mathbb{H})^3}\int \overline{f(z_1)\phi(z_2)\psi(z_3)}\left(\mathcal{D}^{\star(a,\nu_2,\nu_3)}\left(\mathbb{G}_{r,s}^3\right)\right)$$

$$\times (\iota_{111}(z_1, z_2, z_3)) \, y_1^{k_1+s-a'} y_2^{k_2+s-a'} y_3^{k_3+s-a'} \prod_{i=1}^{3} \dfrac{dx_i dy_i}{y_i^2}$$

where $\mathbb{G}^3_{r,s}$ is the Eisenstein series on $\mathbb{H}_3$ defined by

$$\mathbb{G}^3_{r,s} = \sum_{M \in \Gamma^3_\infty \backslash \Gamma^3_0(N)} 1 \mid_{r+s,s} M$$

(2.2)

$$= \sum_{\left(\begin{smallmatrix} * & * \\ C & D \end{smallmatrix}\right) = M \in \Gamma^3_\infty \backslash \Gamma^3_0(N)} \det(CZ+D)^{-r-s} \det(C\bar{Z}+D)^{-s} .$$

In the applications we shall need modified versions of the integral (2.1); it is appropriate to describe these here : we use the well-known fact (see e.g. [25, equation (2.28)]) that holomorphic cusp forms are orthogonal to (C^∞)-automorphic forms in the image of the differential operators δ, therefore we may replace $\left(\mathcal{D}^{\star(a,\nu_2,\nu_3)}\right) \left(\mathbb{G}^3_{r,s}\right) (\iota_{111}(z_1,z_2,z_3)) y_1^{k_1+s-a'} y_2^{k_2+s-a'} y_3^{k_3+s-a'}$ in the integrand of (2.1) by

$$\left(\mathbb{D}_{r+s}(a,\nu_2,\nu_3,0,0,0)\left(\mathbb{G}^3_{r,s}\right)\right)(\iota_{111}(z_1,z_2,z_3)) y_1^{k_1+s} y_2^{k_2+s} y_3^{k_3+s}$$

or by

(2.3) $c_r(s)\mathbb{D}_r(a,\nu_2,\nu_3,0,0,0)\left(\mathbb{E}^3_{r,s}\right)(\iota_{111}(z_1,z_2,z_3)y_1^{k_1} y_2^{k_2} y_3^{k_3}$

where

$$\mathbb{E}^3_{r,s}(Z) = \det(Y)^s \cdot \mathbb{G}^3_{r,s} = \sum_{M \in \Gamma^3_\infty \backslash \Gamma^3_0(N)} \det(Y)^s \mid_r M .$$

The actual computation of $\mathcal{A}(f,\phi,\psi,s)$ is however most conveniently done using the integral in the version of (2.1).

2.1. — The first integration

To understand the integration with respect to z_1, it is better to consider first the integral

(2.4)

$$I(s) = \int_{\Gamma_0(N)\backslash\mathbb{H}} \overline{f(z_1)} \left(\mathcal{D}^{(a,b)}_{r+s}\mathbb{G}^3_{r,s}\right)(\iota_{1,2}(z_1,Z)) y_1^{k_1+s-a'} \frac{dx_1 dy_1}{y_1^2} \times \det(Y)^{s-a'}$$

with $Z = X + iY \in \mathbb{H}_2$. We recall from [2, Thm. 1.1] that the double cosets

$$\Gamma_\infty\backslash\Gamma^3_0(N)/\iota_{1,2}\left(\Gamma_0(N) \times \Gamma^2_0(N)\right)$$

can be parametrized by the following set of representatives

(2.5) $$\{g_m \mid m \in \mathbb{N} \cup \{0\}, m \equiv 0 \bmod N\}$$

where

$$g_m = \begin{pmatrix} & 1_3 & & 0_3 \\ 0 & m & 0 & \\ m & 0 & 0 & 1_3 \\ 0 & 0 & 0 & \end{pmatrix}$$

We split the integral (2.4) into the contributions of the double cosets (2.5) :

(2.6) $$I(s) = \sum_m I_m(s).$$

It is easy to see that the double coset with $m = 0$ decomposes into left cosets as follows

$$\{\iota_{1,2}(\gamma, \delta) \mid \gamma \in \Gamma_\infty \backslash \Gamma_0(N), \delta \in \Gamma^2_\infty \backslash \Gamma^2_0(N)\}.$$

Therefore its contribution to $\mathcal{D}^{\star(a,b)}\mathbb{G}^3_{r+s,s}$ is just

$$\mathcal{D}^{\star(a,b)}_{r+s,s}\left(\sum_{\gamma,\delta} 1 \mid_{r+s,s} \iota_{1,2}(\gamma,\delta)\right)$$

$$= \sum_{\gamma,\delta} \mathcal{D}^{\star(a,b)}_{r+s}(1) \mid^{z_1}_{r+s+a'+b,s-a'} \gamma \mid^{Z}_{r+s+a',s-a',\sigma_b} \delta.$$

It is obvious that

$$\mathcal{D}^{\star(a,b)}_{r+s}(1) = \begin{cases} 0 & b > 0 \\ \varepsilon_3(r+s)\varepsilon_3(r+s+1)\dots\varepsilon_3(r+s+a'-1) & b = 0. \end{cases}$$

Unfolding the integral defining $I_0(s)$ we easily get (by the cuspidality of f) that

$$I_0(s) = 0$$

(we omit the standard calculation).

For fixed $m > 0$, $m \equiv 0 \bmod N$ the left cosets are given by (see [2, Thm. 1.2])
(2.7)
$$\{g_m \iota_{1,2}(\gamma, l(h))g \mid \gamma \in \Gamma_0(N), h \in \Gamma[m]\backslash\Gamma_0(N), g \in C_{2,1}(N)\backslash\Gamma^2_0(N)\},$$

where

$$\Gamma[m] = \Gamma_0(N) \cap \begin{pmatrix} 0 & m^{-1} \\ -m & 0 \end{pmatrix} \Gamma_0(N) \begin{pmatrix} 0 & -m^{-1} \\ m & 0 \end{pmatrix},$$

$l = \iota_{11}$ and $C_{2,1}$ is the standard maximal parabolic subgroup of $Sp(2)$ given by

$$C_{2,1} = \left\{ \begin{pmatrix} & & & \\ 0 & 0 & 0 & \end{pmatrix} \right\}$$

with

$$C_{2,1}(N) := C_{2,1}(\mathbb{Z}) \cap \Gamma_0^2(N).$$

The summation over γ unfolds the integral for $I_m(s)$ to

(2.8)
$$I_m(s) =$$
$$\int_{\mathbb{H}} \overline{f(z_1)} \sum_{h,g} \mathcal{D}_{r+s}^{\star(a,b)} \left(1 \mid_{r+s,s} \boldsymbol{g}_m \right) \mid_{r+s+a',s-a',\sigma_b}^{Z}$$
$$\times l(h)g \cdot y_1^{k_1+s-a'} \frac{dx_1 dy_1}{y^2} \times \det(Y)^{s-a'}.$$

By lemma 4.2 of [6] and (1.4) we have

$$\mathcal{D}^{\star(a,b)} \left(1 \mid_{r+s,s} \boldsymbol{g}_m \right)$$
(2.9)
$$= 2^{-a}(2r + 2s - 2)^{[a]}(r+s)^{[a']}\mathbb{L}_{r+s+a'}^{(b)} \left(1 \mid_{r+s+a',s-a'} \boldsymbol{g}_m \right)$$
$$= A(r+s,b) \cdot (1 - m^2 z_1 Z^*)^{-r-s-a'-b} \cdot (1 - m^2 \bar{z}_1 \bar{Z}^*)^{-s+a'} \cdot (mX_2)^b$$

where for $Z \in \mathbb{H}_2$ we denote by Z^* the entry in the upper left corner of Z and

$$A(s,b) = \left(\frac{-1}{2\pi i} \right)^b \cdot \frac{2^{-a}(2s-2)^{[a]}(s)^{[a']}(2s+a-2)^{[b]}}{b!(s+a'-1)^{[b]}}.$$

This implies

(2.10)
$$I_m(s) = A(r+s,b) \sum_{g,h}$$
$$\left(\int_{\mathbb{H}} \overline{f(z_1)}(1-m^2 z_1 Z^*)^{-r-s-a'-b}(1-m^2 \bar{z}_1 \bar{Z}^*)^{-s+q'} y_1^{k_1+s-a'} \frac{dx_1 dy_1}{y^2} \right.$$
$$\left. \times (mX_2)^b \right) \mid_{r+s+a',s-a',\sigma_b}^{Z} l(h)g \times \det(Y)^{s-a'}.$$

The integral in (2.10) is exactly of the same type as in [2, (1.4), (1.5)]. Using the same notation

$$\mu(k,s) = (-1)^{\frac{k}{2}} \cdot 2^{3-k-2s} \frac{\pi}{k+s-1}$$

as in [2] we get by the same reasoning as there

$$
I_m(s) = A(r+s,b) \cdot \mu(r+a+b, s-a') \cdot m^{-r-2s}
$$

(2.11)
$$
\times \sum_g \left(f^\rho \mid_{r+b+a} \Gamma_0(N) \begin{pmatrix} 0 & -m^{-1} \\ m & 0 \end{pmatrix} \Gamma_0(N) \right)
$$

$$
\times <g<Z>^*> \left(\sigma_b \otimes \overset{k+a}{\det} (j(g,Z)) \right)^{-1} X_2^b \left(\frac{\det \operatorname{Im}(g<Z>)}{\operatorname{Im}(g<Z>^*)} \right)^{s-a'}
$$

with $j(\begin{pmatrix} a & b \\ c & d \end{pmatrix}, z) = cz + d$. This is (essentially) a vector-valued Klingen-type Eisenstein series attached to the modular form

$$
f^\rho \mid_{k_1} \Gamma_0(N) \begin{pmatrix} 0 & -m^{-1} \\ m & 0 \end{pmatrix} \Gamma_0(N)
$$

where

$$
f^\rho(z) = \overline{f(-\bar{z})}.
$$

From now on we assume that f is a normalized newform (eigenform) of level $N_f \mid N$; we write $N = N_f \cdot N^f$ whenever it is convenient. The Fourier coefficients of f are then totally real and we have an Euler product expansion of type
(2.12)
$$
\sum a_f(n) n^{-s} = \left(\prod_{p \mid N_f} \frac{1}{(1 - a_f(p)p^{-s})} \right) \cdot \left(\prod_{p \nmid N_f} \frac{1}{(1 - \alpha_p p^{-s})(1 - \alpha'_p p^{-s})} \right).
$$

Moreover for any prime q dividing N_f we have

$$
a_f(q)^2 = q^{k_1 - 2}
$$

and

$$
f \mid_{k_1} \mathbb{V}_q^N = f \mid_{k_1} \mathbb{V}_q^{N_f} = -a_f(q) q^{1 - \frac{k_1}{2}} f
$$

where $\mathbb{V}_q^N$ denotes the "Atkin-Lehner-involution" given by

$$
\mathbb{V}_q^N = \begin{pmatrix} x & y \\ N & q \end{pmatrix}
$$

with $xq - Ny = q$ and $q \mid x$ (for details we refer to [18]).

Actually we have to work not with the newform f itself, but with $f \mid_{k_1} \begin{pmatrix} D^f & 0 \\ 0 & 1 \end{pmatrix}$ where D^f is a fixed divisor of N^f.

To simplify (2.11) further (for $f|_{k_1} \begin{pmatrix} D^f & 0 \\ 0 & 1 \end{pmatrix}$), we have to study

(2.13)

$$\sum_{m \equiv 0(N)} f|_{k_1} \begin{pmatrix} D^f & 0 \\ 0 & 1 \end{pmatrix} |_{k_1} \Gamma_0(N) \begin{pmatrix} 0 & -m^{-1} \\ m & 0 \end{pmatrix} \Gamma_0(N) \cdot m^{-s}$$

$$= N^{-s} \sum_{m'=1}^{\infty} f|_{k_1} \begin{pmatrix} D^f & 0 \\ 0 & 1 \end{pmatrix} |_{k_1} \Gamma_0(N) \begin{pmatrix} 1 & 0 \\ 0 & m'^2 N \end{pmatrix} \Gamma_0(N) |_{k_1} \begin{pmatrix} 0 & -1 \\ N & 0 \end{pmatrix} \cdot m'^{-s}$$

Now we are essentially in a "local" situation, because we may decompose the "Fricke involution" into Atkin-Lehner involutions :

$$\begin{pmatrix} 0 & -1 \\ N & 0 \end{pmatrix} = \gamma \circ \prod_{q|N} V_q^N$$

with $\gamma \in \Gamma_0(N)$.

We use the following formal identities :

$p \nmid N$:

(2.14)

$$\sum_{l=0}^{\infty} f \mid \Gamma_0(N) \begin{pmatrix} 1 & 0 \\ 0 & p^{2l} \end{pmatrix} \Gamma_0(N) X^l =$$

$$\frac{(1-X)(1-p^2 X^2)}{(1-pX)(1-\alpha_p^2 p^{-k_1+2} X)(1-\alpha_p'^2 p^{-k_1+2} X)}$$

Proof. Standard

(2.15)
$p|N_f$

$$\sum_{l=0}^{\infty} f \mid \Gamma_0(N) \begin{pmatrix} 1 & 0 \\ 0 & p^{2l+1} \end{pmatrix} \Gamma_0(N) \circ V_p^N \cdot X^l = p^{1-\frac{k}{2}} \cdot \frac{a_f(p)}{1-X} \cdot f|_{k_1} V_p^{N_f}$$

$$= \frac{-1}{1-X} f$$

Proof. Standard, using $a_f(p^n) = a_f(p)^n$ and $a_f(p^2) = p^{k_1-2}$

$p|N^f$:

(2.16)

$$\sum_{l=0}^{\infty} f \mid \Gamma_0(N) \begin{pmatrix} 1 & 0 \\ 0 & p^{2l+1} \end{pmatrix} \Gamma_0(N) \circ V_p^N \times X^l$$

$$= p^{1-\frac{k_1}{2}} \cdot \frac{(f|U(p) - p^{\frac{k}{2}} \cdot f|_{k_1} \begin{pmatrix} p & 0 \\ 0 & 1 \end{pmatrix} \cdot X)}{(1-\alpha_p^2 p^{-k_1+2} X)(1-\alpha_p'^2 p^{-k_1+2} X)} |_{k_1} V_p^N.$$

Proof. Standard. Using

$$f|T(p) = f|U(p) + p^{\frac{k_1}{2}-1} \cdot f|_{k_1} \begin{pmatrix} p & 0 \\ 0 & 1 \end{pmatrix}$$

and

$$f|_{k_1} \begin{pmatrix} p & 0 \\ 0 & 1 \end{pmatrix} |_{k_1} \mathbb{V}_p^N = f ,$$

we get for (2.16) :

(2.17)
$$\frac{p^{1-\frac{k_1}{2}} a_f(p)|_{k_1} \begin{pmatrix} p & 0 \\ 0 & 1 \end{pmatrix} - (1+pX)f}{(1 - \alpha_p^2 p^{-k_1+2} X)(1 - \alpha_p'^2 p^{-k_1+2} X)} .$$

In quite the same way we get (still for the case $p|N^f$) :

$$\sum_{l=0}^{\infty} f|_{k_1} \begin{pmatrix} p & 0 \\ 0 & 1 \end{pmatrix} | \Gamma_0(N) \begin{pmatrix} 1 & 0 \\ 0 & p^{2l+1} \end{pmatrix} \Gamma_0(N)|_{k_1} \mathbb{V}_p^N \cdot X^l =$$

$$\frac{p(1+pX) \cdot f - a_f(p) p^{-\frac{k_1}{2}+2} X \cdot f |_{k_1} \begin{pmatrix} p & 0 \\ 0 & 1 \end{pmatrix}}{(1 - \alpha_p^2 p^{-k_1+2} X)(1 - \alpha_p'^2 p^{-k_1+2} X)} |_{k_1} \mathbb{V}_p^N =$$

$$\frac{p(1+pX)f|_{k_1} \begin{pmatrix} p & 0 \\ 0 & 1 \end{pmatrix} - a_f(p) p^{2-\frac{k_1}{2}} f \cdot X}{(1 - \alpha_p^2 p^{-k_1+2} X)(1 - \alpha_p'^2 p^{-k_1+2} X)} .$$

The usual procedure ($X \longmapsto p^{-s}$) yields for (2.13)

(2.19)
$$\sum_{m \equiv 0(N)} f |_{k_1} \begin{pmatrix} D^f & 0 \\ 0 & 1 \end{pmatrix} |_{k_1} \Gamma_0(N) \begin{pmatrix} 0 & -m^{-1} \\ m & 0 \end{pmatrix} \Gamma_0(N) \circ \begin{pmatrix} 0 & -1 \\ N & 0 \end{pmatrix} \cdot m^{-s}$$

$$= N^{-s} \prod_{p \nmid N} \frac{(1 - p^{-s})(1 - p^{2-2s})}{1 - p^{-s+1}} \prod_{p \nmid N_f} \frac{1}{(1 - \alpha_p^2 p^{-s-k_1+2})(1 - \alpha_p^2 p^{-s-k_1+2})}$$

$$\times \prod_{p|N_f} \frac{-1}{(1 - p^{-k-s})} \times \tilde{f}_s$$

with

(2.20)
$$\tilde{f}_s = \sum_{d|N^f} \alpha(d, D, s) f|_{k_1} \begin{pmatrix} d & 0 \\ 0 & 1 \end{pmatrix} ,$$

where

(2.21) $$\alpha(d, D, s) = \prod_{p|N^f} \alpha(d_p, D_p, s)$$

is a multiplicative function given by (2.14)-(2.18). Here we denote by t_p the p-part of a positive rational number t. In the sequel we write $\alpha_p(d, D, s)$ instead of $\alpha(d_p, D_p, s)$.

2.2. — Second Unfolding

To continue the computation of the integral (2.1) we first need to find a good parametrization of $C_{2,1}(N)\backslash\Gamma_0^2(N)$; we shall follow [22] (with the modifications necessary for level $N > 1$). We first remark that two elements of $Sp_2(\mathbb{Z})$ are equivalent modulo $C_{2,1}(\mathbb{Z})$ if and only if their last rows are equal up to sign (the same is true for $\Gamma_0(N)$ and $C_{2,1}(N)$).

For $C_{2,1}(\mathbb{Z})\backslash Sp_2(\mathbb{Z})$ the parametrization given in [22] is as follows :

(2.22) $$\{\iota_{1,1}(1_2, h) \mid h \in Sl_2(\mathbb{Z})_\infty\backslash Sl_2(\mathbb{Z})\} ,$$

(2.23) $$\{d(\mathbb{J})\iota_{1,1}(h, 1_2) \mid h \in Sl_2(\mathbb{Z})_\infty^+\backslash Sl_2(\mathbb{Z})\} ,$$

(2.24) $$\bigcup_{\substack{u,v \in \mathbb{N} \\ u,v \text{ coprime}}} \{d(M) \circ \iota_{1,1}(h, h') \mid h \in Sl_2(\mathbb{Z})_\infty\backslash Sl_2(\mathbb{Z}), h' \in Sl_2(\mathbb{Z})_\infty^+\backslash Sl_2(\mathbb{Z})\}$$

where

$$d : \begin{cases} Gl_2(\mathbb{R}) & \longrightarrow & Sp_2(\mathbb{R}) \\ A & \longmapsto & \begin{pmatrix} (A^t)^{-1} & 0 \\ 0 & A \end{pmatrix} \end{cases}$$

$\mathbb{J} = \begin{pmatrix} 0 & -1 \\ 1 & 0 \end{pmatrix}$ and M is an element of $Sl_2(\mathbb{Z})$ with $M = \begin{pmatrix} * & * \\ u & v \end{pmatrix}$. Among (2.22), (2.23),(2.24) precisely the following elements have their last row congruent to $(0, 0, *, *)$ modulo N :

(2.22') $$\{\iota_{1,1}(1_2, h) \mid h \in \Gamma_\infty\backslash\Gamma_0(N)\}$$

(2.23') $$\{d(\mathbb{J})\iota_{1,1}(h, 1_2) \mid h \in \Gamma_\infty^+\backslash\Gamma_0(N)\}$$

$$(2.24') \quad \left\{ d(M) \circ \iota_{1,1}(h,h') \mid \begin{array}{c} h \in Sl_2(\mathbb{Z})_\infty \backslash Sl_2(\mathbb{Z}),\, h' \in Sl_2(\mathbb{Z})^+_\infty \backslash Sl_2(\mathbb{Z}) \\ M = \begin{pmatrix} * & * \\ u & v \end{pmatrix} \in Sl_2(\mathbb{Z})^+_\infty \backslash Sl_2(\mathbb{Z}) \\ u,v \in \mathbb{N},\, uc \equiv 0(N),\, vc' \equiv 0(N) \end{array} \right\}$$

Here c (and c') denote the lower left entry of h (and h').

At this point we should emphasize that (2.22')-(2.24') do *not* give representatives of $C_{2,1}(N)\backslash \Gamma_0^2(N)$, since these elements are in general not in $\Gamma_0^2(N)$, but they are equivalent modulo $C_{2,1}(\mathbb{Z})$ to such representatives (by using a suitable transformation we shall finally transport them into $\Gamma_0^2(N)$).

To describe (2.24') more appropriately we fix two decompositions

$$N = N_1 \cdot N_2 \qquad \text{and} \qquad N = N_1' \cdot N_2'$$

and consider for the moment only those $h = \begin{pmatrix} * & * \\ c & d \end{pmatrix}$ and $h' = \begin{pmatrix} * & * \\ c' & d' \end{pmatrix}$ with $\gcd(c,N) = N_1$ and $\gcd(c',N) = N_1'$. Then the data

$$u, v, \begin{pmatrix} * & * \\ c & d \end{pmatrix}, \begin{pmatrix} * & * \\ c' & d' \end{pmatrix}$$

describe an element of (2.24') if and only if $N_2 | u$ and $N_2' | v$. These data exist only if $N | N_1 \cdot N_1'$, because we require u,v to be coprime. It is a standard procedure to translate these considerations into more group theoretic terms : we denote by $\tau_{N_1} = \tau_{N_1}^N$ an element of $Sl_2(\mathbb{Z})$ with $\tau_{N_1} = \begin{pmatrix} \alpha & \beta \\ N_1 & N_2 \end{pmatrix}$ and $N_2 | \alpha$. This implies in particular that $(\tau_{N_1})^2 \in \Gamma_0(N)$.

Then (2.24') can also be described by

(2.24")

$$\bigcup_{\substack{N_1 N_2 = N \\ N_1' N_2' = N \\ N | N_1 N_1'}} \left\{ d(M)\iota_{1,1}(h,h') \mid \begin{array}{c} h \in \left(\tau_{N_1}\Gamma_0(N)\tau_{N_1}^{-1}\right)_\infty \backslash \tau_{N_1}\Gamma_0(N) \\ h \in \left(\tau_{N_1'}\Gamma_0(N)\tau_{N_1'}^{-1}\right)^+_\infty \backslash \tau_{N_1'}\Gamma_0(N) \\ u,v \text{ positive, coprime, } N_2 | u,\, N_2' | v \end{array} \right\}$$

By a routine matrix calculation, we see that (with h, h', M, N_1, N_1' as in (2.24"))

$$(2.24''') \qquad \iota_{1,1}(\tau_{(N_1,N'_1)}, \mathbb{1}_2) \circ d(M) \circ \iota_{1,1}(h,h')$$

is indeed in $\Gamma_0^2(N)$, if we require (as we are allowed to do!) that M is of type $M = \begin{pmatrix} r & s \\ u & v \end{pmatrix}$ with $v | r$.

Now we denote by $I_{m,\nu_1,\nu_2}(z_2, z_3, s)$ the $X_2^{\nu_2} X_3^{\nu_3}$–component of the $\mathbb{C}[X_2, X_3]_b$-valued function $I_m(s)$, restricted to $(z_2, z_3) \in \mathbb{H} \times \mathbb{H} \hookrightarrow \mathbb{H}_2$. With f, ϕ, ψ as before (f now again an arbitrary element in $[\Gamma_0(N), k_1]_0$) we consider the double integral

$$(2.25) \qquad \int\!\!\!\int_{(\Gamma_0(N)\backslash\mathbb{H})^2} \overline{\phi(z_2)\psi(z_3)}\mathcal{K}_f^{\nu_2,\nu_3}(z_2, z_3, s)y_2^{k_2}y_3^{k_3}\frac{dx_2 dy_2}{y_2^2}\frac{dx_3 dy_3}{y_3^2}$$

where

$$\mathcal{K}_f(Z, s) =$$
$$\sum_{g \in C_{2,1}(N)\backslash\Gamma_0^2(N)} f(g<Z>^*) \left(\sigma_b \otimes \det{}^{k+a}(j(g, Z))\right)^{-1} X_2^b \cdot \left(\frac{\det \operatorname{Im}(g<Z>)}{\operatorname{Im}(g<Z>^*)}\right)^s$$

is the same Klingen-type Eisenstein series as in (2.11), but with the Hecke operator removed; again $\mathcal{K}_f^{\nu_2,\nu_3}$ denotes the $X_2^{\nu_2} X_3^{\nu_3}$-component of $\mathcal{K}_f$, restricted to $\mathbb{H} \times \mathbb{H}$. We split $\mathcal{K}_f^{\nu_2,\nu_3}$ into three parts according to the three types (2.22),(2.23),(2.24) of left cosets :

$$(2.26) \qquad \mathcal{K}_f^{\nu_2,\nu_3} = \sum_{i=1}^{3} \mathcal{K}_{f,i}^{\nu_2,\nu_3}$$

It is again easy to see that $\mathcal{K}_{f,1}^{\nu_2,\nu_3}$ and $\mathcal{K}_{f,2}^{\nu_2,\nu_3}$ do not contribute to the integral (2.25). Using (2.24") we further split $\mathcal{K}_{f,3}^{\nu_2,\nu_3}$ as

$$(2.27) \qquad \mathcal{K}_{f,3}^{\nu_2,\nu_3} = \sum_{N_1,N_1'} \mathcal{K}_{f,N_1,N_1'}^{\nu_2,\nu_3}$$

We can express $\mathcal{K}_{f,N_1,N_1'}^{\nu_2,\nu_3}$ more explicitly as follows

$$(2.28)$$
$$\mathcal{K}_{f,N_1,N_1'}^{\nu_2,\nu_3}$$
$$= 2\binom{b}{\nu_2} \sum_{h,h',u,v} f \mid_{k_1} \tau_{(N_1,N_1')}(v^2 h(<z_2>) + u^2 h'(<z_3>))$$
$$\times v^{\nu_2} u^{\nu_3} j(h, z_2)^{-k-a-\nu_2} j(h', z_3)^{-k-a-\nu_3}$$
$$\times \left(\frac{\operatorname{Im}(h<z_2>) \cdot \operatorname{Im}(h'<z_3>)}{v^2 \cdot \operatorname{Im}(h<z_2>) + u^2 \cdot \operatorname{Im}(h'<z_3>)}\right)^s$$

where the summation over h, h', u, v is given by (2.24") with N_1 and N_1' fixed.

It is well known how to unfold integrals like

(2.29)
$$\mathcal{I}(f,\phi,\psi,N_1,N_1',s) := \int\int_{(\Gamma_0(N)\backslash\mathbb{H})^2} \overline{\phi(z_2)\psi(z_3)}\mathcal{K}^{\nu_2,\nu_3}_{f,N_1,N_1'}(z_2,z_3,s)y_2^{k_2}y_3^{k_3}\frac{dx_2dy_2}{y_2^2}\frac{dx_3dy_3}{y_3^2}$$

by applying $\tau_{N_1}^{-1}$ and $\tau_{N_1'}^{-1}$: The result is
(2.30)
$$\mathcal{I}(f,\phi,\psi,N_1,N_1',s) =$$

$$2\binom{b}{\nu_2}\int_{\left(\tau_{N_1}\Gamma_0(N)\tau_{N_1}^{-1}\right)_\infty\backslash\Gamma_0(N)}\int_{\left(\tau_{N_1'}\Gamma_0(N)\tau_{N_1'}^{-1}\right)_\infty\backslash\Gamma_0(N)}\overline{\left(\phi\,|_{k_2}\tau_{N_1}^{-1}\right)(z_2)\cdot\left(\psi\,|_{k_3}\tau_{N_1'}^{-1}\right)(z_3)}$$

$$\times\sum_{u,v}f\,|_{k_1}\tau_{(N_1,N_1')}(v^2z_2+u^2z_3)v^{\nu_2}u^{\nu_3}\left(\frac{y_2y_3}{v^2y_2+u^2y_3}\right)^s$$

$$\times y_2^{k_2}y_3^{k_3}\frac{dx_2dy_2}{y_2^2}\frac{dx_3dy_3}{y_3^2}.$$

We do not want to work with Fourier expansions at several cusps, therefore we assume from now on that f, ϕ, ψ are normalized newforms (eigenforms of all Hecke operators) of levels N_f, N_ϕ and N_ψ (all dividing N).

We decompose N_1 and N_2 as

(2.31)
$$N_1 = N_{1,f}\cdot N_1^f, N_2 = N_{2,f}\cdot N_2^f$$

(and the same for N_1', N_2' and also for ϕ and ψ).

We mention here the following facts, which we shall use in the sequel :

*
$$R|N \Longrightarrow \tau_{N_1}^N = \gamma\circ\tau_{(R,N_1)}^R \quad\text{with}\quad \gamma\in\Gamma_0(R).$$

* For any divisor d of N^f we have

$$\begin{pmatrix}d & 0 \\ 0 & 1\end{pmatrix}\circ\tau_{N_1}^N = \gamma\circ\begin{pmatrix}d & 0 \\ 0 & 1\end{pmatrix}\circ\tau_{(dN_f,N_1)}^{dN_f} = \gamma\circ\tau_{N_{1,f}}^{N_f}\circ\begin{pmatrix}(d,N_1^f) & 0 \\ 0 & \frac{d}{(d,N_1^f)}\end{pmatrix}$$

with $\gamma\in\Gamma_0(N_f)$.

•

$$\tau^N_{N_1} = \gamma \circ \left(\prod_{q|N_2} \mathbb{V}^N_q \right) \circ \begin{pmatrix} \frac{1}{N_2} & 0 \\ 0 & 1 \end{pmatrix}$$

with $\gamma \in \Gamma_0(N)$ and (using the same notation as in [18]

$$\mathbb{V}^N_q = \begin{pmatrix} x & y \\ N & q \end{pmatrix}$$

with $xq - Ny = q$ and $q|x$.

• For any newform g of level N and weight k we have (see [18, Theorem 3])

$$g \mid_k \mathbb{V}^N_q = -a_g(q)q^{1-\frac{k}{2}} \cdot g.$$

At this point we introduce - as we already did for N^f in the previous subsection- divisors D^ϕ and D^ψ of N^ϕ and N^ψ; as usual we further factorize them (for given N_1 and N_1') as $D^\phi = D_1^\phi \cdot D_2^\phi$ and $D^\psi = D_1'^\psi \cdot D_2'^\psi$.

Using these facts we get the following Fourier expansions
(2.32)

$$\left(\phi|_{k_2} \begin{pmatrix} D^\phi & 0 \\ 0 & 1 \end{pmatrix} |_{k_2} \tau^{-1}_{N_1} \right)(z_2) = \phi|_{k_2} \begin{pmatrix} D^\phi & 0 \\ 0 & 1 \end{pmatrix} |_{k_2} \tau_{N_1}(z_2)$$

$$= \phi|_{k_2} \tau^{N_\phi}_{N_{1,\phi}} \, |_{k_2} \begin{pmatrix} D_1^\phi & 0 \\ 0 & D_2^\phi \end{pmatrix}(z_2)$$

$$= \phi|_{k_2} \left(\prod_{q|N_{2,\phi}} \mathbb{V}^{N_\phi}_q \right) |_{k_2} \begin{pmatrix} \frac{D_1^\phi}{N_{2,\phi}} & 0 \\ 0 & D_2^\phi \end{pmatrix}(z_2)$$

$$= \left(\prod_{q|N_{2,\phi}} -a(q)q^{-\frac{k_2}{2}+1} \right) \cdot$$

$$\times N_{2,\phi}^{-\frac{k_2}{2}} D_1^{\phi \frac{k}{2}} D_2^{\phi \frac{-k}{2}} \cdot \phi(\frac{D_1^\phi}{D_2^\phi N_{2,\phi}} \cdot z_2)$$

$$= D_1^{\phi \frac{k_2}{2}} D_2^{\phi -\frac{k_2}{2}} \left(\prod_{q|N_{2,\phi}} -a_\phi(q)q^{-k_2+1} \right)$$

$$\times \sum a_\phi(n')e^{2\pi i \frac{D_1^\phi}{D_2^\phi N_{2,\phi}} \cdot n' z_2}.$$

$$\left(\psi|_{k_3}\begin{pmatrix} D^\psi & 0 \\ 0 & 1 \end{pmatrix}|_{k_3}\tau_{N_1'}^{-1}\right)(z_3) = D_1'{}^{\psi\frac{k_3}{2}}D_2'{}^{\psi-\frac{k_3}{2}}$$

(2.33)

$$\left(\prod_{q|N_{2,\psi}'} -a_\psi(q)q^{-k_3+1}\right)\sum a_\psi(n'')e^{2\pi i\frac{D_1'{}^\psi}{D_2'{}^\psi N_{2,\psi}'}n''z_3}$$

and for a divisor d of N^f :

(2.34)

$$f|_{k_1}\begin{pmatrix} d & 0 \\ 0 & 1 \end{pmatrix}\circ\tau_{(N_1,N_1')}^N(z) = f|_{k_1}\tau_{(N_1,N_1')_f}^{N_f}\circ\begin{pmatrix} (d,(N_1,N_1')^f) & 0 \\ 0 & \frac{d}{(d,(N_1,N_1')^f)} \end{pmatrix}$$

$$= f|_{k_1}\tau_{(N_{1,f},N_{1,f}')}^{N_f}\circ\begin{pmatrix} (d,N_1^f,N_1'{}^f) & 0 \\ 0 & \frac{d}{(d,N_1^f,N_1'{}^f)} \end{pmatrix}$$

$$= d^{-\frac{k_1}{2}}\cdot(d,N_1^f,N_1'{}^f)^{k_1}\prod_{q||\mathrm{lcm}(N_{2,f},N_{2,f}')}$$

$$\left(-a_f(q)q^{-k_1+1}\right)\sum a_f(n)e^{2\pi iB\cdot nz}\,.$$

We use here the simple fact that $\frac{N^f}{(N_{1,f},N_{1,f}')} = \mathrm{lcm}(N_{2,f},N_{2,f}')$ and

(2.35)
$$B := \frac{1}{\mathrm{lcm}(N_{2,f},N_{2,f}')}\cdot\frac{(d,N_1^f,N_1'{}^f)^2}{d}\,.$$

Now we are ready to plug these Fourier expansions into the expression (2.30) for

$$\mathcal{I}(f|_{k_1}\begin{pmatrix} d & 0 \\ 0 & 1 \end{pmatrix},\phi|_{k_2}\begin{pmatrix} D^\phi & 0 \\ 0 & 1 \end{pmatrix},\psi|_{k_3}\begin{pmatrix} D^\psi & 0 \\ 0 & 1 \end{pmatrix},N_1,N_1',s)\,.$$

By integration over $x_2 \bmod N_2$ and $x_3 \bmod N_2'$ we see that only those terms $a_f(n)a_\phi(n')a_\psi(n'')$ give non–zero contributions, for which

(2.36)
$$Bnv^2 = \frac{n'}{N_{2,\phi}}\cdot\frac{D_1^\phi}{D_2^\phi}$$

and

(2.37)
$$Bnu^2 = \frac{n''}{N_{2,\psi}'}\cdot\frac{D_1'{}^\psi}{D_2'{}^\psi}$$

Using

$$
\int\limits_0^\infty \int\limits_0^\infty e^{-4\pi Bnv^2 y_2 - 4\pi Bnu^2 y_3} \left(\frac{y_2 y_3}{v^2 y_2 + u^2 y_3} \right)^s y_2^{k_2 - 2} y_3^{k_3 - 2} \, dy_2 \, dy_3
$$

(2.38)

$$
= \frac{\Gamma(s + k_2 + k_3 - 2)\Gamma(s + k_2 - 1)\Gamma(s + k_3 - 1)}{\Gamma(2s + k_2 + k_3 - 2)}
$$

$$
\times (4\pi Bn)^{-s - k_2 - k_3 + 2} \cdot v^{-2s - 2k_2 + 2} \cdot u^{-2s - 2k_3 + 2},
$$

we obtain
(2.39)
$$
\mathcal{I}(f|_{k_1} \begin{pmatrix} d & 0 \\ 0 & 1 \end{pmatrix}, \phi|_{k_2} \begin{pmatrix} D^\phi & 0 \\ 0 & 1 \end{pmatrix}, \psi|_{k_3} \begin{pmatrix} D^\psi & 0 \\ 0 & 1 \end{pmatrix}, N_1, N_1', s) =
$$

$$
2 \binom{b}{\nu_2} \cdot \mathcal{I}_\infty(s) N_2 N_2' B^{-s - k_2 - k_3 + 2} \cdot D_1^{\phi \frac{k_2}{2}} D_2^{\phi - \frac{k_2}{2}} D_1'^{\psi \frac{k_3}{2}} D_2'^{\psi - \frac{k_3}{2}}.
$$

$$
\times d^{-\frac{k_1}{2}} \cdot (d, N_1^f, N_1'^f)^{k_1}
$$

$$
\prod_{q | \mathrm{lcm}(N_{2,f}, N_{2,f}')} (-a_f(q) q^{-k_1 + 1}) \prod_{q | N_{2,\phi}} (-a_\phi(q) q^{-k_2 + 1}) \prod_{q | N_{2,\psi}'} (-a_\psi(q) q^{-k_3 + 1}).
$$

$$
\times \sum_{u,v,n} a_f(n) a_\phi(nv^2 B N_{2,\phi} \cdot \frac{D_2^\phi}{D_1^\phi}) a_\psi(nu^2 B N_{2,\psi}' \frac{D_2'^\psi}{D_1'^\psi})
$$

$$
\times v^{-2s - 2k_2 + 2 + \nu_2} u^{-2s - 2k_3 + 2 + \nu_3} n^{-s - k_2 - k_3 + 2}
$$

with
(2.40)
$$
\mathcal{I}_\infty(s) = (4\pi)^{-s - k_2 - k_3 + 2} \cdot \frac{\Gamma(s + k_2 + k_3 - 2)\Gamma(s + k_2 - 1)\Gamma(s + k_3 - 1)}{\Gamma(2s + k_2 + k_3 - 2)}.
$$

To finish this section, we collect all the information obtained so far; we must take care of the fact that we worked in section (2.2) with the Eisenstein series $\mathcal{K}_f(Z, s)$ rather than with $\mathcal{K}_f(Z, s - a')$ as is required by (2.11). The value of the threefold integral $\mathcal{A}$ as defined by (2.1) is

(2.41)

$$\mathcal{A}(f|_{k_1}\begin{pmatrix} D^f & 0 \\ 0 & 1 \end{pmatrix}, \phi|_{k_2}\begin{pmatrix} D^\phi & 0 \\ 0 & 1 \end{pmatrix}, \psi|_{k_3}\begin{pmatrix} D^\psi & 0 \\ 0 & 1 \end{pmatrix}, s) =$$

$$2\binom{b}{\nu_2} \cdot A(r+s,b) \cdot \mu(r+a+b, s-a') \cdot \mathcal{I}_\infty(s-a') \cdot N^{-2s-r}$$

$$\left(\prod_{p \nmid N} \frac{(1-p^{-2s-r})(1-p^{2-4s-2r})}{1-p^{-2s-r+1}} \right) \left(\prod_{p \nmid N_f} \frac{1}{(1-\alpha_p^2 p^{-2s-r-k_1+2})(1-\alpha_p'^2 p^{-2s-r-k_1+2})} \right)$$

$$\left(\prod_{p | N_f} \frac{-1}{1-p^{-2s-r}} \right) \cdot$$

$$\sum_{d | N^f} \alpha(d, D^f, 2s-r) \cdot d^{\frac{-k_1}{2}} \cdot (d, N_1^f, N_1'^f)^{k_1} (D_1^f)^{\frac{k_2}{2}} \cdot (D_2^\phi)^{-\frac{k_2}{2}} \cdot (D_1'^\psi)^{\frac{k_3}{2}}$$

$$\cdot (D_2'^\psi)^{-\frac{k_3}{2}} \sum_{N_1 N_2 = N, N_1' N_2' = N, N | N_1 N_1'} N_2 \cdot N_2' \cdot B^{-s+a'-k_2-k_3+2}$$

$$\left(\prod_{p | \text{lcm}(N_{2,f}, N_{2,f}')} (-a_f(p) p^{-k_1+1}) \right) \left(\prod_{p | N_{2,\phi}} (-a_\phi(p) p^{-k_2+1}) \right)$$

$$\left(\prod_{p | N_{2,\psi}'} (-a_\psi(p) p^{-k_3+1}) \right)$$

$$\times \sum_{n,u,v} a_f(n) a_\phi(nv^2 B N_{2,\phi} \cdot \frac{D_2^\phi}{D_1^\phi}) a_\psi(nu^2 B N_{2,\psi}' \cdot \frac{D_2'^\psi}{D_1'^\psi})$$

$$\times u^{-2s+a-2k_3+2+\nu_3} v^{-2s+a-2k_2+2+\nu_2} n^{-s+a'-k_2-k_3+2} .$$

We remind the reader that the summation over u and v is subject to the condition $N_2 | u$ and $N_2' | v$.

3. — The Euler factors

Using the multiplicativity properties of the Fourier coefficients of f, ϕ, ψ and that the conditions of summation are of multiplicative type we may now write (2.41) as

(3.1) $\quad 2\binom{b}{\nu_2} \cdot A(r+s,b) \cdot \mu(r+a+b, s-a') \cdot \mathcal{I}_\infty(s-a') \cdot N^{-2s-r} \times \prod_p T_p(s)$

To save notation we write

(3.2) $$T_p = T_p^0 \cdot T_p^1$$

where T_p^1 denotes the last four lines of (2.41). In all cases to be considered, the pair $[(N_1)_p, N_{1p}']$ can take the values $[p, p]$, $[p, 1]$ and $[1, p]$; therefore we split T_p^1 as

(3.3)
$$T_p^1 = C_{pp} + C_{p1} + C_{1p}.$$

If d_p can take both values 1 and p, we further decompose C_{**} as

(3.4)
$$C_{**} = C_{**}^1 + C_{**}^p.$$

according to the cases $d_p = 1$ and $d_p = p$.

Part I : $D_p^f = D_p^\phi = D_p^\psi = 1$

IA : The case $p|N_f$, $p|N_\phi$, $p|N_\psi$

(This is the case also considered by Gross/Kudla[11]).

The conditions imply $d_p = N_p^f = N_p^\phi = N_p^\psi = 1$, $\alpha_p(d, D^f, s) = 1$ and

$$B_p = \begin{cases} 1 & \text{for the case } [p,p] \\ \dfrac{1}{p} & \text{otherwise.} \end{cases}$$

(3.5) $C_{pp}(s) =$
$$\left(\sum_{l=0}^{\infty} \sum_{t=0}^{\infty} a_f(p^l) a_\phi(p^{l+2t}) a_\psi(p^l)(p^l))^{-s+a'-k_2-k_3+2} (p^t))^{-2s+a-2k_2+2+\nu_2} + \right.$$
$$\left. + \sum_{l=0}^{\infty} \sum_{t=1}^{\infty} a_f(p^l) a_\phi(p^l) a_\psi(p^{l+2t})(p^l))^{-s+a'-k_2-k_3+2} (p^t))^{-2s+a-2k_3+2+\nu_3} \right).$$

This equals (we use the fact that $a_\phi(p^2)$ and $a_\psi(p^2)$ are powers of p)

$$\frac{1}{1 - a_f(p)a_\phi(p)a_\psi(p)p^{-s+a'-k_2-k_3+2}} \cdot$$
$$\left\{ \frac{1}{1-a_\phi(p^2)p^{-2s+a-2k_2+2+\nu_2}} + \frac{a_\psi(p^2)p^{-2s+a-2k_3+2+\nu_3}}{1-a_\psi(p)p^{-2s+a-k_3+2+\nu_3}} \right\}$$
$$= \frac{1}{1 - a_f(p)a_\phi(p)a_\psi(p)p^{-s+a'-k_2-k_3+2}} \cdot \left\{ \frac{1}{p^{-2s+a-k_2+\nu_2}} + \frac{p^{-2s+a-k_3+\nu_3}}{1-p^{-2s+a-k_3+\nu_3}} \right\}$$
$$= \frac{1}{1- a_f(p)a_\phi(p)a_\psi(p)p^{-s-3a'-2r-b+2}} \cdot \frac{1+p^{-2s-r}}{1-p^{-2s-r}} \cdot$$

$$(3.6) \quad C_{1,p} = p \cdot p^{s-a'+k_2+k_3-2} \cdot (-a_f(p)p^{-k_1+1}) \cdot (-a_\phi p^{-k_2+1}) \times$$

$$\times \sum_{l=0}^{\infty} \sum_{t=1}^{\infty} a_f(p^l) a_\phi\left(p^l \cdot \frac{1}{p} \cdot p\right) a_\psi\left(p^{l+2t} \cdot \frac{1}{p}\right)(p^t)^{-2s+a-2k_3+2+\nu_3} \cdot (p^l)^{-s+a'-k_2-k_3+2}$$

$$= \frac{1}{1 - a_f(p)a_\phi(p)a_\psi(p)p^{-s+a'-k_2-k_3+2}} \cdot \frac{a_f(p)a_\phi(p)a_\psi(p)p^{3-s-3a'-b-2r}}{1 - a_\psi(p^2)p^{-2s+a-2k_3+2+\nu_3}}$$

$$= \frac{1}{1 - a_f(p)a_\phi(p)a_\psi(p)p^{-s-3a'-2r-b+2}} \cdot \frac{a_f(p)a_\phi(p)a_\psi(p)p^{3-s-3a'-b-2r}}{1 - p^{-2s-r}}.$$

Quite the same computation shows that

$$(3.7) \qquad\qquad\qquad C_{p,1} = C_{1,p}.$$

Hence

$$(3.8) \quad T_p^1 = \frac{1}{1 - a_f(p)a_\phi(p)a_\psi(p)p^{-s-3a'-2r-b+2}} \cdot$$

$$\frac{1 + 2a_f(p)a_\phi(p)a_\psi(p)p^{-s-3a'-b-2r+3} + p^{-2s-r}}{1 - p^{-2s-r}}.$$

Now we write the numerator as

$$(3.9) \quad 1 + 2a_f(p)a_\phi(p)a_\psi(p)p^{-s-3a'-b-2r+3} + p^{-2s-r}$$

$$= \left(1 + a_f(p)a_\phi(p)a_\psi(p)p^{-s+3-3a'-b-2r}\right)^2$$

$$= \frac{(1 - p^{-2s-r})^2}{(1 - a_f(p)a_\phi(p)a_\psi(p)p^{-s+3-3a'-b-2r})^2}.$$

Therefore we obtain

$$(3.10) \quad T_p(s) =$$

$$= \frac{-1}{1 - a_f(p)a_\phi(p)a_\psi(p)p^{-s-3a'-2r-b+2}} \cdot \frac{1}{(1 - a_f(p)a_\phi(p)a_\psi(p)p^{-s+3-3a'-b-2r})^2}.$$

IB : The case $p|N^f$, $p|N_\phi$, $p|N_\psi$ These conditions imply

$$(N_f)_p = N_p^\phi = N_p^\psi = D_p^\phi = D_p^\psi = 1 \qquad \text{and} \qquad d_p \in \{1, p\}$$

$$\alpha_p(d, D^f, s) = \begin{cases} -(1+p^{1-s}) & \text{if } d_p = 1 \\ p^{1-\frac{k_1}{2}} \cdot a_f(p) & \text{if } d_p = p \end{cases}$$

and

$$B_p = \begin{cases} p & \text{if } d_p = (N_1)_p = N'_{1p} = p \\ \dfrac{1}{d_p} & \text{otherwise}. \end{cases}$$

What we get is this :

(3.11) $C^1_{pp}(s) =$

$$= \frac{-(1+p^{1-2s-r})(1+p^{-2s-r})}{(1-\alpha_p a_\phi(p)a_\psi(p)p^{-s-3a'-2r-b+2})(1-\alpha_p a_\phi(p)a_\psi(p)p^{-s-3a'-2r-b+2})(1-p^{-2s-r})}$$

(similar calculation as in (3.5));

(3.12) $C^1_{1p}(s) =$

$$= -(1+p^{1-2s-r}) \cdot p \cdot (-a_\phi(p)p^{-k_2+1}) \times$$

$$\sum_{l=0}^{\infty}\sum_{t=1}^{\infty} a_f(p^l)a_\phi(p^l \cdot p)a_\psi(p^{l+2t})(p^t)^{-2s+a-2k_3+2+\nu_3} \cdot (p^l)^{-s+a'-k_2-k_3+2}$$

$$= \frac{(1+p^{1-2s-r}) \cdot p^{-2s-r}}{(1-\alpha_p a_\phi(p)a_\psi(p)p^{-s-3a'-2r-b+2})(1-\alpha_p a_\phi(p)a_\psi(p)p^{-s-3a'-2r-b+2})(1-p^{-2s-r})} .$$

(3.13) $C^1_{p1} = C^1_{1p} .$

(3.14) $C^p_{pp}(s) =$

$$= p^{1-\frac{k_1}{2}} a_f(p)p^{-\frac{k_1}{2}}p^{k_1}p^{-s+a'-k_2-k_3+2} \times$$

$$\left\{ \sum_{l=0}^{\infty}\sum_{t=0}^{\infty} a_f(p^l)a_\phi(p^{l+2t}p)a_\psi(p^l p)(p^t)^{-2s+a-2k_2+2+\nu_2}(p^l)^{-s+a'-k_2-k_3+2} + \right.$$

$$\left. \sum_{l=0}^{\infty}\sum_{t=1}^{\infty} a_f(p^l)a_\phi(p^l p)a_\psi(p^{l+2t}p)(p^t)^{-2s+a-2k_3+2+\nu_3}(p^l)^{-s+a'-k_2-k_3+2} \right\}$$

$$= \frac{a_f(p)a_\phi(p)a_\psi(p)p^{-s-3a'-2r-b+3}(1+p^{-2s-r})}{(1-\alpha_p a_\phi(p)a_\psi(p)p^{-s-3a'-2r-b+2})(1-\alpha'_p a_\phi(p)a_\psi(p)p^{-s-3a'-2r-b+2})(1-p^{-2s-r})} .$$

(3.15) $C_{1,p}^p$

$$= p^{1-\frac{k_1}{2}}a_f(p)p^{-\frac{k_1}{2}} \cdot p \cdot p^{s-a'+k_1+k_2-2}(-a_\phi(p)p^{-k_2+1}) \times$$

$$\sum_{l=0}^\infty \sum_{t=1}^\infty a_f(p^l)a_\phi(p^l\frac{1}{p} \cdot p)a_\psi(p^{l+2t}\frac{1}{p})(p^t)^{-2s+a-k_3+2+\nu_3}(p^l)^{-s+a'-k_2-k_3+2}$$

$$= \frac{-a_f(p)a_\phi(p)a_\psi(p)p^{-s-2r-3a'-b+3}}{(1-\alpha_p a_\phi(p)a_\psi(p)p^{-s-3a'-2r-b+2})(1-\alpha'_p a_\phi(p)a_\psi(p)p^{-s-3a'-2r-b+2})(1-p^{-2s-r})} \cdot$$

(3.16) $C_{p1}^p = C_{1p}^p \, .$

All the C_{**}^* have the same denominator, the numerator of $\sum C_{**}^*$ being equal to

$$1 - p^{-2s-r} + p^{1-2s-r} - a_f a_\phi a_\psi \left\{ p^{-3s-3r-a-b+3} - p^{-s-a-2r-b+3} \right\} \, .$$

This can be factorized as

$$-(1-p^{-2s-r})(1+\alpha_p a_\phi(p)a_\psi(p)p^{-s-2r-3a'-b+3})$$
$$(1+\alpha'_p a_\phi(p)a_\psi(p)p^{-s-2r-3a'-b+3}) \, .$$

Now we also write the denominator of T_p^0 as product of linear factors in p^{-s} using

$$(1-\alpha_p^2 p^{-k_1+2-2s-r}) = (1-\alpha_p a_\phi a_\psi p^{-s-3a'-b-2r+3})(1+\alpha_p a_\phi a_\psi p^{-s-3a'-b-2r+3})$$

(and the same with α'_p instead of α_p); therefore we get some cancellations and arrive at :

(3.17)

$$T_p = \frac{-1}{(1-\alpha_p a_\phi(p)a_\psi(p)p^{-s-3a'-b-2r+2})(1-\alpha'_p a_\phi(p)a_\psi(p)p^{-s-3a'-b-2r+2})}$$

$$\times \frac{1}{(1-\alpha_p a_\phi(p)a_\psi(p)p^{-s-3a'-b-2r+3})(1-\alpha'_p a_\phi(p)a_\psi(p)p^{-s-3a'-b-2r+3})} \cdot$$

IC : The case $p|N_f$, $p|N^\phi$, $p|N^\psi$

We have

$$(B)_p = \begin{cases} 1 & \text{if} \quad N_1 = N_1' = p \\ \dfrac{1}{p} & \text{otherwise,} \end{cases}$$

and we get

(3.18) $C_{1p} =$

$$= p \cdot p^{s-a'+k_2+k_3-2} \cdot (-a_f(p)p^{-k_1+1}) \times$$

$$\sum_{l=1}^{\infty}\sum_{t=1}^{\infty} a_f(p^l)a_\phi(p^l\tfrac{1}{p})a_\psi(p^{l+2t}\tfrac{1}{p})(p^t)^{-2s+a-2k_3+2+\nu_3}(p^l)^{-s+a'-k_2-k_3+2}$$

$$= -\sum_{l'=0}^{\infty}\sum_{t=1}^{\infty} a_f(p^{l'})a_\phi(p^{l'})a_\psi(p^{l'+2t})(p^t)^{-2s+a-2k_3+2+\nu_3}(p^{l'})^{-s+a'-k_2-k_3+2}.$$

(3.19)

$$C_{p1} = -\sum_{l=0}^{\infty}\sum_{t=1}^{\infty} a_f(p^l)a_\phi(p^{l+2t})a_\psi(p^l)(p^t)^{-2s+a-2k_2+2+\nu_2}(p^l)^{-s+a'-k_2-k_3+2}.$$

3.20) $C_{pp}(s) =$

$$\left(\sum_{l=0}^{\infty}\sum_{t=0}^{\infty} a_f(p^l)a_\phi(p^{l+2t})a_\psi(p^l)\,(p^l)\right)^{-s+a'-k_2-k_3+2}(p^t))^{-2s+a-2k_2+2+\nu_2} +$$

$$+\sum_{l=0}^{\infty}\sum_{t=1}^{\infty} a_f(p^l)a_\phi(p^l)a_\psi(p^{l+2t})\,(p^l))^{-s+a'-k_2-k_3+2}(p^t))^{-2s+a-2k_3+2+\nu_3}\Bigg).$$

Summing up the C_{**} we see that the summands in (3.20) with $t \neq 0$ cancel against (3.18) and (3.19) and we get

$$T_p(s) = -(1-p^{-2s-r})^{-1}\sum_{l=0}^{\infty} a_f(p^l)a_\phi(p^l)a_\psi(p^l)(p^l)^{-s-3a'-b-2r+2}$$

$$= \frac{-1}{(1-a_f(p)\beta_p\gamma_p p^{-s-3a'-b-2r+2})(1-a_f(p)\beta_p\gamma'_p p^{-s-3a'-b-2r+2})} \times$$

$$\times \frac{1}{(1-a_f(p)\beta'_p\gamma_p p^{-s-3a'-b-2r+2})(1-a_f(p)\beta'_p\gamma'_p p^{-s-3a'-b-2r+2})}.$$

I D : The case $p \nmid N$

This case (which is in some sense the most difficult one) was previously considered in [9] for the case of equal weights. It was already noticed in [21] and [22] that the result from [9] carries over to the case of arbitrary weights. We just state the result here :

$$(3.22) \quad T_p = (1 - p^{-2s-r})(1 - p^{2-4s-2r})L_p(f \otimes \phi \otimes \psi, s + 3a' + 2r + b - 2).$$

Part II : $D_p^f = N_p^f$, $D_p^\phi = N_p^\phi$, $D_p^\psi = N_p^\psi$

Only the cases B and C are to be investigated, the case B being the most complicated :

II B : The case $p|N^f$, $p|N_\phi$, $p|N_\psi$

These conditions imply

$$\alpha_p(d, D^f, s) = \begin{cases} -a_f(p)p^{2-\frac{k_1}{2}-s} & \text{if} \quad d_p = 1 \\ p(1 + p^{1-s}) & \text{if} \quad d_p = p \end{cases}$$

$$B_p = \begin{cases} p & \text{if} \quad d_p = (N_1)_p = (N_1')_p = p \\ 1 & \\ \dfrac{1}{d} & \text{otherwise.} \end{cases}$$

$$(3.23) \quad C_{pp}^1 = \alpha_p(1, D^f, 2s + r)\times$$

$$\left\{ \sum_{l=0}^{\infty}\sum_{t=0}^{\infty} a_f(p^l)a_\phi(p^{l+2t})a_\psi(p^l)(p^t)^{-2s+a-2k_2+2+\nu_2}(p^l)^{-s+a'-k_2-k_3+2} + \right.$$

$$\left. \sum_{l=0}^{\infty}\sum_{t=1}^{\infty} a_f(p^l)a_\phi(p^l)a_\psi(p^{l+2t})(p^t)^{-2s+a-2k_3+2+\nu_3}(p^l)^{-s+a'-k_2-k_3+2} \right\}$$

$$= \frac{\alpha_p(1, D^f, s)(1 + p^{-2s-r})}{(1 - \alpha_p a_\phi(p)a_\psi(p)p^{-s-3a'-2r-b+2})(1 - \alpha_p' a_\phi(p)a_\psi(p)p^{-s-3a'-2r-b+2})(1 - p^{-2s-r})}$$

(computation similar to (3.5)).

$$(3.24) \quad C_{1p}^1 = \alpha_p(1, D^f, s) \cdot p \cdot (-a_\phi(p)p^{-k_2+1})\times$$

$$\sum_{l=0}^{\infty}\sum_{t=1}^{\infty} a_f(p^l)a_\phi(p^l p)a_\psi(p^{l+2t})(p^t)^{-2s+a-2k_3+2+\nu_3}(p^l)^{-s+a'-k_2-k_3+2}$$

$$= \frac{-p^{-2s-r}\alpha_p(1, D^f, 2s + r)}{(1 - \alpha_p a_\phi(p)a_\psi(p)p^{-s-3a'-2r-b+2})(1 - \alpha_p' a_\phi(p)a_\psi(p)p^{-s-3a'-2r-b+2})(1 - p^{-2s-r})}.$$

$$(3.25) \quad\quad\quad\quad C_{p,1}^1 = C_{1,p}^1,$$

and

(3.26)

$$\sum C^1_{**} = \frac{-a_f(p)p^{2-\frac{k_1}{2}-s}}{(1-\alpha_p a_\phi(p)a_\psi(p)p^{-s-3a'-2r-b+2})(1-\alpha'_p a_\phi(p)a_\psi(p)p^{-s-3a'-2r-b+2})}.$$

(3.27) $C^p_{pp}(s) =$

$$= \alpha_p(p, D^f, s+2r) \cdot p^{-\frac{k_1}{2}} p^{k_1} p^{-s+a'-k_2-k_3+2} \times$$

$$\sum_{l=0}^{\infty}\sum_{t=0}^{\infty} a_f(p^l)a_\phi(p^{l+2t}p)a_\psi(p^l p)(p^t)^{-2s+a-2k_2+2+\nu_2}(p^l)^{-s+a'-k_2-k_3+2} +$$

$$\sum_{l=0}^{\infty}\sum_{t=1}^{\infty} a_f(p^l)a_\phi(p^l p)a_\psi(p^{l+2t}p)(p^t)^{-2s+a-2k_3+2+\nu_3}(p^l)^{-s+a'-k_2-k_3+2}$$

$$= \frac{\alpha_p(p, D^f, 2s+r)a_\phi(p)a_\psi(p)p^{\frac{k_1}{2}-s-3a'-2r-b+2}(1+p^{-2s-r})}{(1-\alpha_p a_\phi(p)a_\psi(p)p^{-s-3a'-2r-b+2})(1-\alpha_p a_\phi(p)a_\psi(p)p^{-s-3a'-2r-b+2})(1-p^{-2s-r})}$$

(similar computation as in (3.5)).

(3.28) $C^p_{1p}(s) = \alpha_p(p, D^f, 2s+r) \cdot p^{-\frac{k_1}{2}} \cdot p \cdot p^{s-a'+k_2+k_3-2}(-a_\phi p^{-k_2+1}) \times$

$$\sum_{l=0}^{\infty}\sum_{t=1}^{\infty} a_f(p^l)a_\phi(p^l \frac{1}{p}p)a-\psi(p^{l+2t}\frac{1}{p})(p^t)^{-2s+a-2k_3+2+\nu_3}(p^l)^{-s+a'-k_2-k_3+2}$$

$$= \frac{-\alpha_p(p, D^f, 2s+r)a_\phi(p)a_\psi(p)p^{-\frac{k_1}{2}-s-a'-r+2}}{(1-\alpha_p a_\phi(p)a_\psi(p)p^{-s-3a'-2r-b+2})(1-\alpha'_p a_\phi(p)a_\psi(p)p^{-s-3a'-2r-b+2})(1-p^{-2s-r})}.$$

(3.29) $C^p_{p1} = C^p_{1p}$,

and hence :

(3.30)

$$\sum C^p_{**} = \frac{-a_\phi a_\psi(1+p^{1-2s-r})p^{-\frac{k_1}{2}-s-a'-r+3}}{(1-\alpha_p a_\phi(p)a_\psi(p)p^{-s-3a'-2r-b+2})(1-\alpha'_p a_\phi(p)a_\psi(p)p^{-s-3a'-2r-b+2})}.$$

The numerator of $\sum C^*_{**}$ is equal to

(3.31)
$$-\frac{1}{a_\phi a_\psi}p^{-\frac{k_1}{2}+r+3a'+b-1-s}(1+a_f(p)a_\phi(p)a_\psi p^{-s-2r+3-b-3a'}+p^{1-2s-r}).$$

We can therefore apply the same kind of trick as in case I B; denoting by $\epsilon_p(\phi)$ the eigenvalue of the Atkin-Lehner-involution V_p^N acting on ϕ (and similarly for ψ) we end up with

$$T_p(s) =$$
$$\frac{-\epsilon_p(\phi)\epsilon_p(\psi)p^{1-\frac{r}{2}-s}}{(1-\alpha_p a_\phi(p)a_\psi(p)p^{-s-3a'-b-2r+2})(1-\alpha'_p a_\phi(p)a_\psi(p)p^{-s-3a'-b-2r+2})}$$

$$\times \frac{1}{(1-\alpha_p a_\phi(p)a_\psi(p)p^{-s-3a'-b-2r+3})(1-\alpha'_p a_\phi(p)a_\psi(p)p^{-s-3a'-b-2r+3})}.$$

II C : The case $p|N_f$, $p|N^\phi$, $p|N^\psi$

(3.33) $C_{pp}(s) = p^{\frac{k_2}{2}}\cdot p^{\frac{k_3}{2}}\times$

$$\left\{\sum_{l=1}^{\infty}\sum_{t=0}^{\infty}a_f(p^l)a_\phi(p^{l+2t}\frac{1}{p})a_\psi(p^l\frac{1}{p})(p^t)^{-2s+a-2k_2+2+\nu_2}(p^l)^{-s+a'-k_2-k_3+2}+\right.$$

$$\left.+\sum_{l=1}^{\infty}\sum_{t=1}^{\infty}a_f(p^l)a_\phi(p^l\frac{1}{p})a_\psi(p^{l+2t}\frac{1}{p})(p^t)^{-2s+a-2k_3+2+\nu_3}(p^l)^{-s+a'-k_2-k_3+2}\right\}$$

$$= p^{-s+a'+2-\frac{k_2+k_3}{2}}\times$$

$$\left\{\sum_{l=0}^{\infty}\sum_{t=0}^{\infty}a_f(p^{l+1})a_\phi(p^{l+2t})a_\psi(p^l)(p^t)^{-2s+a-2k_2+2+\nu_2}(p^l)^{-s+a'-k_2-k_3+2}+\right.$$

$$\left.\sum_{l=0}^{\infty}\sum_{t=1}^{\infty}a_f(p^{l+1})a_\phi(p^l)a_\psi(p^{l+2t})(p^t)^{-2s+a-2k_3+2+\nu_3}(p^l)^{-s+a'-k_2-k_3+2}\right\}.$$

(3.34) $C_{1p} = p^{-\frac{k_2}{2}+\frac{k_3}{2}}\cdot p\cdot p^{s-a'+k_2+k_3-2}\cdot(-a_f(p)p^{-k_1+1})\times$

$$\sum_{l=0}^{\infty}\sum_{t=1}^{\infty}a_f(p^l)a_\phi(p^l\frac{1}{p}p)a_\psi(p^{l+2t}\frac{1}{p}\frac{1}{p})(p^t)^{-2s+a-2k_3+2+\nu_3}(p^l)^{-s+a'-k_2-k_3+2}$$

$$= -p^{-s+a'+2-\frac{k_2+k_3}{2}}\times$$

$$\sum_{l=0}^{\infty}\sum_{t=0}^{\infty}a_f(p^{l+1})a_\phi(p^l)a_\psi(p^{l+2t})(p^t)^{-2s+a-2k_3+2+\nu_3}(p^l)^{-s+a'-k_2-k_3+2}.$$

(3.35) $C_{p1} = -p^{-s+a'+2-\frac{k_2+k_3}{2}} \times$

$$\sum_{l=0}^{\infty} \sum_{t=0}^{\infty} a_f(p^{l+1}) a_\phi(p^{l+2t}) a_\psi(p^l)(p^t)^{-2s+a-2k_2+2+\nu_3} (p^l)^{-s+a'-k_2-k_2+2}.$$

Hence in the sum of the C_{**} only the "$t = 0$-part" of C_{p1} survives and we obtain

$$T_p = \frac{-\epsilon_p(f)p^{1-s-\frac{r}{2}}}{(1 - a_f(p)\beta_p\gamma_p p^{-s-3a'-b-2r+2})(1 - a_f(p)\beta_p\gamma'_p p^{-s-3a'-b-2r+2})}$$

$$\times \frac{1}{(1 - a_f(p)\beta'_p\gamma_p p^{-s-3a'-b-2r+2})(1 - a_f(p)\beta'_p\gamma'_p p^{-s-3a'-b-2r+2})}.$$

Remark : Although our list of Euler factors is complete, the reader should be aware of the fact that in our integral representation (2.1) we are free to interchange the roles of ϕ and ψ (interchanging the roles of ν_2 and ν_3 at the same time), but f has to be the cusp form of largest weight. Therefore e.g. in case IB, IIB we should also consider the case where p does not divide the level of ϕ or ψ. It will be left to the reader to show by similar computations as above that in those cases the Euler factor will be the same (as should be more or less clear from an adelic point of view).

4. — The Functional equation

The factors $T_{p.} = T_p(s)$ computed in the previous section are for $p \nmid N$ and for $p \mid \gcd(N_f, N_\phi, N_\psi)$ up to a shift in the argument and an elementary factor the known Euler factors of the triple product L-function $L(f, \phi, \psi, s)$ associated to f, ϕ, ψ. We define therefore now :

DEFINITION 4.1. — *The triple product L-function associated to* f, ϕ, ψ *is for* $s > \frac{k_1+k_2+k_3-1}{2}$ *defined as* $L(f, \phi, \psi, s) = \prod_p L_p(f, \phi, \psi, s)$ *where the Euler factor* $L_p(f, \phi, \psi, s)$ *is given by*

(4.1) $L_p(f, \phi, \psi, s + 3a' + 2r + b - s) =$
$$\begin{cases} -T_p(s) & \text{if } p \mid N \\ (1 - p^{-2s-r})^{-1}(1 - p^{2-4s-2r})^{-1}T_p(s) & \text{if } p \nmid N \end{cases}$$

With

(4.2) $L_\infty(f, \phi, \psi, s) = \Gamma_{\mathbb{C}}(s)\Gamma_{\mathbb{C}}(s + 1 - k_1)\Gamma_{\mathbb{C}}(s + 1 - k_2)\Gamma_{\mathbb{C}}(s + 1 - k_3)$

(where $\Gamma_{\mathbb{C}}(s) = (2\pi)^{-s}\Gamma(s)$ as usual) the completed triple product L-function is

$$\Lambda(f, \phi, \psi, s) = L_{\infty}(f, \phi, \psi, s)L(f, \phi, \psi, s).$$

With these notations we have for the integral $\mathcal{A}(f, \phi, \psi, s)$ from 2.1 (with $c_2(s)$ as in (1.19), $\zeta_p(s) = (1 - p^{-s})^{-1}$ for finite primes p and $\zeta_{\infty}(s) = \pi^{-s/2}\Gamma(s/2)$) :

THEOREM 4.2. —

(4.3) $\dfrac{\mathcal{A}(f, \phi, \psi, s)}{c_2(s)}$

$$= \binom{b}{\nu_2} i^{b+k_1} N^{-2s-2} 2^{3-2b-a} \frac{\pi^{2+3a'}}{(s+1)(2s+1)} \frac{L_{\infty}(f, \phi, \psi, s + \frac{k_1+k_2+k_3}{2} - 1)}{zeta_{\infty}(4s+2)\zeta_{\infty}(2s+2)}$$

$$\times \prod_{p\nmid N} \frac{L_p(s + \frac{k_1+k_2+k_3}{2} - 1)}{\zeta_p(4s+2)\zeta_p(2s+2)} \prod_{p|N}(-L_p(f, \phi, \psi, s + \frac{k_1+k_2+k_3}{2} - 1))$$

Proof. This follows from 2.41 and the results of Section 3.

THEOREM 4.3. — The function $\Lambda(f, \phi, \psi, s)$ has a meromorphic continuation to all of $\mathbb{C}$ and satisfies the functional equation
(4.4)
$$\Lambda(f, \phi, \psi, s) = -N^{-4(s+\frac{k_1+k_2+k_3}{2}-1)}(\gcd(N_f, N_\phi, N_\psi))^{-(s+\frac{k_1+k_2+k_3}{2}-1)}$$

$$(\textstyle\prod_{p|N} \epsilon_p)\Lambda(k_1 + k_2 + k_3 - 2 - s)$$

where for $p \mid N$ the number ϵ_p is defined as the product of the eigenvalues under the p-Atkin-Lehner-involution w_p of those forms among f, ϕ, ψ whose level is divisible by p.

Proof. We put $r = 2$ since for this choice the functional equation of the Eisenstein series is under $s \mapsto -s$. It is then not difficult to read off the functional equation of $\mathbb{E}$ from the calculations in [11]; actually things become somewhat simplified since we need the functional equation only up to oldforms. We need the following Lemma :

LEMMA 4.4. — The Eisenstein series $\mathbb{E}_{2,s}$ satisfies the functional equation :

$$(s + 1)(2s + 1)\zeta_{\infty}(4s + 2)\zeta_{\infty}(2s + 2) \prod_{p\nmid N} \zeta_p(4s + 2)\zeta_p(2s + 2)\mathbb{E}_{2,s}(Z)$$

$$= -N^{-9s}(s - 1)(2s - 1)\zeta_{\infty}(-4s + 2)\zeta_{\infty}(-2s + 2)$$

$$\times \prod_{p\nmid N} \zeta_p(-4s + 2)\zeta_p(-2s + 2)\mathbb{E}_{2,-s}\begin{pmatrix} 0_3 & -1_3 \\ N1_3 & 0_3 \end{pmatrix}(Z) + \widetilde{\mathbb{E}_s}$$

where $\widetilde{\mathbb{E}_s}$ is a linear combination of Eisenstein series for groups $\Gamma_0^3(N')$ with N' strictly dividing N or conjugates of these by a matrix $\begin{pmatrix} 0_3 & -1_3 \\ M1_3 & 0_3 \end{pmatrix}$) with $M \mid (N/N')$.

Proof. Let $\mathbb{F}_{r,s}(Z)$ be the Eisenstein series of degree 3 defined in the same way as $\mathbb{E}_{r,s}$ in Section 2, but with the summation running over coprime symmetric pairs (C, D) with $\gcd(\det C, N) = 1$; one has

$$(4.6) \qquad \mathbb{E}_{r,s}|_r \begin{pmatrix} 0 & -1 \\ N & 0 \end{pmatrix} = N^{-(3r/2)-3s} \mathbb{F}_{r,s}.$$

The calculation of the local intertwining operators $M(s) = M_p(s)$ in sections 5 and 6 of [11] gives (in the notations of that article) $M_p(s)\Phi_p(s)$ for $p = \infty$ or $p \nmid N$ and allows for $p \nmid N$ to express $M_p(s)\Phi_p^0(s)$ explicitly as a linear combination of the sections $\Phi_p^0(-s), \Phi_p^3(-s), (\Phi_p)_K(-s), (\Phi_p)_{K'}(-s)$. An elementary calculation shows in this case that the coefficient at $\Phi_p^0(-s)$ is zero and that the coefficient at $\Phi_p^3(-s)$ equals

$$p^{-6s-3} \frac{\zeta_p(2s-1)\zeta_p(4s-1)}{\zeta_p(-4s+2)\zeta_p(-2s+2)}.$$

We let $Z = X + iY$ and $g = (g_\infty, 1, \dots) \in Sp_3(\mathbb{A})$ with $g_\infty = \begin{pmatrix} 1_3 & X \\ 0_3 & 1_3 \end{pmatrix} \begin{pmatrix} Y^{1/2} & 0_3 \\ 0_3 & Y^{-1/2} \end{pmatrix}$. Then it is well known that with

$$\Phi^0 = \Phi_\infty^{-2} \times \prod_{p|N} \Phi_p^0 \times \prod_{p \nmid N} (\Phi_p)_K$$

and analogously defined Φ^3 one has $\mathbb{E}_{2,s}(\bar{Z}) = (\det Y)^{-1} E(g, s, \Phi^0)$ and $\mathbb{F}_{2,s}(\bar{Z}) = (\det Y)^{-1} E(g, s, \Phi^3)$ (see Proposition 7.5 of [11]), and analogous formulae are true for all the global sections Φ for which the p-adic components Φ_p are one of the $\Phi_p^0, \Phi_p^3, (\Phi_p)_K, (\Phi_p)_{K'}$. The assertion of the lemma follows upon using the duplication formula for the gamma-function for the contribution from the infinite place and then applying the functional equation of the Riemann zeta function.

We can now finish the proof of Theorem 4.3. From Theorem 4.2, Lemma 4.4 and the fact that the integrand in $\frac{A(f,\phi,\psi,s)}{c_2(s)}$ contains by 1.18 a differential operator independent of s we see (using the results of Part II in Section 3) that

$$(4.7) \qquad \begin{aligned} \Lambda(s + \tfrac{k_1+k_2+k_3}{2} - 1) &= -N^{-4s}(\gcd(N_f, N_\phi, N_\psi))^{-s} \\ &\quad (\textstyle\prod_{p|N} \epsilon_p)\Lambda(-s + \tfrac{k_1+k_2+k_3}{2} - 1) \end{aligned}$$

where for $p \mid N$ the number ϵ_p is defined as in the assertion of Theorem 4.3. Putting $s' = s + \frac{k_1 + k_2 + k_3}{2} - 1$ we obtain the desired functional equation under $s' \mapsto k_1 + k_2 + k_3 - 2 - s'$.

5. — Computation of the central critical value

Following the strategy of [12] we now evaluate the integral 2.2 at the point $s = 0$ using a variant of Siegel's theorem, i. e. , expressing the value at $s = 0$ of the Eisenstein series G as a linear combination of theta series. The setting for this is basically the same as in [2, 4, 6]. Let M_1, M_2, M_3 be relatively prime square free integers such that M_1 has an odd number of prime divisors. By $D = D(M_1)$ we denote the quaternion algebra over $\mathbb{Q}$ ramified at ∞ and the primes dividing M_1 and by $R = R(M_1, M_2)$ an Eichler order of level $M = M_1 M_2$ in D, i. e. , the completion R_p is a maximal order for $p \nmid M_2$ and conjugate to $\{\begin{pmatrix} a & b \\ c & d \end{pmatrix} \in M_2(\mathbb{Z}_p) \mid c \equiv 0 \bmod p\}$ for $p \mid M_2$, where we identify $D \otimes \mathbb{Q}_p$ with $M_2(\mathbb{Q}_p)$ for $p \nmid M_1$. By $\mathrm{gen}(M_1, M_2, M_3)$ we denote the genus of $\mathbb{Z}$-lattices with quadratic form (quadratic lattices) of $R(M_1, M_2)$ equipped with the norm form of the quaternion algebra scaled by M_3. The genus theta series of degree n of $\mathrm{gen}(M_1, M_2, M_3)$ is then

$$\Theta^{\mathrm{gen},(n)}_{M_1,M_2,M_3}(Z) = \sum_{\{K\} \in \mathrm{gen}(M_1,M_2,M_3)} \frac{\Theta^{(n)}(K,Z)}{|O(K)|}$$

where the summation is over a set of representatives of the classes in $\mathrm{gen}(M_1, M_2, M_3)$, $O(K)$ is the (finite) group of orthogonal units of the quadratic lattice K, Z is a variable in the Siegel upper half space $\mathbb{H}_n$ and

$$\Theta^{(n)}(K,Z) = \sum_{\boldsymbol{x}=(x_1,\ldots,x_n) \in K^n} \exp(2\pi i \, \mathrm{tr}(q(\boldsymbol{x})Z))$$

with $q(\boldsymbol{x}) = (\frac{1}{2}(M_3 \, \mathrm{tr}(x_i \bar{x}_j)))_{i,j}$.

We consider a double coset decomposition

$$D^{\times}_{\mathbb{A}} = \bigcup_{i=1}^{h=h(M_1,M_2)} D^{\times}_{\mathbb{Q}} y_i R^{\times}_{\mathbb{A}}$$

of the adelic multiplicative group of D with $R^{\times}_{\mathbb{A}} = D^{\times}_{\infty} \times \prod_{p \neq \infty} R^{\times}_p$ and representatives y_i with $n(y_i) = 1$ and $(y_i)_{\infty} = 1$. Then the lattices $I_{ij} = y_i R y_j^{-1}$ (with the norm form scaled by M_3) exhaust $\mathrm{gen}(M_1, M_2, M_3)$ (with some classes possibly occurring more than once) and it is easily seen that with $R_i = I_{ii}$ and $e_i = |R_i^\times|$ we have

(5.1)
$$\sum_{i,j=1}^{h} \frac{\Theta^{(n)}(I_{ij}, Z)}{e_i e_j} = 2^\omega \Theta_{M_1,M_2,M_3}^{\text{gen},(n)}(Z)$$

where $\omega = \omega(M_1, M_2)$ is the number of prime divisors of $M_1 M_2$. With these notations we have from [2] (Theorem 3.2 and Corollary 3.2) and [6], (p.229) :

LEMMA 5.1. — *The value at $s = 0$ of $G_{2,s}(Z)$ is*

$$\sum_{M_1 M_2 M_3 | N} \alpha_{M_1,M_2,M_3} \Theta_{M_1,M_2,M_3}^{\text{gen},(3)}(Z)$$

with

$$\alpha_{M_1,M_2,M_3} = (-1)^{1+\omega(M_1,M_2)}(M_1 M_2)^{-3} M_3^{-6} 8\pi^4 \zeta^{(N)}(2)^{-2}.$$

In order to compute the value at $s = 0$ of the differentiated Eisenstein series from Section 2 we have to compute

(5.2)
$$\left(\mathcal{D}_2^{*(a,\nu_2,\nu_3)} \Theta^{(3)}(K, -)\right)(\iota_{111}(z_1, z_2, z_3))$$

for the individual theta series appearing in the sum in Lemma 5.1. We denote by U_μ the space of homogenous harmonic polynomials of degree μ in 4 variables and identify an element of U_μ with a polynomial on D_∞ by evaluating it at the component vector of an element of D_∞ with respect to an orthonormal basis relative to the quaternion norm on D. Similarly, for $\nu \in \mathbb{N}$ let $U_\nu^{(0)}$ be the space of homogeneous harmonic polynomials of degree ν on $\mathbb{R}^3$ and view $P \in U_\nu^{(0)}$ as a polynomial on $D_\infty^{(0)} = \{x \in D_\infty | \text{tr}(x) = 0\}$. The representations τ_ν of $D_\infty^\times / \mathbb{R}^\times$ of highest weight (ν) on $U_\nu^{(0)}$ given by $(\tau_\nu(y))(P)(x) = P(y^{-1}xy)$ for $\nu \in \mathbb{N}$ give all the isomorphism classes of irreducible rational representations of $D_\infty^\times / \mathbb{R}^\times$. By $\langle\langle \ , \ \rangle\rangle_\nu^{(0)}$ we denote the invariant scalar product in the representation space $U_\nu^{(0)}$, by $\langle\langle \ , \ \rangle\rangle_\mu$ the invariant scalar product in the $SO(D_\infty, \text{norm}) =: H_{\mathbb{R}}^+$-space U_μ. We notice that the invariant scalar products $\langle\langle , \rangle\rangle_{\mu_i}$ on the U_{μ_i} can be normalized in such a way that they take rational values on the subspaces of polynomials with rational coefficients and that these subspaces generate the U_{μ_i}. Indeed, consider the Gegenbauer polynomial $C^{(\mu_i)}(x, x') = $ obtained from

$$C_1^{(\mu_i)}(t) = 2^{\mu_i} \sum_{j=0}^{[\frac{\mu_i}{2}]} (-1)^j \frac{1}{j!(\mu_i - 2j)!} \frac{(\mu_i - j)!}{2^{2j}} t^{\mu_i - 2j}$$

by

$$\tilde{C}^{(\mu_i)}(x_1, x_2) = 2^{(\mu_i)}(\text{norm}(x_1)\text{norm}(x_2))^{\mu_i/2}C_1^{(\mu_i)}\left(\frac{\text{tr}(x_1\overline{x_2})}{2\sqrt{\text{norm}(x_1)\text{norm}(x_2)}}\right)$$

and normalize the scalar product on U_{μ_i} such that $C^{(\mu_i)}$ is a reproducing kernel, i.e.,

$$\langle\langle C^{(\mu_i)}(x, x'), Q(x)\rangle\rangle_{\mu_i} = Q(x')$$

for all $Q \in U_{\mu_i}$. Then the $C^{(\mu_i)}(\cdot, x')$ with rational x' are rational, generate U_{μ_i}, and the reproducing property implies that they have rational scalar products whith each other. The same argument applies to the $U_\nu^{(0)}$.

It is well known that the group of proper similitudes of the quadratic space (D, norm) is isomorphic to $(D^\times \times D^\times)/Z(D^\times)$ via

$$(x_1, x_2) \mapsto \sigma_{x_1, x_2} \text{ with } \sigma_{x_1, x_2}(y) = x_1 y x_2^{-1}$$

and that under this isomorphism $SO(D, \text{norm})$ is the image of

$$\{(x_1, x_2) \in D^\times \times D^\times \mid n(x_1) = n(x_2)\}.$$

Moreover, the $SO(D_\infty, \text{norm}) =: H_{\mathbb{R}}^+$-space $U_\nu^{(0)} \otimes U_\nu^{(0)}$ is isomorphic to the $H_{\mathbb{R}}^+$-space $U_{2\nu}$ and the isomorphism can be normalized in such a way that it preserves rationality and is compatible with the invariant scalar products on both spaces (which are assumed to be normalized as above).

Denoting by S the Gram matrix of the quadratic lattice K we know from Section 1 that (5.2) is of the form

$$\sum_{x=(x_1,x_2,x_3)\in(\mathbb{Z}^4)^3} P(S^{\frac{1}{2}}x_1, S^{\frac{1}{2}}x_2, S^{\frac{1}{2}}x_3) \exp(\pi i(S[x_1]z_1 + S[x_2]z_2 + S[x_3]z_3))$$

where $P \in \otimes_{i=1}^3 U_{k_i-2}$ is a harmonic polynomial of degree $\mu_i = k_i - 2$ on $\mathbb{R}^4$ in each of the variables and is invariant under the (diagonal) action of $H_{\mathbb{R}} = O(D \otimes \mathbb{R}, \text{norm})$. Moreover, P is independent of S and has rational coefficients (up to a factor of π^{3a+2b}). The $H_{\mathbb{R}}$-invariant trilinear form T on $U_{\mu_1} \otimes U_{\mu_2} \otimes U_{\mu_3}$ defined by taking the scalar product with the invariant polynomial $\pi^{-3a-2b}P$ (as remarked in Section 1 this is up to scalars the unique invariant trilinear form) is hence rational (i. e. takes rational values on tensor products of polynomials with rational coefficients).

If all the μ_i are even (as is the case in our situation) then the decomposition of $H_{\mathbb{R}}^+$ and of U_μ as $U_{\mu/2}^{(0)} \otimes U_{\mu/2}^{(0)}$ from above gives furthermore that T factors as $T^{(0)} \otimes T^{(0)}$, where the unique (up to scalars) $D_\infty^\times$-invariant trilinear form $T^{(0)}$ on $U_{\mu_1/2}^{(0)} \otimes U_{\mu_2/2}^{(0)} \otimes U_{\mu_3/2}^{(0)}$ has the same rationality properties as T. Of course both T and T_0 are just the ordinary multiplication if all the μ_i are 0.

For any positive definite symmetric 4×4-matrix S we define the U_{μ_i}-valued theta series $\tilde{\Theta}_S^{(\mu_i)}$ by

$$(5.3) \qquad \tilde{\Theta}_S^{(\mu_i)}(z)(x') = \sum_{x \in \mathbb{Z}^4} C^{(\mu_i)}(S^{1/2}x, S^{1/2}x') \exp(\pi i S[x]z).$$

We notice that if K is a quaternary quadratic lattice with Gram matrix S the right hand side of 5.3 does not depend on the choice of basis of K with respect to which the Gram matrix is computed (because of the invariance of $C^{(\mu_i)}$ under the (diagonal) action of the orthogonal group); we may therefore write it as $\tilde{\Theta}^{(\mu_i)}(K)$ as well.

We denote by $\tilde{\Theta}_{M_1, M_2, M_3}^{(\mu_i)}$ the weighted average of the $\tilde{\Theta}_S^{(\mu_i)}$ over the Gram matrices of representatives of the classes in the genus $\mathrm{gen}(M_1, M_2, M_3)$ as above. We find

$$(5.4) \qquad \begin{aligned} (\mathcal{D}_2^{*(a, \nu_2, \nu_3)} \Theta(S, -))(\iota_{111}(z_1, z_2, z_3)) = \\ \pi^{3a+2b} T(\tilde{\Theta}^{(\mu_1)}(S, z_1) \otimes \tilde{\Theta}^{(\mu_2)}(S, z_2) \otimes \tilde{\Theta}^{(\mu_3)}(S, z_3)) \end{aligned}$$

(where T is as above). For the value at $s = 0$ of (2.1) we obtain therefore

$$(5.5) \qquad \begin{aligned} \pi^{3a+2b} \sum_{M_1 M_2 M_3 | N} \alpha_{M_1, M_2, M_3} T(\langle \tilde{\Theta}_{M_1, M_2, M_3}^{(\mu_1)}(z_1), f(z_1) \rangle \otimes \\ \langle \tilde{\Theta}_{M_1, M_2, M_3}^{(\mu_2)}(z_2), \phi(z_2) \rangle \otimes \langle \tilde{\Theta}_{M_1, M_2, M_3}^{(\mu_3)}(z_3), \psi(z_3) \rangle) \end{aligned}$$

where by $\langle , \rangle$ we denote the Petersson product.

In order to evaluate this expression further we use Eichler's correspondence. We fix M_1, M_2 and an Eichler order $R(M_1, M_2) \subseteq D(M_1)$ as above and set $M = M_1 M_2$. For an irreducible rational representation (V_τ, τ) (with $\tau = \tau_\nu$ as above) of $D_{\mathbb{R}}^\times / \mathbb{R}^\times$ we denote by $\mathcal{A}(D_{\mathbb{A}}^\times, R_{\mathbb{A}}^\times, \tau)$ the space of functions $\varphi : D_{\mathbb{A}}^\times \to V_\tau$ satisfying $\varphi(\gamma x u) = \tau(u_\infty^{-1})\varphi(x)$ for $\gamma \in D_{\mathbb{Q}}^\times$ and $u = u_\infty u_f \in R_{\mathbb{A}}^\times$. It has been discovered by Eichler that these functions are in correspondence with the elliptic modular forms of weight $2 + 2\nu$ and level $M = M_1 M_2$. This correspondence can be described as follows (using the double coset decomposition $D_{\mathbb{A}}^\times = \overset{h}{\underset{i=1}{\cup}} D^\times y_i R_{\mathbb{A}}^\times$ from above) : recall from Section 5 of [2] and Section 3 of [4] that for each $p \mid M_1 M_2$ we have an involution $\widetilde{w_p}$ on the space $\mathcal{A}(D_{\mathbb{A}}^\times, R_{\mathbb{A}}^\times, \tau_\nu)$ given by right translation by a suitable element $\pi_p \in R_p^\times$ of norm p normalizing R_p. This space then splits into common eigenspaces of all these (pairwise commuting) involutions. On $\mathcal{A}(D_{\mathbb{A}}^\times, R_{\mathbb{A}}^\times, \tau_\nu)$ we have furthermore for $p \nmid N$ Hecke operators $\tilde{T}(p)$ whose action on these functions is expressed by the Brandt-matrices $(B_{ij}^{(\nu)}(p))$ (whose entries are endomorphisms of V_{τ_ν}). They commute with the involutions $\widetilde{w_p}$. On the space $\mathcal{A}(D_{\mathbb{A}}^\times, R_{\mathbb{A}}^\times, \tau_\nu)$ we have moreover the natural inner product $\langle\ ,\ \rangle_\nu$ defined by integration, it is explicitly given by

$$\langle \varphi, \rho \rangle_\nu = \sum_{i=1}^{h} \frac{\langle\langle\varphi(y_i), \rho(y_i)\rangle\rangle_\nu^{(0)}}{e_i}.$$

By abuse of language we call (in the case $\nu = 0$) forms cuspidal, if they are orthogonal to the constant functions with respect to this inner product.

We denote for p dividing M_2 the p-essential part by $\mathcal{A}_{p,\text{ess}}(D_{\mathbb{A}}^\times, R_{\mathbb{A}}^\times, \tau)$ consisting of functions φ that are orthogonal to all $\rho \in \mathcal{A}(D_{\mathbb{A}}^\times, (R_{\mathbb{A}}')^\times, \tau)$ for orders $R' \supseteq R$ for which the completion R_p' strictly contains R_p. It is invariant under the $\tilde{T}(p)$ for $p \nmid M_1 M_2$ and the $\tilde{w}_p$ for $p \mid M_1 M_2$ and hence has a basis of common eigenfunctions of all the $\tilde{T}(p)$ for $p \nmid N$ and all the involutions $\tilde{w}_p$ for $p \mid N$. Moreover by the results of [7, 15, 17, 23] we know that in the space $\mathcal{A}_{\text{ess}}(D_{\mathbb{A}}^\times, R_{\mathbb{A}}^\times, \tau)$ of forms that are p-essential for all p dividing $M = M_1 M_2$ strong multiplicity one holds, i.e., each system of eigenvalues of the $\tilde{T}(p)$ for $p \nmid M$ occurs at most once, and the eigenfunctions are in one to one correspondence with the newforms in the space $S^{2+2\nu}(M)$ of elliptic cusp forms of weight $2 + 2\nu$ for the group $\Gamma_0(M)$ that are eigenfunctions of all Hecke operators (if τ is the trivial representation and R is a maximal order one has to restrict here to cuspidal forms on the quaternion side in order to obtain cusp forms on the modular forms side). This correspondence (Eichler's correspondence) preserves Hecke eigenvalues for $p \nmid M$, and if φ corresponds to $g \in S^{2+2\nu}(M)$ then the eigenvalue of g under the Atkin-Lehner involution w_p is equal to that of φ under $\tilde{w}_p$ if D splits at p and equal to minus that of φ under $\tilde{w}_p$ if D_p is a skew field. From (3.13) of [6] we know that if g having first Fourier coefficient 1 corresponds in this way to φ with $\langle \varphi, \varphi \rangle_\nu = 1$ then $\langle g, \tilde{\Theta}^{(\mu)}(K) \rangle = \langle g, g \rangle (\varphi(y_i) \otimes \varphi(y_j))$ holds.

It is not difficult (see also [13]) to extend this correspondence to not necessarily new forms $g \in S^{2+2\nu}(N)$ in the following way :

LEMMA 5.2. — *Let $N = M_1 M_2$ be a decomposition as above. Call $\tilde{g} \in S^{2+2\nu}(N)$ an M'-new form if it is orthogonal to all oldforms coming from $g' \in S^{2+2\nu}(M')$ for $M' \mid N$. Then Eichler's correspondence from above extends to a one to one correspondence between the set of all M_1-new eigenforms of the Hecke operators for $p \nmid N$ in $S^{2+2\nu}(N)$ that are eigenfunctions of all the w_p for $p \mid M_2$ with the set of all Hecke eigenfunctions in $\mathcal{A}(D_{\mathbb{A}}^\times, R_{\mathbb{A}}^\times, \tau)$ that are eigenfunctions of all the involutions $\tilde{w}_p$. This correspondence is compatible with the Hecke action and the eigenvalues under the respective involutions as above; it maps newforms of level $M' \mid N$ (with $M_1 \mid M'$) to forms that* are p-essential precisely for the $p \mid (M'/M_1)$. The correspondence can be explicitly given (in a nonlinear way) by the first Yoshida lifting sending $\tilde{\varphi}$ to the form

$$\sum_{ij}^{h} \frac{\langle\langle\tilde{\varphi}(y_i) \otimes \tilde{\varphi}(y_j), \tilde{\Theta}^{(2\nu)}(I_{ij})\rangle\rangle}{e_i e_j},$$

it then sends $\tilde{\varphi}$ with $\langle\tilde{\varphi}, \tilde{\varphi}\rangle_\nu = 1$ to $\tilde{g}$ having first Fourier coefficient one. Moreover, in this normalization it satisfies the scalar product relations from above, i. e. , if $\tilde{g}$ corresponds to $\tilde{\varphi}$ then $\langle\tilde{g}, \tilde{\Theta}^{(2\nu)}(I_{ij})\rangle\langle\tilde{\varphi}, \tilde{\varphi}\rangle_\nu^2 = \langle\tilde{g}, \tilde{g}\rangle(\tilde{\varphi}(y_i) \otimes \tilde{\varphi}(y_j))$ holds.

Proof. Let M' be a divisor of N and $g \in S^{2+2\nu}(M')$ and let ϵ be a function from the set S of prime divisors of N/M' (whose cardinality we denote by $\omega(M/M')$) to $\{\pm 1\}^{\omega(M/M')}$. Then the function $g^\epsilon := \sum_{S' \subseteq S} g| \prod_{p \in S} \epsilon(p) w_p)$ in $S^{2+2\nu}(N)$ is an eigenfunction of the Atkin-Lehner involutions w_p for the $p \mid (N/M')$ with eigenvalues $\epsilon(p)$. Similarly, fix a maximal order $\tilde{R} \subseteq D$ and an Eichler order $R = R(M_1, M_2) \subseteq \tilde{R}$, let $M_2' \mid M_2$ and let $R(M_1, M_2')$ be the Eichler order of level $M_1 M_2'$ in D containing R and contained in $\tilde{R}$. Let ϵ be a function on the set of prime divisors of M_2/M_2' as above. Then to $\varphi \in \mathcal{A}_{\mathrm{ess}}(D_{\mathbb{A}}^\times, (R(M_1, M_2'))_{\mathbb{A}}^\times, \tau)$ we construct as above a unique φ^ϵ having the same Hecke eigenvalues for $p \nmid N$ and the same $\tilde{w}_p$-eigenvalues for $p \mid M_1 M_2'$ as φ such that φ^ϵ is an eigenfunction of the $\tilde{w}_p$ for $p \mid (M_2/M_2')$ with eigenvalues $\epsilon(p)$.

Given an M_1-new Hecke eigenform $\tilde{g}$ in $S^{2+2\nu}(N)$ that is an eigenfunction of all the w_p for $p \mid M_2$ with eigenvalues ϵ_p we then associate to it the newform g of some level $M' = M_1 M_2' \mid M_1 M_2$ that has the same Hecke eigenvalues for $p \nmid N$ so that $\tilde{g} = g^\epsilon$ with $\epsilon(p) = \epsilon_p$ and apply Eichler's correspondence to get a $\varphi \in \mathcal{A}_{\mathrm{ess}}(D_{\mathbb{A}}^\times, (R(M_1, M_2'))_{\mathbb{A}}^\times, \tau_\nu)$. From [2] we know that this can be normalized such that g is obtained from φ by Yoshida's lifting. We then pass to the eigenfunction φ^ϵ of all $\widetilde{w_p}$ with the same eigenvalues as g for $p \mid M_2$. The scalar product relation (up to normalization) follows then in the same way as in [6], using the uniqueness of the given set of $\tilde{w}_p$-eigenvalues and the fact that application of w_p for $p \mid M_2$ transforms $\tilde{\Theta}^{(2\nu)}(I_{ij})$ to $\tilde{\Theta}^{(2\nu)}(I_{i'j})$, where $y_{i'}$ represents the double coset of $y_i \pi_p^{-1}$ (this is an easy generalization of Lemma 9.1 a) of [2], see also [3]). The same argument shows that Yoshida's lifting realizes this correspondence (using the well known fact that it gives the right Hecke eigenvalues for the $p \nmid N$), and using the expression of g^ϵ as a Yoshida lifting we find the correct normalization of the scalar product relation as in [6].

We will need a version of the scalar product relation in Lemma 5.2 also for the case of newforms of level strictly dividing the level of the theta series involved.

LEMMA 5.3. — *Let M_1, M_2' with $M = M_1 M_2'$ dividing N be as before and let g be a normalized newform of level M and weight $k = 2 + 2\nu$ as in the previous Lemma; let S be the set of prime divisors of N/M. Put $M_2 = N/M_2'$ and let $R' = R(M_1, M_2')$ and $R = R(M_1, M_2) \subseteq R'$ be Eichler orders of levels M, N respectively in $D(M_1)$ and consider a double coset decomposition $D_A^\times = \cup_{i=1}^h D_\mathbb{Q}^\times y_i R_A^\times$ and corresponding quadratic lattices (ideals in D) I_{ij} relative to R as before. Let $\varphi \in \mathcal{A}_{ess}(D_A^\times, (R_A')^\times, \tau_\nu)$ be the essential form corresponding to g under Eichler's correspondence (with $\langle \varphi, \varphi \rangle_\nu = 1$). Then*

$$(5.6) \quad \langle g, \tilde{\Theta}^{(2\nu)}(I_{ij}) \rangle = \langle g, g \rangle \sum_{S' \subseteq S} \left(\prod_{p \in S'} \widetilde{(w_p)} \varphi \right)(y_i) \otimes \left(\prod_{p \in S'} \widetilde{(w_p)} \varphi \right)(y_j).$$

Proof. Let ϵ be a function from the set of prime divisors of N/M to $\{\pm 1\}^{\omega(N/M)}$ and let $g^\epsilon, \varphi^\epsilon$ be as in the proof of Lemma 5.2. Then g^ϵ corresponds to $(\varphi^\epsilon)/\sqrt{\langle \varphi^\epsilon, \varphi^\epsilon \rangle_\nu}$ under Yoshida's correspondence, so we get

$$(5.7) \quad \langle g^\epsilon, \tilde{\Theta}^{(2\nu)}(I_{ij}) \rangle = \frac{\langle g^\epsilon, g^\epsilon \rangle}{\langle \varphi^\epsilon, \varphi^\epsilon \rangle_\nu} \varphi^\epsilon(y_i) \otimes \varphi^\epsilon(y_j)$$

by Lemma 5.2. For any $S' \subseteq S$ the Petersson product $\langle g, g|(\prod_{p \in S'} \epsilon(p)(w_p)) \rangle$ is the same as the product of g with the image of $g|(\prod_{p \in S'} \epsilon(p)(w_p))$ under the trace operator from modular forms for $\Gamma_0(N)$ to modular forms for $\Gamma_0(M)$, hence equal to $\prod_{p \in S'} \epsilon(p) p^{-\nu} a_g(p) \langle g, g \rangle$, since for $p \nmid M$ the map $g \mapsto g|w_p$ composed with the trace just gives the (renormalized) Hecke operator. The same argument applies to $\langle \varphi, (\prod_{p \in S'} \epsilon(p) \widetilde{(w_p)} \varphi) \rangle$ and gives the same factor of comparison with $\langle \varphi, \varphi \rangle$ since g and φ have the same (renormalized) Hecke eigenvalues. Thus we have $\langle g^\epsilon, g^\epsilon \rangle \langle \varphi^\epsilon, \varphi^\epsilon \rangle^{-1} = \langle g, g \rangle \langle \varphi, \varphi \rangle^{-1}$, and summing up the identities (5.7) for all the functions ϵ gives the assertion.

LEMMA 5.4. — *Let f, ϕ, ψ as in Section 2 be newforms of square free levels N_f, N_ϕ, N_ψ with $N = \text{lcm}(N_f, N_\phi, N_\psi)$ and with weights $k_i = 2 + 2\mu_i$. Then in 5.5 the summand for M_1, M_2, M_3 is zero unless $M_3 = 1$, $N = M_1 M_2$ and $M_1 \mid \gcd(N_f, N_\phi, N_\psi)$ hold.*

Proof. Since by Lemma 5.2 we can express f, ϕ, ψ as Yoshida-liftings an easy generalization of Lemma 9.1 b) of [2] shows that f is orthogonal to all $\tilde{\Theta}^{(\mu_1)}(K)$ for K in $\text{gen}(M_1, M_2, M_3)$ for which N_f does not divide $M_1 M_2$ and analogously for ϕ, ψ. This establishes the vanishing of all summands for which $N \neq M_1 M_2$.

If there is a $p \mid M_1$ not dividing $\gcd(N_f, N_\phi, N_\psi)$ then say $p \nmid N_f$. The Petersson product of f with $\tilde{\Theta}^{(2\mu_1)}(K)$ for K in $\text{gen}(M_1, M_2, 1)$ is then the same as that of f with the form obtained by applying the trace operator from modular forms on $\Gamma_0(N)$ to modular forms on $\Gamma_0(N/p)$ to the theta series. But it is easily checked that this trace operator annihilates the theta series of $K \in \text{gen}(M_1, M_2, M_3)$ if $p \mid M_1$, see [8]; the same argument is applied to ϕ, ψ which shows the last part of the assertion.

LEMMA 5.5. — *Let* $\tilde{f}, \tilde{\phi}, \tilde{\psi}$ *be cusp forms of weights* k_1, k_2, k_3 *for* $\Gamma_0(N)$ *with square free* N *as in Section 2, assume them to be eigenforms of the Hecke operators for* $p \nmid N$ *and of all the Atkin-Lehner involutions* w_p *with eigenvalues* $\epsilon_f(p), \epsilon_\phi(p), \epsilon_\psi(p)$ *but not necessarily newforms. Let* $M_1 M_2 = N$ *(with* M_1 *as always having an odd number of prime factors) and let* $R = R(M_1, M_2)$ *be an Eichler order in* $D = D(M_1)$. *Let* $\varphi_f, \varphi_\phi, \varphi_\psi$ *be the forms in* $\mathcal{A}(D_\mathbb{A}^\times, R_\mathbb{A}^\times, \tau_{\nu_i})$ *corresponding to* f, ϕ, ψ *under the correspondence of Lemma 5.2 (with* $k_i = 2 + 2\mu_i$ *for* $i = 1, \ldots, 3$). *Then the summand for* $M_1, M_2, M_3 = 1$ *in 5.5 is*

$$2^{-\omega(N)} \alpha_{M_1, M_2, 1} \langle f, f \rangle \langle \phi, \phi \rangle \langle \psi, \psi \rangle \sum_{i=1}^{h} \left(\frac{T_0(\varphi_f(y_i) \otimes \varphi_\phi(y_i) \otimes \varphi_\psi(y_i))}{e_i} \right)^2 .$$

The latter expression is zero unless for all $p \mid N$ *one has* $p \mid M_1$ *if and only if* $\epsilon_f(p) \epsilon_\phi(p) \epsilon_\psi(p) = -1$.

Proof. The first part of the assertion is an immediate consequence of Lemma 5.2 and the decomposition of T as $T^{(0)} \otimes T^{(0)}$. For the second part we notice that the expression $\sum_{i=1}^{h} \left(\frac{T_0(\varphi_f(y_i) \otimes \varphi_\phi(y_i) \otimes \varphi_\psi(y_i))}{e_i} \right)$ does not change if an involution $\widetilde{w_p}$ is applied to all three functions $\varphi_f, \varphi_\phi, \varphi_\psi$ since this only permutes the order of summation. On the other hand each summand is multiplied with the product of the eigenvalues of $\varphi_f, \varphi_\phi, \varphi_\psi$ under $\widetilde{w_p}$, which in view of the relation between the w_p eigenvalues and the $\widetilde{w_p}$-eigenvalues of corresponding functions proves the assertion.

Although the scalar product relation in Lemma 5.2 is not true if one omits the condition that $\tilde{g}$ is an eigenfunction of all the involutions w_p, the next Lemma shows that by an amusing newforms argument an only slightly changed version of Lemma 5.5 (which is based on this scalar product relation) remains true without this condition.

LEMMA 5.6. — *Let* f, ϕ, ψ *be normalized newforms of levels* N_f, N_ϕ, N_ψ *of weights* $k_i = 2 + 2\mu_i$ *(*$i = 1, \ldots, 3$*) as in Section 2 and let* $\Lambda_1, \Lambda_2, \Lambda_3$ *be pairwise disjoint subsets of the set of primes divisors of* $N^f = N/N_f, N^\phi, N^\psi$ *respectively such that* $\Lambda_1 \cup \Lambda_2 \cup \Lambda_3$ *is the set of all primes dividing precisely one of the integers* N_f, N_ϕ, N_ψ. *For* $\kappa = 1, \ldots, 3$ *let*

$$\widetilde{w_{\Lambda_\kappa}} = \prod_{p \in \Lambda_\kappa} \widetilde{w_p}.$$

Let a decomposition $N = M_1 M_2$ as before be given, fix a maximal order $\tilde{R}$ in
$D = D(M_1)$ and an Eichler order $R = R(M_1, M_2) \subseteq \tilde{R}$ of level $M_1 M_2 = N$
in D and consider a double coset decomposition $D_A^\times = \cup_{i=1}^h D_Q^\times y_i R_A^\times$ and
corresponding quadratic lattices (ideals in D) I_{ij} relative to R as before. For
each $M_2' \mid M_2$ let $R(M_1 M_2') = R(M_1, M_2')$ be the unique Eichler order of level
$M_1 M_2'$ contained in $\tilde{R}$ and containing R. Let $\varphi_1 \in A_{\mathrm{ess}}(D_A^\times, (R(N_f)_A^\times, \tau_{\mu_1}))$
be the form corresponding to f under Eichler's correspondence and define
φ_2, φ_3 analogously with respect to ϕ, ψ. Then

$$T\big(\langle f, \tilde{\Theta}^{(\mu_1)}\rangle \langle \phi, \tilde{\Theta}^{(\mu_2)}\rangle \langle \psi, \tilde{\Theta}^{(\mu_3)}\rangle\big)$$
$$= 2^{-\omega(\gcd(N_f, N_\phi, N_\psi))}\langle f, f\rangle \langle \phi, \phi\rangle \langle \psi, \psi\rangle$$

(5.8)

$$\times \big(T_0(\sum_{i=1}^h \frac{1}{e_i} \widetilde{w_{\Lambda_1}}\varphi_1(y_i) \otimes \widetilde{w_{\Lambda_2}}\varphi_2(y_i) \otimes \widetilde{w_{\Lambda_3}}\varphi_3(y_i)))^2.$$

Proof. This is an immediate consequence of Lemmas 5.2 and 5.3 : upon
inserting the scalar product relations from these lemmata into the left hand
side of (5.8) we obtain a sum of terms of the type

$$\big(T_0((\sum_{i=1}^h \widetilde{w_{\Lambda_1'}}\varphi_1(y_i) \otimes \widetilde{w_{\Lambda_2'}}\varphi_2(y_i) \otimes \widetilde{w_{\Lambda_3'}}\varphi_3(y_i)))\big)^2$$

with arbitrary subsets Λ_κ' of the sets of primes dividing $N/N_f, N/N_\phi, N/N_\psi)$
respectively. Let p be a prime dividing two of the levels, say $p \mid N_f, p \mid N_\phi$.
Then since applying $\widetilde{w_p}$ to all three of the φ_κ only changes the order of
summation, the involution $\widetilde{w_p}$ for $p \in \Lambda_3'$ may be pulled over to φ_2, φ_3 which
are eigenfunctions of $\widetilde{w_p}$. The terms with the set Λ_3' and those with $\Lambda_3' \setminus \{p\}$
give therefore the same contribution. Let now p be a prime dividing only one
of the levels, say $p \mid N_f$. If $p \notin \Lambda_2' \cup \Lambda_3'$ then the component of $\varphi_2 \otimes \varphi_3$ in the
τ_{μ_1}-isotypic component of $\tau_{\nu_2} \otimes \tau_{\nu_3}$ is an oldform with respect to p, hence
orthogonal to the p-essential form φ_1. Since $T_0(\varphi_1 \otimes \varphi_2 \otimes \varphi_3)$ is proportional
to the scalar product of φ_1 with this component of $\varphi_1 \otimes \varphi_2$ such a term
gives no contribution; the same argument applies if $p \in \Lambda_2' \cap \Lambda_3'$ holds. If
p is in precisely one of Λ_2', Λ_3' then the same argument as in the first case
shows that p may be shifted to either one of these sets without changing the
contribution of the term. Taking together both cases we find that all terms
appearing are of the shape

$$\big(T_0((\sum_{i=1}^h \widetilde{w_{\Lambda_1}}\varphi_1(y_i) \otimes \widetilde{w_{\Lambda_2}}\varphi_2(y_i) \otimes \widetilde{w_{\Lambda_3}}\varphi_3(y_i)))\big)^2$$

with each term appearing $2^{\omega(N/\gcd(N_f, N_\phi, N_\psi))}$ times, which in view of 5.1
implies the assertion.

Collecting all the information obtained we arrive at the main theorem :

THEOREM 5.7. — *The value of the triple product L-function $L(f, \phi, \psi, s)$ at the central critical value $s = \frac{k_1 + k_2 + k_3}{2} - 1$ is*

(5.9)

$$(-1)^{a'} 2^{5+4a+3b-\omega(\gcd(N_f, N_\phi, N_\psi))} \pi^{5+9a'+4b} \frac{(a'+1)^{[b]}}{2^{[a+b]} 2^{[a']} (\nu_2+1)^{[a']} (\nu_3+1)^{[a']}}$$

$$\times \langle f, f \rangle \langle \phi, \phi \rangle \langle \psi, \psi \rangle \left(T_0 \left(\sum_{i=1}^{h} \frac{1}{e_i} \widetilde{w_{\Lambda_1}} \varphi_1(y_i) \otimes \widetilde{w_{\Lambda_2}} \varphi_2(y_i) \otimes \widetilde{w_{\Lambda_3}} \varphi_3(y_i) \right) \right)^2$$

where the notation is as in Lemma 5.6 and T_0 is (as explained in the beginning of this section) the up to scalars unique rational invariant trilinear form on the representation space $U_{\mu_1/2}^{(0)} \otimes U_{\mu_2/2}^{(0)} \otimes U_{\mu_3/2}^{(0)}$ (with $\mu_i = k_i - 2$) and takes values in the coefficient fields of f, ϕ, ψ respectively on the polynomials $\widetilde{w_{\Lambda_1}} \varphi_\kappa(y_i)$ (for $\kappa = 1, \ldots, 3$).

It should be noted that the rational quantity on the right hand side can be interpreted as the height pairing of a diagonal cycle with itself in the same way as in [11]. One has just to replace (for $\kappa = 1, \ldots, 3$) the group $\mathrm{Pic}(X)$ of [11] with the group $\mathrm{Pic}(V_\kappa)$ from [14] obtained by attaching to each y_i in the double coset decomposition of $D_\mathbb{A}^\times$ used above the space of $R_i^\times$-invariant polynomials in $U_{\mu_\kappa}^{(0)}$. Our functions φ_κ may then be interpreted as elements of $\mathrm{Pic}(V_\kappa)$. One may then form the tensor product of these three groups and obtain an analogue of the diagonal cycle Δ from [11] by using our Gegenbauer polynomials from above and proceed as in loc. cit.

Manuscrit reçu le 26 juin 1995

References

[1] S. Böcherer. — *Über die Fourier-Jacobientwicklung der Siegelschen Eisensteinreihen II*, Mathem. Z. **189** (1989),81-110.

[2] S. Böcherer, R. Schulze-Pillot. — *Siegel modular forms and theta series attached to quaternion algebras*, Nagoya Math. J. **121** (1991), 35-96.

[3] S. Böcherer, R. Schulze-Pillot. — *Siegel modular forms and theta series attached to quaternion algebras II*, Preprint 1995.

[4] S. Böcherer, R. Schulze-Pillot. — *Mellin transforms of vector valued theta series attached to quaternion algebras*, Math. Nachr. **169** (1994), 31-57.

[5] S. Böcherer, R. Schulze-Pillot. — *Vector valued theta series and Waldspurger's theorem*, Abh. Math. Sem. Hamburg **64** (1994), 211-233.

[6] S. Böcherer, T. Satoh,T. Yamazaki. — *On the pullback of a differential operator and its application to vector valued Eisenstein series*, Comm. Math. Univ. S. Pauli **41** (1992), 1-22.

[7] M. Eichler. — *The basis problem for modular forms and the traces of the Hecke operators*, p. 76-151 in Modular functions of one variable *I*, Lecture Notes Math. **320**, Berlin-Heidelberg-New York 1973.

[8] J. Funke. — *Spuroperator und Thetareihen quadratischer Formen*, Diplomarbeit Köln 1994.

[9] P. Garrett. — *Decomposition of Eisenstein series : Rankin triple products*, Annals of Math. **125** (1987), 209-235.

[10] P. Garrett, M. Harris. — *Special values of triple product L-Functions*, Am. J. of Math. **115** (1993), 159-238.

[11] B. Gross, S. Kudla. — *Heights and the central critical values of triple product L-functions*, Compositio Math. **81** (1992), 143-209.

[12] M. Harris, S. Kudla. — *The central critical value of a triple product L-function*, Annals of Math. 133 (1991), 605-672

[13] K. Hashimoto. — *On Brandt matrices of Eichler orders*, Preprint 1994.

[14] R. Hatcher. — *Heights and L-series*, Can. J. math. **62** (1990), 533-560.

[15] H. Hijikata, H. Saito. — *On the representability of modular forms by theta series*, p. 13-21 in Number Theory, Algebraic Geometry and Commutative Algebra, in honor of Y. Akizuki, Tokyo 1973.

[16] T. Ibukiyama. — *On differential operators on automorphic forms and invariant pluriharmonic polynomials*, Preprint 1990.

[17] H. Jacquet, R. Langlands. — *Automorphic forms on $GL(2)$*, Lect. Notes

in Math. **114**, Berlin–Heidelberg–New York 1970.

[18] W. Ll. — *Newforms and functional equations*, Math. Ann. **212** (1975), 285–315.

[19] P. LITTELMANN. — *On spherical double cones*, J. of Algebra **166** (1994), 142–157.

[20] H. MAAβ. — *Siegel's modular forms and Dirichlet series* Lect. Notes Math. **216** Berlin, Heidelberg, New York : Springer 1971.

[21] T. ORLOFF. — *Special values and mixed weight triple products (with an appendix by Don Blasius)*, Invent. Math. **90** (1987), 169–180.

[22] T. SATOH. — *Some remarks on triple L-functions*, Math. Ann. **276** (1987), 687–698.

[23] H. SHIMIZU. — *Theta series and automorphic forms on GL_2*, J. of the Math. Soc. of Japan **24** (1972), 638–683.

[24] G. SHIMURA. — *The special values of the zeta functions associated with cusp forms*, Comm.Pure Appl. Math. **29** (1976), 783–804.

[25] G. SHIMURA. — *The arithmetic of differential operators on symmetric domains*, Duke Math. J. **48** (1981), 813–843.

S. Böcherer and R. Schulze-Pillot
Fakultät für Mathematik und Informatik
Universität Mannheim
Seminargebäude A5
D–68131 MANNHEIM

Number Theory
Paris 1993–94

Abelian extensions of complete discrete valuation fields

Ivan B. Fesenko

Introduction

In 1968 Y. Ihara [Ih] proposed to study class field theory of the p-adically complete field $\widehat{\mathbb{Q}(t)}_p$ that is the quotient field of $\varprojlim \mathbb{Z}[t]/p^i\mathbb{Z}[t]$. This field in modern terminology is a two-dimensional local-global field. Ihara considered cyclic extensions of degree p of this field. He suggested that its class field theory could "explain arithmetically" the map $j \to \Pi_j$ which associates to each $j \in \mathbb{F}_p^{\text{sep}}, j \neq 0; 1$ ($p \neq 2; 3$) the subgroup $\Pi_j \subset \mathbb{Q}_p^*$ generated by α'/α where $(1 - \alpha u)(1 - \alpha' u)$ is the numerator of the zeta function of E with E running all elliptic curves defined over finite extensions of $\mathbb{F}_p(j)$ with modulus j.

Later that work of Ihara stimulated two completely different series of works on abelian extensions of complete discrete valuation fields with very general residue field produced by H. Miki ([M2], 1977, also [M1]) and K. Kato ([Ka1–Ka7], 1977–1982).

The first direction describes some types of abelian extensions of a complete discrete valuation field with imperfect residue field via study of the group of principal units without using cohomological methods. In the second direction abelian extensions of an n-dimensional local fields are described in terms of the Milnor K_n-groups. The latter theory is based in particular on Galois cohomology groups calculations. Independently A. N. Parshin from different motivations suggested and then developed higher local theory in positive characteristic by using quotients of Milnor K-groups endowed with some proper topology and applied it to a description of abelian coverings of two-dimensional arithmetic schemes ([P1–P5], 1975–78, 85, 90).

The aim of this work is to sketch the present–day scenery of local class field theories including [Kur], [Ko1–Ko2], [Sp], [F1–F5].

I am grateful to Y. Ihara for sending me a copy of [Ih] together with its translation into English.

Parts of the first part of the work were prepared during a seminar on higher class field theory in Max–Planck–Institut für Mathematik organized by W. Raskind and the author in spring 1994. I am thankful to W. Raskind, Y. Koya and M. Spieß. Special thanks to T. Fimmel for his stimulating questions and discussions. An essential part of this work was written while I was staying in the University of Sydney in August 1994. Its hospitality is gratefully acknowledged. I am grateful to K. F. Lai and other participants of my lectures there for stimulating atmosphere.

Higher local theories

First we introduce main objects describing abelian extensions of multi-dimensional local fields : Milnor K-groups and topological K-groups ($1°$ – $6°$). We follow [P1–P5], [Ka3–Ka4], [F1,F2,F4]. Then we consider higher local class field theories ($7°$ – $10°$).

1. — Multidimensional fields

Given a two-dimensional smooth projective scheme X over a finite field of characteristic p one can attach to a point $x \in X$ and a curve $y \subset X$ passing smoothly through x the quotient field of the completion $(\widehat{\mathcal{O}_{X,x}})_y$ of the localization at y of the completion $\widehat{\mathcal{O}_{X,x}}$ of the localization at x. This is a two-dimensional local field over a finite field (which is itself considered as a 0-dimensional local field). More generally, an n-dimensional local field F is a complete discrete valuation field with residue field being $(n-1)$-dimensional. Due to classical structure theorems F is noncanonically isomorphic to $k_m((t_{m+1}))\ldots((t_n))$ where k_m is a coefficient field corresponding to the $(n-m)$-th residue field of F and either $m = 0$ or $\mathrm{char}(k_m) = 0, \mathrm{char}(k_{m-1}) = p$. In the latter case the group of principal units of F with respect to the discrete valuation of rank $n - m$ is divisible, and so is not of interest for class field theory. The field k_m for $m \neq 0$ is called a mixed characteristic field, it is a natural higher analog of a p-adic fields. Lifting prime elements from F and residue fields $k_{n-1}, \ldots, k_1$ to the field F one obtains an ordered system of local parameters $t_n, \ldots, t_1$ (t_n is a prime element of F).

If a field M has mixed characteristic and $\mathcal{O}_M$ is its ring of integers with respect to the discrete structure of rank m and t_m is a main local parameter, then the quotient field $M\{\{t\}\}$ of $\varprojlim\mathcal{O}_M[[t]][t^{-1}]/t_m^i\mathcal{O}_M[[t]][t^{-1}]$ is an $(m+1)$-dimensional mixed characteristic field. In general a mixed characteristic field is a finite extension of a field like $\mathbb{Q}_p\{\{t_1\}\}\ldots\{\{t_{m-1}\}\}$ ([FV, sect. 5 Ch. II], for more details see [Zh]).

It occurs that the Milnor K-groups are not the most suitable objects to be related with abelian extensions of an n-dimensional local field F.

It is more convenient to work with quotients $K_m^M(F)/\Lambda_m(F)$ of Milnor K-groups endowed with a special topology ($\Lambda_m(F)$ is the intersection of all neighbourhoods of zero). Arithmetical homomorphisms from Milnor K-groups (like a reciprocity map) usually factorize through such quotients.

2. — Topology on the multiplicative group

It is natural to expect compatibility of theories of a field and its residue field (lifting of extensions of the residue field as unramified extensions of the field and the border homomorphism in K-theory). Thus, a genuine topology on a multidimensional field has to take into account topologies of its residue fields. Denote by $\mathcal{O}_0$ the field in F corresponding to the last finite residue field k_0 when $\mathrm{char}(F) = p$ and the image in F of the ring of Witt vectors of k_0 corresponding to the homomorphism $k_0 \to k_{m-1}$ (which is uniquely determined, see [FV, sect. 5 Ch. II]) when $\mathrm{char}(F) = 0$. The ring $\mathcal{O}_0$ contains the set of canonical liftings $\mathcal{R}$ from k_0, so called multiplicative representatives.

For a 2-dimensional local field F with local parameters t_2, t_1 define a basis of open neighborhoods of 1 as $1 + t_2^i \mathcal{O}_F + t_1^j \mathcal{O}_0[[t_1]]$ (e.g. [Ka3]). Then every element $\alpha \in F^*$ can be expanded as a convergent with respect to the just defined topology product

$$\alpha = t_2^{a_2} t_1^{a_1} \theta \prod (1 + \theta_{i,j} t_2^i t_1^j)$$

with $\theta \neq 0, \theta_{i,j} \in \mathcal{R}, a_1, a_2 \in \mathbb{Z}$.

In the multidimensional case one can define topology by induction on dimension. Let F be an n-dimensional local field with $\mathrm{char}(k_{n-1}) = p$. Fix a lifting (and thus a set of representatives S) of k_{n-1} in F compatible with the residue morphism : $k_0 \to \mathcal{O}_0$, the residues $\bar{t}_i \in k_{n-1}, 1 \leqslant i \leqslant n-1$, of local parameters go to t_i in F. Given the topology on the additive group k_{n-1}, introduce the following topology on the additive group F. First, an element $\alpha \in F$ is said to be a limit of a sequence of elements $\alpha_v \in F, v \to +\infty$, if and only if given any writing $\alpha_v = \sum_i \theta_{v,i} t_n^i, \alpha = \sum_i \theta_i t_n^i$ with $\theta_* \in S$, for every set $\{U_i, \quad -\infty < i < +\infty\}$ of neighbourhoods of zero in k_{n-1} and every i_0 for almost all v the residue of $\theta_{v,i} - \theta_i$ belongs to U_i for all $i < i_0$. Second, a subset U in F is called open if and only if for every $\alpha \in U$ and every sequence $\alpha_v \in F$ having α as a limit almost all α_v belong to U. This determines a topology τ on F. Then α is a limit of α_v if and only if the sequence α_v converges to α with respect to the topology τ.

By induction on dimension one verifies that limit is uniquely determined, each Cauchy sequence with respect to the topology τ converges in F, the limit of the sum of two convergent sequences is the sum of their limits.

If a subgroup $A = \{\alpha = \sum \theta_i t_n^i : \alpha \in F, \theta_i \in S_i \subset S\}$ is open, then all sets of residues $\overline{S_i}$ are open subgroups in k_{n-1}.

Note that the topology τ on the additive group is different from that introduced by Parshin in [P4] for $n \geqslant 2$: for example, the set $W = F \setminus \{t_2^a t_1^{-c} + t_2^{-a} t_1^c : a, c \geqslant 1\}$ in $F = \mathbb{F}_p((t_1))((t_2))$ is open in the just defined topology, i.e. for each convergent sequence $x_v \to x \in W$ almost all x_v belong to W. If for some open subgroups U_i in the additive group of $\mathbb{F}_p((t_1))$ such that $U_i = \mathbb{F}_p((t_1))$ for $i \geqslant a$ the group $\{x = \sum a_i t_2^i : x \in F, a_i \in U_i\}$ were contained in W, then for any positive c such that $t_1^c \in U_{-a}$ we would have $t_2^a t_1^{-c} + t_2^{-a} t_1^c \in W$, a contradiction. However, a sequence of elements in F coverges to an element of F with respect to τ if and only if it converges with respect to the topology introduced by Parshin. The group F is not a topological group for $n \geqslant 2$ with respect to τ. For example, if $W' + W' \subset W$, then W' is not open with respect to τ.

If $\operatorname{char}(k_{n-1}) = p$, then define the topology τ on F^* as the product of the induced from F topology on the group of principal units $V_F = 1 + (t_n, \ldots, t_1)\mathcal{O}_F$, the discrete topologies on the cyclic groups generated by t_i and the cyclic group of non-zero multiplicative representatives of k_0 in F.

A general principle on higher dimensions states that there are two essential breaks in objects and methods that are to be involved : from 1 to 2 and from 2 to >2. For a 2-dimensional local field its multiplicative group F^* is a topological group and it has a countable basis of open subgroups. In the case of at least three dimensional field both assertions don't hold. For example, if $W'^2 \cdot W' \subset 1 + W t_3 + \mathcal{O}_L t_3^2$ for $L = F((t_3))$, then W' is not open in $1 + \mathcal{O}_L t_3$.

If $\operatorname{char}(k_{n-1}) = p$, then every element $\alpha \in F^*$ can be expanded into a convergent product :

$$\alpha = t_n^{a_n} \ldots t_1^{a_1} \theta \prod (1 + \theta_{i_n, \ldots, i_1} t_n^{i_n} \ldots t_1^{i_1}).$$

Local parameters $t_n, \ldots, t_1$, multiplicative representatives θ and principal units $1 + \theta t_n^{i_n} \ldots t_1^{i_1}$ are *topological generators* of F^* : each element $\alpha \in F^*$ can be expanded into a convergent product of some of them. F^* is a semi-topological group and sequential group (the multiplication is sequentially continuous).

This topology on the multiplicative group is different from that introduced by Parshin in [P4] and then refered to in [F1-F4]. It is easy to check that for $n \geqslant 3$ each open subgroup A in F^* with respect to the topology introduced in [P4] possesses the property : $1 + t_n^2 \mathcal{O}_F \subset (1 + t_n^3 \mathcal{O}_F)A$. However, the subgroup in $1 + t_n \mathcal{O}_F$ topologically generated by $1 + \theta t_n^{i_n} \ldots t_1^{i_1}$ with $(i_n, \ldots, i_1) \neq (2, 1, \ldots, 0)$, $i_n \geqslant 1$ (i.e. the sequential closure of the subgroup generated by these elements), is open in τ and doesn't satisfy the above-mentioned property.

If $\mathrm{char}(F) = \mathrm{char}(k_m) = 0$, $\mathrm{char}(k_{m-1}) = p$, then define the topology τ on F^* as the product of the trivial (anti-discrete) topology on the divisible part of F^*, the discrete topology on the cyclic groups generated by t_i with $i > m$ and the just defined topology on k_m^*.

For class field theory sequential continuity seems to be more important than continuity. This is a hidden phenomenon in dimension 1 and 2, where continuity is the same as sequential continuity.

For other definitions and details see [Ka3], [F1,F2,F4], [MZh].

3. — Topology on K–groups

Let λ be the finest topology on $K_m^{\mathrm{M}}(F)$ for which the map $(F^*)^m \to K_m^{\mathrm{M}}(F)$ is sequentially continuous with respect to the product of the introduced above topology on F^* and for which the subtraction in $K_m^{\mathrm{M}}(F)$ is sequentially continuous. Put

$$K_m^{\mathrm{top}}(F) = K_m^{\mathrm{M}}(F)/\Lambda_m(F)$$

with the quotient topology where $\Lambda_m(F)$ is the intersection of all neighborhoods of 0 with respect to λ (hence is a subgroup). When $m = 1$, $K_1^{\mathrm{top}}(F) = F^*$ algebraically and topologically.

Note that for two principal units $\epsilon, \eta \in F^*$ the following holds in $K_2^{\mathrm{M}}(F)$:

$$\{\epsilon, \eta\} = \{1 - \epsilon, 1 - (1 - \epsilon^{-1})(1 - \eta)\} + \{\eta, 1 - (1 - \epsilon^{-1})(1 - \eta)\}$$

with the principal unit $1 - (1 - \epsilon^{-1})(1 - \eta)$ of higher order than that of ϵ, η. This allows one to continue this process in $K_2^{\mathrm{top}}(F)$ and finally to write $\{\epsilon, \eta\}$ as a sum of symbols in the form $\{\rho_i, t_i\}$ with principal units ρ_i and local parameters t_i. Note also that $\{\theta, \theta'\} = \{\theta, \epsilon\} = 0$ for $(q-1)$–th roots of unity θ, θ' and a principal unit ϵ. Therefore, every element x of $K_m(F)$ can be written as a sum of a fixed number of terms in the form $\{\alpha_i\} \cdot \{$some local parameters$\}$ with $\alpha_i \in F^*$ plus an element of $\Lambda_m(F)$.

This definition of topological K-groups is somewhat implicit, but it is the most convenient for an initial study of them. Note that if there is a symbolic sequentially continuous homomorphism from the tensor product of m copies of F^* to a group G on which the topology is defined by means of a set of subgroups, then it induces a continuous homomorphism from $K_m^{\mathrm{top}}(F)$ to G.

Later in 5° we will show that $\Lambda_m(F)$ coincides with $\cap_{l \geqslant 1} l K_m^{\mathrm{M}}(F)$ for $m \leqslant n$ and is a divisible group.

The structure of topological K-groups of multidimensional local fields is almost completely known (in contrast to the Milnor K-groups). The simplest way to describe it is to introduce at first Artin–Schreier–Witt–Parshin, Vostokov and higher tame pairings, and then to apply *explicit* formulas defining them. The role of the Artin–Schreier–Witt pairings in the one-dimensional case is known from the theory of class formation in characteristic p [KS]. Recall that the Vostokov pairing [V2] has appeared as a multidimensional variant of his explicit formulas [V1] for the Hilbert norm residue symbol in number local fields in case the residue field being of odd characteristic. A general philosophy due to Shafarevich [Sh] as reflection of similarities between Riemann surfaces and algebraic number fields is to find an explicit formula for the p^r-th Hilbert symbol, then forget about class field theory, and using the pairing correctly defined by the explicit formula develop independently and explicitly class field theory for Kummer extensions. For higher local fields both Artin–Schreier–Witt–Parshin and Vostokov pairings are first applied to determine structure of the quotient group $K_n^{\text{top}}(F)/p^r$ (in characteristic p for arbitrary r). Then they serve as implements to construct Artin–Schreier–Witt and Kummer extensions which correspond via these pairings to open subgroups of finite index in $K_n^{\text{top}}(F)$. Coincidence of both pairings with the corresponding pairings induced from class field theory enables one to deduce existence theorem.

4. — Pairings of the topological K–groups

Let F be an n-dimensional local field of characteristic p. For $\alpha_1, \ldots, \alpha_n \in F^*$, and a Witt vector $(\beta_0, \ldots, \beta_r) \in W_r(F)$ put

$$(\alpha_1, \ldots, \alpha_n, (\beta_0, \ldots, \beta_r)]_r = \text{Tr}_{k_0/\mathbb{F}_p}(\gamma_0, \ldots, \gamma_r)$$

where k_0 is the last finite residue field of F and the i-th ghost component $\gamma^{(i)}$ of $(\gamma_0, \ldots, \gamma_r)$ is defined as $\text{res}_{k_0}(\beta^{(i)}\alpha_1^{-1}d\alpha_1 \wedge \ldots \wedge \alpha_n^{-1}d\alpha_n)$.

This is a sequentially continuous symbolic in the first n coordinates map. Hence it defines the *Artin–Schreier–Witt–Parshin pairing* :

$$K_n^{\text{top}}(F)/p^r \times W_r(F)/(\text{Frob} - 1)W_r(F) \to W_r(\mathbb{F}_p)$$

where Frob is the Frobenius map.

Let F be an n-dimensional mixed characteristic local field and let a primitive p^r-th root of unity ζ be contained in F. Let $X_1, \ldots, X_n$ be independent indeterminates over the quotient field of $\mathcal{O}_0$. For an element

$$\alpha = t_n^{a_n} \ldots t_1^{a_1} \theta \prod (1 + \theta_{i_n, \ldots, i_1} t_n^{i_n} \ldots t_1^{i_1})$$

of F^*, with $\theta \neq 0, \theta_{i_n, \ldots, i_1} \in \mathcal{R}$ put

$$\alpha(X) = X_n^{a_n} \ldots X_1^{a_1} \theta \prod (1 + \theta_{i_n, \ldots, i_1} X_n^{i_n} \ldots X_1^{i_1}).$$

The formal power series $\alpha(X) \in \mathcal{O}_0((X_1)) \ldots ((X_n))$ depends on the choice of local parameters and the choice of the decomposition of α. Denote $z(X) = \zeta(X)$, $s(X) = z(X)^{p^r} - 1$. Define the action of the operator Δ on θ's and on X_i as raising to the p-th power. For $\alpha \in F^*$ put $l(\alpha) = p^{-1} \log \alpha(X)^{p-\Delta}$.

Now for elements $\alpha_1, \ldots, \alpha_{n+1} \in F^*$ define $\Phi(\alpha_1, \ldots, \alpha_{n+1})$ as

$$\sum_{i=1}^{n+1} (-1)^{n+1-i} l(\alpha_i) \left(\frac{d\,\alpha_1}{\alpha_1} \wedge \ldots \wedge \frac{d\,\alpha_{i-1}}{\alpha_{i-1}} \wedge p^{-1} \frac{d\,\alpha_{i+1}^{\Delta}}{\alpha_{i+1}^{\Delta}} \wedge \ldots \wedge p^{-1} \frac{d\,\alpha_{n+1}^{\Delta}}{\alpha_{n+1}^{\Delta}} \right).$$

Let μ_{p^r} denote the cyclic group generated by ζ. Define a map

$$V_r \colon (F^*)^{n+1} \to \mu_{p^r}$$

as

$$V_r(\alpha_1, \ldots, \alpha_{n+1}) = \zeta^{\gamma}, \qquad \gamma = \mathrm{Tr}_{\mathcal{O}_0/\mathbb{Z}_p} \mathrm{res} \Phi(\alpha_1, \ldots, \alpha_{n+1})/s(X).$$

This is a very deep result [V2] that V_r doesn't depend on the attaching formal power series to elements of F (for another proof involving syntomic cohomologies of Fontaine–Messing see [Ka8]). The map V_r is sequentially continuous and symbolic. It defines the *Vostokov pairing* :

$$K_n^{\mathrm{top}}(F)/p^r \times F^*/p^r \to \mu_{p^r}.$$

Let F be an n-dimensional local field and let the cardinality of the group μ of all roots of unity of order prime to p be $q-1$ (in other words, $k_0 = \mathbb{F}_q$). For an element $\alpha \in F^*$ and its writing as above put $v^{(j)}(\alpha) = a_j$ for $1 \leqslant j \leqslant n$. For elements $\alpha_1, \ldots, \alpha_{n+1}$ of F^* define $\mathfrak{c}(\alpha_1, \alpha_2, \ldots, \alpha_{n+1})$ as the $(q-1)$-th root of unity whose residue is equal to the residue of $\alpha_1^{b_1} \ldots \alpha_{n+1}^{b_{n+1}} (-1)^b$ in the last residue field k_0, where $b = \sum_{s, i<j} v^{(s)}(\alpha_i) v^{(s)}(\alpha_j) b_{i,j}^s$ and b_j is the determinant of the matrix obtained by omitting the j-th row with the sign $(-1)^{j-1}$ from the matrix $A = (v^{(i)}(\alpha_j))$, and $b_{i,j}^s$ is the determinant of the matrix obtained by omitting the i-th and j-th columns and s-th line from A. The map $\mathfrak{c}$ is well defined and is sequentially continuous and symbolic. Therefore, it induces the *tame symbol* — the pairing

$$K_n^{\mathrm{top}}(F)/(q-1) \times F^*/(q-1) \to \mu.$$

The tame symbol is the composition

$$K_{n+1}^{\mathrm{top}}(F) \xrightarrow{\partial} K_n^{\mathrm{top}}(k_{n-1}) \xrightarrow{\partial} \ldots \xrightarrow{\partial} K_1(k_0) = \mathbb{F}_q^*,$$

where ∂ are border homomorphism in K-theory, c.f. [FV, Ch. IX].

5. — Structure of K^{top}-groups

Note that for two principal units $\epsilon, \eta \in V_F$ the following holds in $K_2^M(F)$:

$$\{\epsilon, \eta\} = \{1 - \epsilon, 1 - (1 - \epsilon^{-1})(1 - \eta)\} + \{\eta, 1 - (1 - \epsilon^{-1})(1 - \eta)\}$$

with the principal unit $1 - (1 - \epsilon^{-1})(1 - \eta)$ of higher order than that of ϵ, η. This allows one to continue this process in $K_2^{top}(F)$ and finally to write $\{\epsilon, \eta\}$ as a sum of symbols in the form $\{\rho_i, t_i\}$ with principal units ρ_i and local parameters t_i. Note also that $\{\theta, \theta'\} = \{\theta, \epsilon\} = 0$ for $(q-1)$–th roots of unity θ, θ' and a principal unit ϵ. Therefore, $K_n^{top}(F)$ is *topologically generated* by the symbols

(1) $\{t_n, \ldots, t_1\}$,

(2) $\{\theta, t_n, \ldots, t_{i+1}, t_{i-1}, \ldots, t_1\}$ with a nonzero multiplicative representative θ, $1 \leqslant i \leqslant n$,

(3) $\{1 + \theta t_n^{i_n} \ldots t_1^{i_1}, t_n, \ldots, t_{i+1}, t_{i-1}, \ldots, t_1\}$ with a multiplicative representative θ, $1 \leqslant i \leqslant n$.

Topological relations among these generators (modulo p^r for each r in the case of $\mathrm{char}(F) = p$, modulo p^r in the case $\mathrm{char}(F) = 0$ and a primitive p^r–th root of unity belongs to F, modulo p if $\mathrm{char}(F) = 0$ and a primitive p–th root of unity doesn't belong to F) are revealed using the Artin–Schreier–Witt–Parshin, Vostokov and tame pairings, see [P1,P3,P4], [F1–F2]. Simulateneously one verifies that all the pairings are nondegenerate.

In particular, $K_n^{top}(F)$ is isomorphic to the product of the cyclic group generated by $\{t_1, \ldots, t_n\}$, n copies of the cyclic groups of order $q - 1$, and the subgroup $V K_n^{top}(F)$ generated by principal units V_F.

In the case of $\mathrm{char}(F) = p$ an extended theorem of Parshin [P4] claims that there exists an isomorphism and homeomorphism $\psi : \prod_J \mathcal{E}_J \to V K_m^{top}(F)$ with the sequential topology on $\prod \mathcal{E}_J$. Here J consists of $j_1, \ldots, j_{m-1}$ and runs all $(m-1)$-elements subsets of $\{1, \ldots, n\}$, $m \leqslant n+1$. $\mathcal{E}_J$ is the subgroup of V_F generated by $1 + \theta t_n^{i_n} \cdots t_1^{i_1}$, $\theta \in \mathcal{O}_0$, with restrictions that p doesn't divide $\gcd(i_1, \ldots, i_n)$ and the smallest index l for which i_l is prime to p doesn't belong to J. This provides an explicit and satisfactory description of the topology on $K_n^{top}(F)$ in the positive characteristic case. Note that in this theorem one should the topology on the multiplicative group defined in $2°$ instead of that defined in [P4] is used.

Let τ be the finest topology on the Milnor K_m^M-group such that for every $r \geqslant 1$ and every $\alpha_j^{(i)} = \lim_{v \to +\infty} \alpha_{j,v}^{(i)}$ in F^*, $1 \leqslant j \leqslant m$, $1 \leqslant i \leqslant r$, the sum $\sum_i \{\alpha_{1,v}^{(i)}, \ldots, \alpha_{m,v}^{(i)}\}$ converges to $\sum_{1 \leqslant i \leqslant r} \{\alpha_1^{(i)}, \ldots, \alpha_m^{(i)}\}$. Denote by $T_m(F)$ the intersection of all neighbourhoods of zero with respect to τ. The topology τ is finer than λ defined in $3°$. In the same way as in $3°$ every element x of

$K_m^M(F)$ can be written as a sum of a fixed number of element in the form $\{\alpha_i\} \cdot \{\text{some local parameters}\}$ with $\alpha_i \in F^*$ plus an element of $T_m(F)$.

One has $\cap_{l \geqslant 1} l K_m^M(F) \subset T_m(F)$, because (1) $\cap_{l \geqslant 1} l K_m^M(F) \subset V K_m^M(F)$ where $V K_m^M(F)$ is the group generated by principal units V_F of F, (2) lx can be written as the sum of symbols $\{\alpha^l\} \cdot \{\text{some local parameters}\}$ and an element of $T_m(F)$, (3) for any $\alpha_v \in V_F$ the sequence $\alpha_v^{p^v}$, $v \to +\infty$, converges to 1.

By induction on dimension one can prove that for $m \leqslant n$ the following holds.

(1) Let $r \geqslant 1$, $m \geqslant 2$, and let U be a neighbourhood of 1 in $1 + t_n \mathcal{O}_F$ with respect to the discrete valuation of rank 1. Then for $\alpha_1 \in V_F$, $\alpha_2, \ldots, \alpha_m \in F^*$ there exist elements $\beta_J \in V_F$ which sequentially continuously depend on $\alpha_1, \ldots, \alpha_m$ such that $\{\alpha_1, \ldots, \alpha_m\}$ can be written as $\sum \{\beta_J, t_{j_1}, \ldots, t_{j_{m-1}}\} \bmod p^r V K_m^M(F) + \{U\} K_{m-1}^M(F)$. Here $\{U\} K_{m-1}^M(F)$ is the subgroup of $K_m^M(F)$ generated by U.

(2) The quotient topology on $V K_m^M(F)/(p^r V K_m^M(F) + \{U\} K_{m-1}^M(F))$ of the product of the topologies on V_F^d, $d = \binom{N}{m-1}$ via the surjective homomorphism $(\beta_J) \to \sum \{\beta_J, t_{j_1}, \ldots, t_{j_{m-1}}\} \bmod p^r V K_m^M(F) + \{U\} K_{m-1}^M(F)$ is equivalent to the quotient topology of τ.

(3) In characteristic p the topology τ is equivalent to λ, $T_m(F) = \Lambda_m(F)$ and the space $V K_m^M(F)/T_m(F)$ with the quotient topology of τ is homeomorphic to $\prod_J \mathcal{E}_J$, $(\beta_J) \to \sum \{\beta_J, t_{j_1}, \ldots, t_{j_{m-1}}\} \bmod T_m(F)$.

(4) $T_m(F) = \cap_{l \geqslant 1} l K_m^M(F)$ and is a divisible group.

(5) The homomorphism $(V_F)^d \to V K_m^M(F)/T_m(F)$, $(\beta_J) \to \sum \{\beta_J, t_{j_1}, \ldots, t_{j_{m-1}}\} \bmod T_m(F)$ is surjective and the quotient topology of the product of the topologies on $(V_F)^d$ is equivalent to the quotient topology of τ on $K_m^M(F)/T_m(F)$.

(6) $T_m(F)$ coincides with $\Lambda_m(F)$ and is the intersection of all open subgroups of finite index in $K_m^M(F)$. The sequence $0 \to \Lambda_m(F) \to K_m^M(F) \to K_m^{\text{top}}(F) \to 0$ splits.

The proof goes as follows. We can assume that $\text{char}(k_{n-1}) = p$. Let $\alpha = 1 + \theta t_n^i \bmod t_n^{i+1} \mathcal{O}_F$ with $i > 0$ and let

$$\bar{\theta} = -\sum \bar{\theta}_{i_{n-1}, \ldots, i_1}^{p^r} \bar{t}_{n-1}^{i_{n-1}} \ldots \bar{t}_1^{i_1}$$

for the residue of θ and t_i and $0 \leqslant i_1, \ldots, i_{n-1} < p^r$. Then for $\beta \in V_F$

$$\{\alpha, \beta\} = \{1 - \sum \theta_{i_{n-1}, \ldots, i_1}^{p^r} t_n^i t_{n-1}^{i_{n-1}} \ldots t_1^{i_1}, \beta\} + \sum_v \{\gamma_v, \delta_v\}$$

with $\gamma_v \in 1 + t_n^{i+1} \mathcal{O}_F$, $\delta_v \in \mathcal{O}_F$ sequentially continuously depending on α, β. Furthermore, putting $\eta = \sum \theta_{i_{n-1}, \ldots, i_1}^{p^r} t_n^i t_{n-1}^{i_{n-1}} \ldots t_1^{i_1}$ we get that $-\{1 - \eta, \beta\}$

can be written $\bmod p^r K_2^M(F)$ as

$$\{1 - \eta(\beta - 1)/(1 - \eta), \beta\} + i\{1 - \eta(\beta - 1)/(1 - \eta), t_n\} + \ldots + \ldots$$
$$+ i_1\{1 - \eta(\beta - 1)/(1 - \eta), t_1\}.$$

If $\alpha, \beta \in V_F, \notin 1 + t_n \mathcal{O}_F$, then one can apply the induction assumption for k_{n-1} and the sequentially continuous lifting from k_{n-1} to F. This implies (1). Then (2) follows from definitions. To prove (3) it is sufficient to notice that there is a surjective map $\varphi : \prod_J \mathcal{E}_J \rightarrow V K_m^M(F)/\mathrm{T}_m(F)$ analogous to Parshin's map ψ. Since the composition of φ and the surjective map $V K_m^M(F)/\mathrm{T}_m(F) \rightarrow V K_m^M(F)/\Lambda_m(F)$ coincides with ψ, $\mathrm{T}_m(F) = \Lambda_m(F)$. As in (2) one deduces that φ is a homeomorphism.

To deduce (4) one applies (2) and then in characteristic p proves that $\cap_U p K_m^M(F) + \{U\} K_{m-1}^M(F) = p K_m^M(F)$ using injectivity of the differential symbol

$$d_F \colon K_m^M(K)/p \rightarrow \Omega_K^m, \qquad d_F\{\alpha_1, \ldots, \alpha_m\} = \frac{d\,\alpha_1}{\alpha_1} \wedge \ldots \wedge \frac{d\,\alpha_m}{\alpha_m}$$

which is a part of Bloch-Kato-Gabber's theorem [BK].

That implies $\mathrm{T}_m(F) = \cap_{l \geqslant 1} l K_m^M(F)$. From (3) it follows that $K_m^{\mathrm{top}}(F)$ doesn't have a nontrivial p-torsion, hence $\mathrm{T}_m(F)$ is a p-divisible group. If $\mathrm{char}(F) = 0$ and a primitive p-th root of unity belongs to F, then the p-torsion of $K_m^M(F)/\mathrm{T}_m(F)$ is generated by the p-torsion in F^* [F1,F2] (the proof uses Kato's theorem which claims that the homomorphism from $K_m^M(F)/p$ to $H^m(F, \mathbb{Z}/p)$ is an isomorphism). Then $\mathrm{T}_m(F)$ is a divisible group.

For $\alpha_1 \in V_F, \alpha_2, \ldots, \alpha_m \in F^*$ there exist elements $\beta_J \in V_F$ which sequentially continuously depend on $\alpha_1, \ldots, \alpha_m$ such that $\{\alpha_1, \ldots, \alpha_m\}$ can be written as $\sum\{\beta_J, t_{j_1}, \ldots, t_{j_{m-1}}\} \bmod \mathrm{T}_m(F)$. Then, due to the definition of the topology of F^*, the preimage in $(V_F)^d$ of an open set in $K_m^M(F)/\mathrm{T}_m(F)$ is open there.

Since the intersection of all open subgroups in $(V_F)^d$ is trivial, (6) follows from (5) and the observation that a subgroup in $K_m^M(F)$ is open in τ if and only if it is open in λ.

Let ρ be the finest topology on $K_m^M(F)$ for which the map from $(F^*)^m$ to $K_m^M(F)$ is sequentially continuous and the intersection of all neighbourhoods of zero in $K_m^M(F)$ contains $\cap_{l \geqslant 1} l K_m^M(F)$ (this topology was used in [F1,F2]). The previous statements imply that ρ is finer than τ and the intersection of all neighbourhoods of zero in $K_m^M(F)$ with respect to ρ coincides with $\cap_{l \geqslant 1} l K_m^M(F)$. On the level of subgroups all three topologies λ, ρ and τ coincide : a subgroup in $K_m^M(F)$ is open in λ if and only if it is open in ρ if and only if it is open in τ.

For $\operatorname{char}(F) = 0$ I. B. Zhukov found (applying higher class field theory) a complete algebraic description of $K_n^{\mathrm{top}}(F)$ in several cases (private communication). In particular, if $T_p K_n^{\mathrm{top}}(F)$ is the topological closure of the p-torsion in $K_n^{\mathrm{top}}(F)$ and F has a local parameter t_n algebraic over $\mathbb{Q}_p$, then $VK_n^{\mathrm{top}}(F)/T_p K_n^{\mathrm{top}}(F)$ possesses a topological basis of the form $\{\epsilon, t_{n-1}, \ldots, t_1\}$ with ϵ running free $\mathbb{Z}_p$-generators of the group of principal units of the algebraic closure of $\mathbb{Q}_p$ in F modulo its p-primary torsion.

For another approach to topologies on K_m^{M} see [Ka6].

6. — Norm

It follows easily from 5° that for a cyclic extension L/F of a prime degree $K_n^{\mathrm{top}}(L)$ is generated by L^* over the image of $K_{n-1}^{\mathrm{top}}(F)$.

In characteristic p there is a very simple way to define the norm mapping on topological K-groups :

(1) for a cyclic extension L/F of a prime degree introduce $N_{L/F}$: $K_n^{\mathrm{top}}(L) \to K_n^{\mathrm{top}}(F)$ as induced by the norm on K_1;

(2) for an arbitrary abelian extension L/F define the norm decomposing L/F in cyclic extension of prime degree.

Correctness of this definition follows from an application of the Artin–Schreier–Witt–Parshin and tame pairings. The norm on $K_n^{\mathrm{top}}(L)$ is dual to the map induced by the fields embedding $F \to L$, for details (one should replace K_2 by K_1 there) see [P4].

For an arbitrary multidimensional local field define the norm on $K^{\mathrm{top}}(F)$ as induced from the norm on Milnor K-groups. Compatibility of the just defined norm with induced from the Milnor K-groups follows then from 5°.

Hilbert Satz 90 plays a very significant role in K-theory. For a general type of fields and arbitrary cyclic extension it is still only known for K_2. If F is a higher local field, then Hilbert Satz 90 holds for K_n^{top}. The proof uses the description of the torsion in K_n^{top} in 5° and small Hilbert Satz 90 :
if L/F is of a prime degree l with a generator σ, then the sequence

$$K_n^{\mathrm{top}}(F)/l \oplus K_n^{\mathrm{top}}(L)/l \xrightarrow{\; i_{F/L} \oplus (1-\sigma) \;} K_n^{\mathrm{top}}(L)/l \xrightarrow{\; N_{L/F} \;} K_n^{\mathrm{top}}(F)/l$$

is exact, where $i_{F/L}$ is induced by the fields embedding. The latter theorem is verified by explicit calculations in K_n^{top}/l-groups whose structure is completely known due to the tame and Vostokov pairings (adjoin if necessary a primitive l–th root of unity ζ and then return without problems as $|F(\zeta) : F|$ is prime to l). Similar calculations show that the index of the norm group $N_{L/F} K_n^{\mathrm{top}}(L)$ in $K_n^{\mathrm{top}}(F)$ is $|L : F|$ when $|L : F|$ is prime [F1, F2].

Now we review 4 approaches to higher class field theory : of K. Kato [Ka1-Ka7], Y. Koya [Ko1-Ko2], A. N. Parshin [P1-P5] and the author [F1,F2,F4].

7. — Kato's approach

For a field F, K. Kato introduced remarkable groups $H^m(F)$ as follows.

(1) $H^m(F) = \varinjlim H^m(F, \mu_l^{\otimes(m-1)})$ for a field of characteristic 0, where μ_l is the group of all l-th roots of unity in F^{sep}, $\mu_l^{\otimes(m-1)}$ is the $(m-1)$-th tensor power, $l \geqslant 1$ and the homomorphisms of the inductive system are induced by the canonical injections $\mu_l^{\otimes(m-1)} \to \mu_{l'}^{\otimes(m-1)}$ when l divides l';

(2) $H^m(F) = \varinjlim H^m(F, \mu_l^{\otimes(m-1)}) \oplus \varinjlim H_{p^r}^m(F)$ for $\mathrm{char}(F) = p > 0$, where l runs all positive integers prime to p, r runs all positive integers. Here $H_{p^r}^m(F) = W_r(F) \otimes \underbrace{(F^* \otimes \cdots \otimes F^*)}_{m-1 \text{ times}}/J$, where J is the subgroup generated by the following three types of elements :

a) $(\mathrm{Frob}(y) - y) \otimes \beta_1 \otimes \ldots \otimes \beta_{m-1}$ with $y \in W_r(F)$, $\beta_i \in F^*$;

b) $\underbrace{(0, \ldots, 0, \beta_1, 0, \ldots, 0)}_{i \text{ times}}(\beta_1) \otimes \beta_1 \otimes \ldots \otimes \beta_{m-1}$ with $0 \leqslant i < r$;

c) $y \otimes \beta_1 \otimes \ldots \otimes \beta_{m-1}$ with $\beta_i = \beta_j$ for some $i \neq j$.

Equivalently one can put $H_{p^r}^m(F)$ to be $H^1(F, W_r\Omega_{F,\log}^{m-1})$ where $W_r\Omega_{F,\log}^{m-1}$ is the logarithmic part of the De Rham–Witt complex.

For any field F the group $H^1(F)$ is isomorphic to the group of all continuous homomorphisms $\mathrm{Gal}(F^{\mathrm{ab}}/F) \to \mathbb{Q}/\mathbb{Z}$ and $H^2(F)$ is isomorphic to $\mathrm{Br}(F)$.

For an n-dimensional local field F a celebrated theorem of Kato claims that there exists a canonical homomorphism $H^{n+1}(F) \simeq \mathbb{Q}/\mathbb{Z}$. This is an analog of the classical theorem describing the Brauer group of a local field with finite residue field. The proof of Kato's theorem is easy for the prime to p part where it follows from typical arguments involving the Hochshild–Serre spectral sequence. The proof is more difficult for the p part and relies in particular on relations among quotients of Milnor K-groups, Galois cohomology groups and subquotient modules in the module of differentials of fields of positive characteristic. Some ingredients are the theorem of Kato on the residue symbol $K_n^M(F)/p \to H^n(F, \mathbb{Z}/p)$ mentioned in 5° and the study of the cohomological residue $H^{n+1}(F, \mu_l^{\otimes(n)}) \to H^n(k_{n-1}, \mu_l^{\otimes(n-1)})$, see [Ka3–Ka4], [R]. In fact, many results established by Kato in [Ka3–Ka4] hold for arbitrary complete discrete valuation fields.

Using the canonical pairing

$$H^1(F) \times K_n^M(F) \to H^1(F) \times H^n(F) \to H^{n+1}(F) \simeq \mathbb{Q}/\mathbb{Z}$$

one obtains the *higher local reciprocity map*

$$\Psi_F^M \colon K_n^M(F) \to \mathrm{Gal}(F^{\mathrm{ab}}/F).$$

It describes finite abelian extensions L/F in a sense that Ψ_F^M induces an isomorphism of $K_n^M(F)/N_{L/F}K_n^M(L)$ onto $\mathrm{Gal}(L/F)$.

Kato has also proved existence theorem which describes norm subgroups [Ka6]. His approach is different from that of 10°.

8. — Koya's approach

The previous theory can be treated as a generalization of Tate's approach in classical class field theory. For one-dimensional fields the notion of formation of classes seems to be the most standard way to define the reciprocity map. There is an important obstruction for its generalization to higher-dimensional fields. Already for (> 1)-dimensional fields the Galois descent for K_n^{top} fails : if L/F is a finite Galois extension, then $i_{F/L}\colon K_n^{\mathrm{top}}(F) \to K_n^{\mathrm{top}}(L)$ induced by $F \to L$ isn't in general injective, and $i_{F/L}K_n^{\mathrm{top}}(F)$ doesn't in general coincide with the $\mathrm{Gal}(L/F)$-invariant elements of $K_n^{\mathrm{top}}(L)$ (the same is true for Milnor K-groups).

For instance, consider $K = \mathbb{Q}_p(\zeta)$ with a primitive p-th root of unity ζ. Let ω be a p-primary element in K, i.e. a principal unit such that $K_1 = K(\sqrt[p]{\omega})$ is the unramified extension of degree p over K. Let t be a transcendental element over K and let $F = K\{\{t\}\}$. Cõnsider the totally ramified (with respect to the 2–dimensional structure) extension $L = F(\sqrt[p]{t})$ of degree p. Put $F_1 = FK_1$, $L_1 = LF_1$. Then, according to properties of K_2 of a local field (see, e.g., [FV, Ch. IX]) for a prime element $\pi = 1 - \zeta$ of K the symbol $i_{K/K_1}\{\omega, \pi\}$ is a divisible element in $K_2^M(K_1)$, since $i_{K/K_1}\{\omega, \pi\}$ belongs to $pK_2^M(K_1)$. Hence $i_{F/F_1}(i_{K/F}\{\omega, \pi\}) = 0$ in $K_2^{\mathrm{top}}(F_1)$, but from explicit calculations it follows that $i_{K/F}\{\omega, \pi\} \notin N_{L/F}K_2^{\mathrm{top}}(L)$. Since $\sigma\{\sqrt[p]{t}, \pi\} - \{\sqrt[p]{t}, \pi\}$ is a multiple of $\{\zeta, 1 - \zeta\} = 0$ for a generator σ of L/F, the symbol $\{\sqrt[p]{t}, \pi\}$ is σ-invariant but it doesn't belong to $i_{F/L}K_2^{\mathrm{top}}(F)$.

Y. Koya found a class formation approach to higher class field theory using bounded complexes of Galois modules and their modified hypercohomology groups $\widehat{\mathbb{H}}$ instead of respectively Galois modules and their modified (Tate) cohomology groups [Ko1-Ko2]. His generalized Tate–Nakayama theorem claims that if for a finite group G and a bounded complex $\mathbb{A}$ of G-modules there is an element $a \in \widehat{\mathbb{H}}^2(G, \mathbb{A})$ such that for every prime l and Sylow l-group G_l of G the group $\widehat{\mathbb{H}}^1(G_l, \mathbb{A})$ is trivial and the group $\widehat{\mathbb{H}}^2(G_l, \mathbb{A})$ is generated by the symbol $\mathrm{res}_{G/G_l}(a)$ of order $|G_l|$, then for every subgroup H of G and $i \in \mathbb{Z}$ the cup-product with $\mathrm{res}_{G/H}(a)$ induces an isomorphism of the Tate cohomology group $\widehat{H}^{i-2}(H, \mathbb{Z})$ onto $\widehat{\mathbb{H}}^i(H, \mathbb{A})$.

Koya's generalized axioms of class formation for a profinite group G and a bounded complex $\mathbb{A}$ of G-modules are the following :

(1) $\widehat{\mathbb{H}}^i(U, \mathbb{A}) = 0$ for every open subgroup U in G and $i = 1$;

(2) for every open subgroup U in G there is a canonical isomorphism

$$\mathrm{inv}_U \colon \widehat{\mathbb{H}}^2(U, \mathbb{A}) \to \mathbb{Q}/\mathbb{Z}$$

such that $\mathrm{inv}_V \circ \mathrm{res}_{U/V} = |U : V| \mathrm{inv}_U$ for every pair of open subgroups $V \subset U$ in G.

For a 2-dimensional local field F it follows from results of S. Saito [S] and Koya [Ko1] that the shifted Lichtenbaum complex $\mathbb{Z}(2)_F[2]$ satisfies the generalized axioms of class formation. Then for every open normal subgroup $U = \mathrm{Gal}(F^{\mathrm{sep}}/L)$ of $G_F = \mathrm{Gal}(F^{\mathrm{sep}}/F)$ the finite group $\mathrm{Gal}(L/F)$ and the complex $\mathbb{A} = \tau_{\leqslant 0}(\mathbb{R}\Gamma(U, \mathbb{Z}(2)_F[2]))$ satisfy the assumption of the generalized Tate–Nakayama theorem. The isomorphism of this theorem for $i = 0$ implies the isomorphism $K_2^{\mathrm{M}}(F)/N_{L/F}K_2^{\mathrm{M}}(L) \to \mathrm{Gal}(L/F)^{\mathrm{ab}}$. This is a more general result than that in 7°, because it provides certain information about nonabelian extensions.

Since a Lichtenbaum complex $\mathbb{Z}(n)$ for $n > 2$ has not yet been constructed, one can't extend this approach for higher dimensions directly. Recently M. Spieß [Sp] proved that for an n-dimensional local field F the shifted complex $\mathbb{Z}'(n)_F[n]$, where $\mathbb{Z}'(n)$ either is the decomposable part of motivic cohomology $\mathbb{G}_m \otimes^{\mathbb{L}n} [-n]$ studied by B. Kahn in [Kn] or is $\tau_{\geqslant 0} \mathcal{Z}^n(F^{\mathrm{sep}}, 2n - .)$ with $\mathcal{Z}^{\cdot}(F^{\mathrm{sep}}, \cdot)$ being the Bloch complex [B], satisfies Koya's class formation axioms. Thus there is an isomorphism

$$K_n^{\mathrm{M}}(F)/N_{L/F}K_n^{\mathrm{M}}(L) \to \mathrm{Gal}(L/F)^{\mathrm{ab}}.$$

9. — Parshin's approach

A. N. Parshin was the first who suggested a K-theoretic approach to higher class field theory. His celebrated theory for multidimensional local fields of characteristic p [P1–P5] is more accessible than that of Kato. In fact, we need to say only few things in addition to those in 2°–4°, 6° in order to describe the reciprocity map in characteristic p.

Recall that for an n-dimensional local field F of characteristic p the torsion subgroup $TK_n^{\mathrm{top}}(F)$ in $K_n^{\mathrm{top}}(F)$ is isomorphic to $(k_0^*)^n$, $k_0 = \mathbb{F}_q$ (see 5°). Put $\tilde{K}(F) = K_n^{\mathrm{top}}(F)/TK_n^{\mathrm{top}}(F)$. Let L/F be a cyclic extension of degree p^m with a generator σ. Then the sequence

$$0 \longrightarrow \tilde{K}(F) \xrightarrow{i_{L/F}} \tilde{K}(L) \xrightarrow{1-\sigma} \tilde{K}(L) \xrightarrow{N_{L/F}} \tilde{K}(F)$$

is exact. The proof is a la [KS] : denote the limit

$$\varinjlim W_r(F)/(\mathrm{Frob} - 1)W_r(F)$$

with respect to usual maps $(\alpha_0, \ldots, \alpha_{r-1}) \to (0, \alpha_0, \ldots, \alpha_{r-1})$ by $W(F)$. Then the Artin–Schreier–Witt–Parshin pairing via extended Pontryagin duality [P5, §1,§2] provides a duality between $\tilde{K}(F)$ and $W(F)$. Therefore, the exactness of the above sequence can be easily deduced from from the exactness of the sequence

$$W(F) \longrightarrow W(L) \xrightarrow{1-\sigma} W(L) \xrightarrow{\mathrm{Tr}_{L/F}} W(F) \longrightarrow 0.$$

Note that the above sequence for $\tilde{K}$-groups is exact even for Milnor K-groups of an arbitrary field F of characteristic p due to a deep theorem of O. T. Izhboldin [Izh].

Now it is straightforward to show that $\tilde{K}$ is class formation in the category of p-extensions of F and in this way the p part of class field theory follows. Nondegeneracy of the tame pairing of 4° and the Kummer theory provide the prime to p part map $K_n^{\mathrm{top}}(F) \to \mathrm{Gal}(F(\sqrt[q-1]{F^*})/F)$, that of the Artin–Schreier–Witt–Parshin pairing (the p part map $K_n^{\mathrm{top}}(F) \to \mathrm{Gal}(F^{\mathrm{abp}}/F)$ where F^{abp} is the maximal abelian p-extension of F). There is the third map which transforms the symbol $\{t_1, \ldots, t_n\}$ for a system of local parameters $t_n, \ldots, t_1$ in F to the lifting of the Frobenius automorphism on $F \otimes_{\mathbb{F}_q} \mathbb{F}_q^{\mathrm{sep}}$. All three maps are compatible, and their stitching is the *reciprocity map*

$$K_n^{\mathrm{top}}(F) \to \mathrm{Gal}(F^{\mathrm{ab}}/F).$$

Thus, the whole construction of the reciprocity map in the Parshin theory is cohomology free. In contrast to the Milnor K-groups used in the previous theories, the group $K_n^{\mathrm{top}}(F)$ describing abelian extensions is completely known, see 5°– 6°.

10. — Explicit approach

Finally we describe main ideas of the approach to local class field theories in [F1,F2,F4]. This approach is a generalization of two explicit constuctions of the reciprocity map and its inverse one for classical local fields due to M. Hazewinkel [H1–H3] and J. Neukirch [N1–N2], essential ingredients are the tame, Artin–Schreier–Witt–Parshin, Vostokov pairings and topological K-groups.

For an n-dimensional local field F denote by $v_F \colon K_n^{\mathrm{top}}(F) \to \mathbb{Z}$ the composition

$$K_n^{\mathrm{top}}(F) \xrightarrow{\partial} K_{n-1}^{\mathrm{top}}(k_{n-1}) \xrightarrow{\partial} \cdots \xrightarrow{\partial} K_0(k_0) = \mathbb{Z},$$

where ∂ are the border homomorphism in K-theory. An element Π_F of $K_n^{\mathrm{top}}(F)$ which is mapped to 1 is called prime. Its role in higher class field theory is in many respects similar to the role of a prime element of a classical

local field. Given a system of local parameters $t_n, \ldots, t_1$ a prime element can
be written as $\{t_1, \ldots, t_n\} + \epsilon$ with $\epsilon \in \ker v_F$. Let $\widetilde{F}$ be the maximal purely
unramified extension of F i.e. the unramified extension with respect to n
structure corresponding to k_0^{sep}/k. The Galois group of $\widetilde{F}/F$ has a canonical
generator — the lifting of the Frobenius automorphism from G_{k_0} which is
called by the same name.

The inverse map to the reciprocity map can be explicitly described as
follows : let L/F be a finite Galois extension, attach to an automorphism
σ the element $N_{\Sigma/F}\Pi_\Sigma \bmod N_{L/F}K_n^{\text{top}}(L)$ where Π_Σ is any prime element
of $K_n^{\text{top}}(\Sigma)$ and Σ is the fixed field of a lifting of the σ on $\text{Gal}(\widetilde{L}/F)$ such
that its restriction on $\text{Gal}(\widetilde{F}/F)$ is a positive integer power of the Frobenius
automorphism. This is a direct generalization of the Neukirch definition
in the classical case. The main result is that the map just defined doesn't
depend on the choice of lifting of σ and the choice of a prime element, and
induces an isomorphism of groups

$$\mathcal{Y}_{L/F} \colon \text{Gal}(L/F)^{\text{ab}} \to K_n^{\text{top}}(F)/N_{L/F}K_n^{\text{top}}(L)$$

[F1,F2,F4].

The proof is essentially based on Hilbert Satz 90 and the norm index
calculation for extensions of a prime degree in 6°. It is convenient to
introduce the second map acting in inverse direction from topological K-
groups to the Galois group as a generalization of the Hazewinkel description
of the reciprocity map for classical local fields. However, in complete extent
this can be done only in characteristic p.

Put $K_n^{\text{top}}(\widetilde{F}) = \varinjlim K_n^{\text{top}}(F')$ with F' running finite subextensions of F'
in $\widetilde{F}$.

For the fields of positive characteristic the Galois descent for the topological K_n-groups holds (see, for example, [F2]). Given a finite Galois extension
L/F linearly disjoint with $\widetilde{F}/F$ denote by $V(L|F)$ the subgroup in $K_n^{\text{top}}(\widetilde{L})$
generated by elements $\sigma\alpha - \alpha$ with $\sigma \in \text{Gal}(\widetilde{L}/\widetilde{F})$, $\alpha \in VK_n^{\text{top}}(\widetilde{L})$. Then the
sequence

$$1 \longrightarrow \text{Gal}(\widetilde{L}/\widetilde{F}) \xrightarrow{c} K_n^{\text{top}}(\widetilde{L})/V(L|F) \xrightarrow{N_{\widetilde{L}/\widetilde{F}}} K_n^{\text{top}}(\widetilde{F}) \longrightarrow 0$$

is exact where $c(\sigma) = \sigma\Pi_L - \Pi_L$ modulo $V(L|F)$ doesn't depend on the
choice of Π_L [F4].

This allows one to define for a finite Galois extension L/F linearly
disjoint with $\widetilde{F}/F$ a generalization of the Hazewinkel homomorphism

$$\Psi_{L/F} \colon K_n^{\text{top}}(F)/N_{L/F}K_n^{\text{top}}(L) \to \text{Gal}(L/F)^{\text{ab}}$$

as follows. Given an element $\epsilon \in \ker v_F$ write it as $N_{\widetilde{L}/\widetilde{F}}\eta$ with $\eta \in K_n^{\mathrm{top}}(\widetilde{L})$. Then for a lifting $\varphi \in \mathrm{Gal}(\widetilde{L}/F)$ of the Frobenius automorphism the element $\varphi\eta - \eta$ belongs to the kernel of $N_{\widetilde{L}/\widetilde{F}}$ and according to the description of this kernel can be written as $\tilde{\sigma}\Pi_L - \Pi_L$ modulo $V(L|F)$ with $\tilde{\sigma} \in \mathrm{Gal}(\widetilde{L}/\widetilde{F})$. The generalized Hazewinkel map attaches the automorphism $\tilde{\sigma}^{-1}|_{L\cap F^{\mathrm{ab}}}$ to $\epsilon \bmod N_{L/F}K_n^{\mathrm{top}}(L)$. It is a well defined homomorphism. The generalized Neukirch and Hazewinkel maps are inverse to each other and thus are isomorphisms. The whole theory here is cohomology free similar to the Parshin theory.

In the case of characteristic zero all essential problems are concentrated in p-extensions. There is a class of p-extensions which are very close to extensions in positive characteristic (so-called $\wp$-extensions (or Artin–Schreier towers) which are towers of subsequent cyclic extensions of degree p generated at each step by roots of an Artin–Schreier polynomial $\wp(X) - a = X^p - X - a$. One can prove that for a cyclic $\wp$-extension L/F linearly disjoint with $\widetilde{F}/F$ a weak Galois descent holds : the homomorphism v_F maps the $\mathrm{Gal}(\widetilde{L}/\widetilde{F})$-invariant elements of $K_n^{\mathrm{top}}(\widetilde{L})$ onto $|L : F|\mathbb{Z}$ [F4, sect. 3]. A generalized Hazewinkel map $\Psi_{L/F}$ for an arbitrary extension in the case of characteristic zero doesn't exist, see 8°. However, it can be defined by the same rule as in the positive characteristic case above for a finite Galois p-extension L/F linearly disjoint with $\widetilde{F}/F$ which is an Artin–Schreier tree (AST) that means that every cyclic intermediate subextension in L/F is a $\wp$-extension. Then for an AST-extension L/F the composition $\Psi_{L/F} \circ \mathcal{Y}_{L\cap F^{\mathrm{ab}}/F}$ is identity. AST-extensions are "dense" in the class of all p-extensions : for a finite Galois p-extension linearly disjoint with $\widetilde{F}/F$ there exists a p-extension Q/F linearly disjoint with $\widetilde{F}/F$ such that $\widetilde{Q}\cap\widetilde{L} = \widetilde{F}$ and any intermediate cyclic extension in LQ/Q is an AST-extension. This allows one to prove that $\mathcal{Y}_{L/F}$ is an isomorphism [F4]. For another proof when three assertions : Hilbert Satz 90, $|K_n^{\mathrm{top}}(F) : N_{L/F}K_n^{\mathrm{top}}(L)| = |L : F|$, $\mathcal{Y}_{L/F}$ is an isomorphism are verified for a cyclic extension L/F by simultaneous induction on degree, see [F1].

Now for an n-dimensional local field F passing to the projective limit for $\Psi_{L/F}$ when L/F runs all abelian subextensions in F^{ab}/F we obtain the *reciprocity map*

$$\Psi_F^{\mathrm{top}} : K_n^{\mathrm{top}}(F) \to \mathrm{Gal}(F^{\mathrm{ab}}/F).$$

It is compatible with the reciprocity maps defined in 7°–10°.

The reciprocity map Ψ_F^{top} is injective and its image is dense in $\mathrm{Gal}(F^{\mathrm{ab}}/F)$. The maximal divisible subgroup $\Lambda_n(F)$ of $K_n^{\mathrm{M}}(F)$ coincides with the intersection of all open subgroups of finite index in $K_n^{\mathrm{M}}(F)$ by 5° ; the latter is the kernel of Ψ_F^{M} due to existence theorem : the lattice of open subgroups of finite index in $K_n^{\text{top}}(F)$ is in an order reversing bijection with the lattice of the finite abelian extensions L/F, $L \to N_{L/F}K_n^{\text{top}}(L)$ [F1,F2].

Using the description of the topology on the Milnor K-groups one can verify that for a finite Galois extension M/F the preimage of an open subgroup of finite index in $K_n^{\text{top}}(F)$ is an open subgroup of finite index in $K_n^{\text{top}}(M)$ and $N_{M/F}K_n^{\text{top}}(M)$ is an open subgroup of finite index in $K_n^{\text{top}}(F)$. Then it is sufficient to prove existence theorem for a prime index. The abelian extension attached to an open subgroup is constructed then as corresponding to the annihilator of the open subgroup via the Artin–Schreier–Witt–Parshin, Vostokov and tame pairings (again, if necessarily, adjoining a root of unity and then descending).

For another approach to existence theorem see [Ka6].

There are several works on class field theory of a local field with a global residue field in terms of K_2-idele groups : [Ka5], [Ko3], [Kuc]. In the next part we describe totally ramified abelian p-extensions of a complete discrete valuation field with arbitrary residue field of characteristic p in terms of the group of principal units.

Totally ramified abelian p-extensions and the group of principal units

Let F be a complete discrete valuation field with residue field $\overline{F}$ of characteristic p. In this part we deal with reciprocity maps describing abelian totally ramified p-extensions of F in terms of subquotients of the group of principal units of F [F3,F5]. We indicate then their relations with Miki's [M2] and Kurihara's results [Kur2].

11. — Perfect residue fields

Let F be a local field with a perfect residue field $\overline{F}$ of characteristic $p > 0$. Let $\wp(X)$ denote as above the polynomial $X^p - X$. Put $\kappa = \dim_{\mathbb{F}_p} \overline{F}/\wp(\overline{F})$. We will assume that $\kappa \neq 0$, the case $\kappa = 0$ when the field $\overline{F}$ is algebraically p-closed may be treated as a limit of class field theories of local fields with nonalgebraically p-closed residue field tending to $\overline{F}$.

To describe the maximal abelian extension F^{ab}/F one must study abelian prime to p-extensions and abelian p-extensions. Totally tamely ramified abelian extensions over F are easily described by the Kummer theory, since any such extension L/F is generated by adjoining a root $\sqrt[l]{\pi}$ for a suitable prime element π in F and in this case a primitive l-th root of unity belongs to F.

The description of the maximal unramified abelian p-extension follows from the Witt theory. Thus, the nontrivial part is study of abelian totally ramified p-extensions of F. A variant of their description in terms of constant pro-quasi-algebraic groups as a generalization of geometric Serre's class field theory was furnished by M. Hazewinkel ([H1–H2]). We describe another approach to abelian totally ramified p-extensions which is cohomology free and of explicit nature [F3].

Let $\widetilde{F}$ denote the maximal abelian unramified p-extension of F. The Witt theory shows that $\mathrm{Gal}(\widetilde{F}/F) \simeq \mathbb{Z}_p^{\kappa}$. Let L/F be a Galois totally ramified p-extension, then $\mathrm{Gal}(L/F)$ can be identified with $\mathrm{Gal}(\widetilde{L}/\widetilde{F})$. Let

$$\mathrm{Gal}(L/F)^{\sim} = \mathrm{Hom}_{\mathbb{Z}_p}\big(\mathrm{Gal}(\widetilde{F}/F), \mathrm{Gal}(L/F)\big)$$

denote the group of continuous homomorphisms from the profinite group $\mathrm{Gal}(\widetilde{F}/F)$ which is a $\mathbb{Z}_p$-module ($a \cdot \sigma = \sigma^a$, $a \in \mathbb{Z}_p$) to the discrete $\mathbb{Z}_p$-module $\mathrm{Gal}(L/F)$. This group is isomorphic (non-canonically) with $\mathrm{Gal}(L/F)^{\oplus \kappa}$. Now let L/F be of finite degree. Let $\chi \in \mathrm{Gal}(L/F)^{\sim}$ and Σ_{χ} be the fixed field of all $\tau_{\varphi} \in \mathrm{Gal}(\widetilde{L}/F)$, where $\tau_{\varphi}|_{\widetilde{F}} = \varphi, \tau_{\varphi}|_{L} = \chi(\varphi)$ and φ runs a topological $\mathbb{Z}_p$-basis of $\mathrm{Gal}(\widetilde{F}/F)$. Then $\Sigma_{\chi} \cap \widetilde{F} = F$, i.e., Σ_{χ}/F is a totally ramified p-extension. Let U_F and $U_{1,F}$ be the groups of units and the group of principal units of F respectively. Let π_{χ} be a prime element of Σ_{χ}. Put

$$\Upsilon_{L/F}(\chi) = N_{\Sigma_{\chi}/F}\pi_{\chi}N_{L/F}\pi_L^{-1} \bmod N_{L/F}U_L,$$

where π_L is a prime element in L. The group $U_F/N_{L/F}U_L$ is mapped isomorphically onto the group $U_{1,F}/N_{L/F}U_{1,L}$ (multiplicative representatives are mapped to 1). So we obtain the map

$$\Upsilon_{L/F} : \mathrm{Gal}(L \cap F^{\mathrm{ab}}/F)^{\sim} \to U_{1,F}/N_{L/F}U_{1,L}$$

which is well defined and is another generalization of the Neukirch map of [N1–N2].

Note that there is an analog of the exact sequence in 10° :

$$1 \longrightarrow \mathrm{Gal}(\widetilde{L}/\widetilde{F})^{\mathrm{ab}} \longrightarrow U_{1,\widetilde{L}}/V(L|F) \xrightarrow{\; N_{\widetilde{L}/\widetilde{F}} \;} U_{1,\widetilde{F}} \longrightarrow 1,$$

where $V(L|F)$ is generated by $\epsilon^{\sigma-1}$ with $\epsilon \in U_{1,\widetilde{L}}$. Now define the generalized Hazewinkel map $\Psi_{L/F}$ as follows. Let $\epsilon \in U_{1,F}$ and $\phi \in \mathrm{Gal}(\widetilde{F}/F)$. Write $\epsilon = N_{\widetilde{L}/\widetilde{F}}\eta$ with $\eta \in U_{1,\widetilde{L}}$. Let $\varphi \in \mathrm{Gal}(\widetilde{L}/F)$ be a lifting of ϕ. Then $\eta^{-1}\varphi(\eta) \equiv \pi_L\sigma(\pi_L^{-1}) \bmod V(L|F)$ for a suitable $\sigma \in \mathrm{Gal}(\widetilde{L}/\widetilde{F})^{\mathrm{ab}}$ where π_L is a prime element in L. Set $\chi(\phi) = \sigma|_{L \cap F^{\mathrm{ab}}}$. Then $\chi \in (\mathrm{Gal}(L \cap F^{\mathrm{ab}}/F))^{\sim}$. Put $\Psi_{L/F}(\epsilon) = \chi$. The main result is that $\Upsilon_{L/F}$ and $\Psi_{L/F}$ are inverse isomorphisms with natural functorial properties [F3]. Thus, the quotients

$U_{1,F}/N_{L/F}U_{1,L}$ still as in classical theories describe abelian extensions. However they are roughly κ times larger than the Galois group of L/F.

Passing to the projective limit one obtains the *reciprocity map*

$$\Psi_F : U_{1,F} \longrightarrow \mathrm{Hom}_{\mathbb{Z}_p}\big(\mathrm{Gal}(\widetilde{F}/F), \mathrm{Gal}(F^{\mathrm{abp}}/\widetilde{F})\big),$$

where $U_{1,F}$ is the group of principal units, F^{abp} is the maximal abelian p-extension of F.

By using extended theory of additive polynomials one can describe for a fixed prime element π of F those open subgroups of finite index in $U_{1,F}$ (normic subgroups) which are norm groups $N_{L/F}U_{1,L}$ for finite abelian totally ramified p-extensions L/F such that π belongs to $N_{L/F}L^*$. Existence theorem in the perfect residue field case claims that for a fixed prime π in F the lattice of abelian extensions L/F such that $\pi \in N_{L/F}L^*$ is in order reversing bijection with the lattice of normic subgroups in $U_{1,F}$ [F3].

We note that there is a synthesis of the theories of 10° and 12° : a description of totally ramified with respect to n-dimensional structure abelian p-extensions of an n-dimensional local field with last residue field being perfect of characteristic p [F4].

12. — General residue field case

Let F be a complete discrete valuation field with arbitrary residue field $\overline{F}$ which isn't separably p-closed.

Denote again by $\widetilde{F}$ the maximal unramified abelian p-extension of F, i.e. the unramified extension corresponding to the maximal abelian p-extension $\overline{F}^{\mathrm{abp}}$ of the residue field $\overline{F}$.

Let L/F be a totally ramified finite Galois p-extension. Note that if $\epsilon = N_{\widetilde{L}/\widetilde{F}}\beta$ with $\beta \in U_{\widetilde{L}}$, then one can write $\beta = \theta\eta$ with $\theta \in U_L, \eta \in U_{1,\widetilde{L}}$ and then $\epsilon' = N_{\widetilde{L}/\widetilde{F}}\eta \in U_{1,F} \cap N_{\widetilde{L}/\widetilde{F}}U_{1,\widetilde{L}}$ is uniquely defined mod $N_{L/F}U_{1,L}$. Thus, the quotient group $(U_F \cap N_{\widetilde{L}/\widetilde{F}}U_{\widetilde{L}})/N_{L/F}U_L$ is mapped isomorphically onto $(U_{1,F} \cap N_{\widetilde{L}/\widetilde{F}}U_{1,\widetilde{L}})/N_{L/F}U_{1,L}$ by $\epsilon \to \epsilon'$.

In the same way as in the perfect residue field case introduce the generalized Neukirch map

$$\Upsilon_{L/F}: \mathrm{Gal}(L \cap F^{\mathrm{ab}}/F)^{\sim} \to (U_{1,F} \cap N_{\widetilde{L}/\widetilde{F}}U_{1,\widetilde{L}})/N_{L/F}U_{1,L}.$$

Assume that the residue field of F is not perfect. Denote by $\mathcal{F} = \mathrm{P}(F)$ a complete discrete valuation field which is an extension of F such that $e(\mathcal{F}|F) = 1$ and the residue field of $\mathcal{F}$ is the perfection $\overline{F}^{\mathrm{perf}} = \cup_{i \geqslant 1}\overline{F}^{p^{-i}}$ of the residue field of F ($\mathcal{F}$ isn't uniquely defined). In the same way define $\mathfrak{F} = \mathrm{P}(\widetilde{F})$.

For $\sigma \in \mathrm{Gal}(L/F)$ put $c(\sigma) = \pi_L^{-1}\sigma\pi_L \bmod V(L|F)$, where π_L is a prime element in L, and $V(L|F)$ is the subgroup of $U_{1,\widetilde{L}}$ generated by the elements $\epsilon^{-1}\sigma(\epsilon)$ with $\epsilon \in U_{1,\mathfrak{L}}$, $\sigma \in \mathrm{Gal}(L/F)$, $\mathfrak{L} = L\mathfrak{F}$. Then the sequence

$$1 \longrightarrow \mathrm{Gal}(L/F)^{\mathrm{ab}} \xrightarrow{\ c\ } U_{1,\widetilde{L}}/V(L|F) \xrightarrow{\ N_{\widetilde{L}/\widetilde{F}}\ } N_{\widetilde{L}/\widetilde{F}}U_{1,\widetilde{L}}$$

analogous to $10°$ and $11°$ is exact.

Now we introduce a reciprocity map acting in inverse direction with respect to $\Upsilon_{L/F}$. Let $\epsilon \in U_{1,F} \cap N_{\widetilde{L}/\widetilde{F}}U_{1,\widetilde{L}}$ and $\phi \in \mathrm{Gal}(\widetilde{F}/F)$. Let $\eta \in U_{1,\widetilde{L}}$ be such that $N_{\widetilde{L}/\widetilde{F}}\eta = \epsilon$. Then for an extension $\varphi \in \mathrm{Gal}(\widetilde{L}/F)$ of ϕ one can write $\eta^{-1}\varphi(\eta) \equiv c(\sigma^{-1})$ for a suitable $\sigma \in \mathrm{Gal}(L/F)^{\mathrm{ab}}$, where π_L is a prime element in L. Set $\chi(\phi) = \sigma$. Then $\chi \in (\mathrm{Gal}(L \cap F^{\mathrm{ab}}/F))^{\sim}$. Put $\Psi_{L/F}(\epsilon) = \chi$. The generalized Hazewinkel map $\Psi_{L/F}$: $(U_{1,F} \cap N_{\widetilde{L}/\widetilde{F}}U_{1,\widetilde{L}})/N_{L/F}U_{1,L} \longrightarrow (\mathrm{Gal}(L \cap F^{\mathrm{ab}}/F))^{\sim}$ is well defined and a homomorphism [F5]. The composition $\Psi_{L/F} \circ \Upsilon_{L\cap F^{\mathrm{ab}}/F}$ is identity.

Put $\mathcal{L} = L\mathcal{F}$. The maps $\Upsilon_{\mathcal{L}/\mathcal{F}}$ and $\Psi_{\mathcal{L}/\mathcal{F}}$ defined in $11°$ are compatible with their descendants for L/F : the diagram

$$
\begin{array}{ccccc}
\mathrm{Gal}(L\cap F^{\mathrm{ab}}/F)^{\sim} & \xrightarrow{\ \Upsilon_{L/F}\ } & (U_{1,F}\cap N_{\widetilde{L}/\widetilde{F}}U_{1,\widetilde{L}})/N_{L/F}U_{1,L} & \xrightarrow{\ \Psi_{L/F}\ } & (\mathrm{Gal}(L\cap F^{\mathrm{ab}}/F))^{\sim} \\
\downarrow & & \downarrow{\scriptstyle \lambda_{L/F}} & & \downarrow \\
\mathrm{Gal}(\mathcal{L}\cap\mathcal{F}^{\mathrm{ab}}/\mathcal{F})^{\sim} & \xrightarrow{\ \Upsilon_{\mathcal{L}/\mathcal{F}}\ } & U_{1,\mathcal{F}}/N_{\mathcal{L}/\mathcal{F}}U_{1,\mathcal{L}} & \xrightarrow{\ \Psi_{\mathcal{L}/\mathcal{F}}\ } & (\mathrm{Gal}(\mathcal{L}\cap\mathcal{F}^{\mathrm{ab}}/\mathcal{F}))^{\sim}
\end{array}
$$

($\lambda_{L/F}$ is induced by the inclusion) is commutative. Since $\Psi_{\mathcal{L}/\mathcal{F}}$ is injective, we deduce that $\lambda_{L/F}$ is surjective and

$$\ker \Psi_{L/F} = \ker \lambda_{L/F} = (U_{1,F} \cap N_{\widetilde{L}/\widetilde{F}}U_{1,\widetilde{L}} \cap N_{\mathcal{L}/\mathcal{F}}U_{1,\mathcal{L}})/N_{L/F}U_{1,L}.$$

In other words, $\Psi_{L/F}$ induces the isomorphism

$$(U_{1,F} \cap N_{\widetilde{L}/\widetilde{F}}U_{1,\widetilde{L}})/(U_{1,F} \cap N_{\widetilde{L}/\widetilde{F}}U_{1,\widetilde{L}} \cap N_{\mathcal{L}/\mathcal{F}}U_{1,\mathcal{L}}) \to (\mathrm{Gal}(L \cap F^{\mathrm{ab}}/F))^{\sim}.$$

In contrast to all previous class field theories a new problem comes on to the stage. The objects which describe abelian extensions in this case are not very simple especially because of the term $N_{\widetilde{L}/\widetilde{F}}U_{1,\widetilde{L}}$. And for a finite abelian totally ramified p-extension L/F there is no a priori as in other class field theories induction on degree

$$(*) \qquad N_{M/F}U_{1,M} \cap N_{\widetilde{E}/\widetilde{F}}U_{1,\widetilde{E}} = N_{M/F}(U_{1,M} \cap N_{\widetilde{E}/\widetilde{M}}U_{1,\widetilde{E}})$$

for every subextension $M/F \subset E/F$ in L/F . One can show that the property (*) holds if and only if $U_{1,F} \cap N_{\widetilde{L}/\widetilde{F}}U_{1,\widetilde{L}} \cap N_{\mathcal{L}/\mathcal{F}}U_{1,\mathcal{L}} = N_{L/F}U_{1,L}$ and if and only if $\Psi_{L/F}$ and $\Upsilon_{L/F}$ are isomorphisms [F5].

If L/F is a cyclic extension, then the property (*) holds, thus $\Psi_{L/F}$ is an isomorphism [F5].

If the residue field of F is imperfect, one can show that $\Psi_{L/F}$ is an isomorphism in the following cases : (1) L is the compositum of cyclic extensions M_i over F, $1 \leqslant i \leqslant m$, such that all the breaks of $\mathrm{Gal}(M_i/F)$ with respect to the upper numbering are not greater than every break of $\mathrm{Gal}(M_{i+1}/F)$ for all $1 \leqslant i \leqslant m - 1$; (2) $\mathrm{Gal}(L/F)$ is the product of cyclic groups of order p and a cyclic group.

Merits of the theory just exposed with respect to higher local class field theories are more simple structure of the objects in comparison to K-groups and more independence of a concrete type of the residue field. The main demerit is that only totally ramified abelian extensions are covered, and not abelian extensions with inseparable residue field extension.

Miki in [M2] has shown without explicit introduction of reciprocity maps that for a totally ramified cyclic extension L/F of degree m and for a finite abelian unramified extension E/F of exponent m the group $(F \cap N_{EL/E}U_{EL})/N_{L/F}U_L$ is canonically isomorphic to the character group of $\mathrm{Gal}(E/F)$.

In the description of the Galois group of a totally ramified p-extension L/F one can take instead the maximal unramified abelian p-extension $\widetilde{F}/F$ any its subextension $\hat{F}/F$ whose Galois group is a free abelain profinite p-group. Then the group $\mathrm{Hom}_{\mathbb{Z}_p}\bigl(\mathrm{Gal}(\hat{F}/F), \mathrm{Gal}(L/F)\bigr)$ becomes smaller (what's nice), but $N_{\hat{L}/\hat{F}}U_{1,\hat{L}} \cap U_{1,F}$ isn't the possible largest subgroup and isn't too far from $N_{L/F}U_{1,L}$ (which is bad). One can consruct in the same way as above the homomorphism

$$\hat{\Upsilon}_{L/F} \colon \mathrm{Hom}_{\mathbb{Z}_p}\bigl(\mathrm{Gal}(\hat{F}/F), \mathrm{Gal}(L/F)\bigr) \to (U_{1,F} \cap N_{\hat{L}/\hat{F}}U_{1,\hat{L}})/N_{L/F}U_{1,L}$$

for an abelian totally ramified p-extension L/F.

Assume now that the residue field $\overline{F}$ is a formal power series field of $n - 1$ indeterminates over a finite field k_0. Denote a lifting in F of a system of local parameters of $\overline{F}$ by $t_{n-1}, \ldots, t_1$. Then $\pi_F, t_{n-1}, \ldots, t_1$ form a system of local parameters of F as of an n-dimensional local field over k. Denote by $\widehat{F}$ the maximal abelian unramified p-extension of F corresponding to the $\mathbb{Z}_p$-extension of k_0. Let L/F be a finite Galois totally ramified p-extension with respect to the discrete valuation of rank 1. Then the following diagram is commutative

$$
\begin{array}{ccc}
\mathrm{Hom}_{\mathbb{Z}_p}\!\left(\mathrm{Gal}(\widetilde{F}/F),\mathrm{Gal}(L/F)\right) & \xrightarrow{\ \Upsilon_{L/F}\ } & (U_{1,F}\cap N_{\widetilde{L}/\widetilde{F}}U_{1,\widetilde{L}})/N_{L/F}U_{1,L} \\
\downarrow & & \downarrow \\
\mathrm{Gal}(L/F) & \xrightarrow{\ \hat{\Upsilon}_{L/F}\ } & (U_{1,F}\cap N_{\hat{L}/\hat{F}}U_{1,\hat{L}})/N_{L/F}U_{1,L} \\
\downarrow & & \downarrow \\
\mathrm{Gal}(L/F) & \xrightarrow{\ \mathcal{Y}_{L/F}\ } & K_n^{\mathrm{top}}(F)/N_{L/F}K_n^{\mathrm{top}}(L)
\end{array}
$$

where the homomorphism $\mathcal{Y}_{L/F}$ is the generalized Neukirch map in $10°$; the right one is induced by $\epsilon \to \{t_1,\dots,t_{n-1},\epsilon\}$.

13. — Some applications The description of the kernel of $\Psi_{L/F}$ for cyclic extensions has numerous applications. First, one can show that for an abelian totally ramified p-extension E/F the norm groups $N_{L/F}U_{1,L}$ are in bijection with subextensions L/F of the extension E/F.

A deeper result is that for a complete discrete valuation field F with non-separable-p-closed residue field the norm group $N_{L/F}L^*$ is uniquely determined by an abelian totally ramified p-extension L/F [F5] : for abelian totally ramified p-extensions L_1, L_2 over F the equality of their norm groups $N_{L_1/F}L_1^* = N_{L_2/F}L_2^*$ holds if and only if $L_1 = L_2$. This generalizes the classical assertion to the most possible extent.

In the case of imperfect residue field, one needs additional information in comparison with the perfect residue field case about the structure of norm subgroups. Existence theorem seems to be very difficult even to formulate. This is natural in view of the description of the norm groups in multidimensional class field theory where one uses all power of topological K-groups. However, for cyclic extensions of the fields with the absolute ramification index 1 there is a satisfactory description of norm groups.

For a complete discrete valuation field F of characteristic 0 with residue field $\overline{F}$ of characteristic $p > 2$, absolute ramification index 1 and a fixed prime element π introduce the function

$$\mathcal{E}_{n,\pi}\colon W_n(\overline{F}) \to U_{1,F}/U_{1,F}^{p^n}$$

by the formula

$$\mathcal{E}_{n,\pi}\left((a_0,\dots,a_{n-1})\right) = \prod_{0\leqslant i\leqslant n-1} E(\tilde{a}_i^{p^{n-1-i}}\pi)^{p^i} \bmod U_{1,F}^{p^n}.$$

Here $E(X) = \exp(X + X^p/p + X^{p^2}/p^2 + \ldots)$ is the Artin–Hasse function, and $\tilde{a}_i$ is a lifting of $a_i \in \overline{F}$ in the ring of integers of F. Then cyclic totally ramified extensions L/F of degree p^n, such that a fixed prime element π of F belongs to $N_{L/F}L^*$, are in one-to-one correspondence with subgroups

$$\mathcal{E}_{n,\pi}(\wp W_n(\overline{F})\mathrm{Frob}(a_0, \ldots, a_{n-1}))U_{1,F}^{p^n}$$

in $U_{1,F}$, where $(a_0, \ldots, a_{n-1})$ is invertible in $W_n(\overline{F})$, $\wp = \mathrm{Frob} - 1$, and Frob is the Frobenius map [F5]. This was first discovered by Kurihara [Kur2] for $\pi = p$. He proved that there is an exact sequence

$$1 \to H^1(F, \mathbb{Z}/p^n)_{nr} \to H^1(F, \mathbb{Z}/p^n) \to W_n(\overline{F}) \to 1$$

with nice functorial properties, i.e. there is a canonical connection between Witt vectors and cyclic p-extensions (in the case of $e < p - 1$ any cyclic p-extension has separable residue field extension, see [M1]). The approach of Kurihara is based on the study of the sheaf of the etale vanishing cycles on the special fiber of a smooth scheme over the ring of integers of F and of filtrations on Milnor's K-groups of local rings.

The class field theories described above seem still to demonstrate "a vivid and lively picture of the great and beautiful edifice of class field theories", at least in the local case.

Manuscrit reçu le 7 août 1995

References

[B] S. BLOCH. — *Algebraic cycles and higher K-theory*, Advances in Math. **61**, (1986), 267–304.

[BK] S. BLOCH, K. KATO. — *p-adic etale cohomology*, Inst. Hautes Études Sci. Publ. Math. **63**, (1986), 107–152.

[F1] I.B. FESENKO. — *Class field theory of multidimensional local fields of characteristic 0, with the residue field of positive characteristic*, Algebra i Analiz **3** (1991), n° 3, 165–196; English transl. in St. Petersburg Math. J. **3** (1992), n° 3, 649–678.

[F2] I.B. FESENKO. — *Multidimensional local class field theory*, II Algebra i Analiz **3** (1991), n° 5, 168–190; English transl. in St. Petersburg Math. J. **3** (1992), 1103–1126.

[F3] I.B. FESENKO. — *Local class field theory : perfect residue field case*, Izvestija Russ. Acad. Nauk. Ser. Mat. **57** (1993), n° 4, 72–91; English transl. in Russ. Acad. Scienc. Izvest. Math. **43** (1994), 65–81.

[F4] I.B. FESENKO. — *Abelian local p-class field theory*, Math. Ann. **301** (1995), 561–586.

[F5] I.B. FESENKO. — *On general local reciprocity maps*, to appear.

[FV] I.B. FESENKO, S.K. VOSTOKOV. — *Local Fields and Their Extensions : A Constructive Approach*, AMS, Providence, R.I., 1993.

[H1] M. HAZEWINKEL. — *Abelian extensions of local fields*, Thesis, Amsterdam Univ., 1969.

[H2] M. HAZEWINKEL. — *Corps de classes local*, H. Demazure, P. Gabriel, Groupes Algébriques, T. 1, North Holland, Amsterdam, 1970.

[H3] M. HAZEWINKEL. — *Local class field theory is easy* , Adv. Math. **18** (1975), 148–181.

[Ih] Y. IHARA. — *Problems on some complete p-adic function fields* (in Japanese), Kokyuroku of the RIMS Kyoto Univ. **41**, (1968), 7–17.

[Izh] O.T. IZHBOLDIN. — *On the torsion subgroup of Milnor K-groups*, Dokl. Akad. Nauk SSSR **294**, (1987), n° 1, 30–33; English transl. in Soviet Math. Dokl. **37**, (1987).

[Ka1] K. KATO. — *A generalization of local class field theory by using K-groups*, I, Proc. Japan Acad. **53**, (1977), 140–143.

[Ka2] K. KATO. — *A generalization of local class field theory by using K-groups*, II, Proc. Japan Acad. **54**, (1978), 250–255.

[Ka3] K. KATO. — *A generalization of local class field theory by using K-groups*, I, J. Fac. Sci. Univ. Tokyo Sect. IA Math. **26**, (1979), 303–376.

[Ka4] K. KATO. — *A generalization of local class field theory by using K-groups*, II, J. Fac. Sci. Univ. Tokyo Sect. IA Math. **27**, (1980), 603–683.

[Ka5] K. KATO. — *A generalization of local class field theory by using K-groups*, III, J. Fac. Sci. Univ. Tokyo Sect. IA Math. **29**, (1982), 31–43.

[Ka6] K. KATO. — *The existence theorem for higher local class field theory*, Preprint IHES, 1980.

[Ka7] K. KATO. — *Galois cohomology of complete discrete valuation fields*, Algebraic K-theory, Part II (Oberwolfach, 1980), Lecture Notes in Math. vol. 967, Springer, Berlin and New York, 1982, 215–238.

[Ka8] K. KATO. — *The explicit reciprocity law and the cohomology of Fontaine–Messing*, Bull. Soc. Math. France **119**, (1991), 397–441.

[Kn] B. KAHN. — *The decomposable part of motivic cohomology and bijectivity of the norm residue homomorphism*, Contemp. Math. **126**, (1992), 79–87.

[Ko1] Y. KOYA. — *A generalization of class formation by using hypercohomology*, Invent. Math. **101**, (1990), 705–715.

[Ko2] Y. KOYA. — *A generalization of Tate–Nakayama theorem by using hypercohomology*, Proc. Japan Acad., Ser. A **69** (1993), n° 3, 53–57.

[Ko3] Y. KOYA. — *Class field theory of higher semi-global fields*, Preprint (1995).

[Kuc] J. KUCERA. — *Über die Brauergruppe von Laurentreihen- und rationalen Funktionenkörpern und deren Dualität mit K-Gruppen*, Dissertation Univ. of Heidelberg (1994).

[Kur1] M. KURIHARA. — *On two types of complete discrete valuation fields*, Comp. Math. **63**, (1987), 237–257.

[Kur2] M. KURIHARA. — *Abelian extensions of an absolutely unramified local field with general residue field*, Invent. Math. **93**, (1988), 451–480.

[KS] Y. KAWADA, I. SATAKE. — *Class formations*, II J. Fac. Sci. Univ. Tokyo Sect. IA Math.**7** (1956), 353–389.

[M1] H. MIKI. — *On $\mathbb{Z}_p$-extensions of complete p-adic power series fields and function fields*, J. Fac. Sci. Univ. Tokyo Sect. IA Math. **21**, (1974), 377–393.

[M2] H. MIKI. — *On unramified abelian extensions of a complete field under a discrete valuation with arbitrary residue field of characteristic $p \neq 0$ and its application to wildly ramified $\mathbb{Z}_p$-extensions*, J. Math. Soc. **29** (1977), n° 2, 363–371.

[MZh] A.I. MADUNTS, I.B. ZHUKOV. — *Multidimensional complete fields : topology and other basic constructions*, Trudy St. Petersburg Mat. Obshchestva **3**, (1994); English transl. in Proceed. St. Petersburg Math. Society, AMS Translation Series, 2 **166**, (1995), 1–34.

[N1] J. NEUKIRCH. — *Neubegründung der Klassenkörpertheorie*, Math. Z. **186**, (1984), 557–574.

[N2] J. NEUKIRCH. — *Class Field Theory*, Springer, Berlin etc., 1986.

[P1] A.N. PARSHIN. — *Class fields and algebraic K-theory*,Uspekhi Mat. Nauk **30**, (1975), n 1, 253–254; English transl. in Russian Math. Surveys.

[P2] A.N. PARSHIN. — *On the arithmetic of two dimensional schemes, I, Distributions and residues)*, Izv. Akad. Nauk SSSR Ser. Mat. **40**, (1976), n° 4, 736–773; English transl. in Math. USSR-Izv. **10**, (1976).

[P3] A.N. PARSHIN. — *Abelian coverings of arithmetic schemes*, Dokl. Akad. Nauk SSSR **243** (1978), n° 4, 855–858; English transl. in Soviet Math. Dokl. **19**, (1978).

[P4] A.N. PARSHIN. — *Local class field theory*, Trudy Mat. Inst. Steklov. **165** (1985), 143–170; English transl. in Proc. Steklov Inst. Math. **1985**, n° 3, 157–185.

[P5] A.N. PARSHIN. — *Galois cohomology and Brauer group of local fields*, Trudy Mat. Inst. Steklov. **183** (1990), 159–169; English transl. in Proc. Steklov Inst. Math. **1991**, n° 4, 191–201.

[R] W. RASKIND. — *Abelian class field theory of arithmetic schemes*, Proc. Symp. Pure Math. Amer. Math. Soc., Providence, R.I. **58**, I, (1995), 85–187.

[S] Sh. SAITO. — *Arithmetic duality on two-dimensional henselian rings*, Arithmetic duality on two-dimensional henselian rings, Preprint Univ. Tokyo, (1988).

[Sh] I.R. SHAFAREVICH. — *A general reciprocity laws*, Mat. Sb. **26 (68)** (1950), 113–146; English transl. in Amer. Math. Soc. Transl. (2) **4**, (1956), 73–106.

[Sp] M. SPIEẞ. — *Class formations and higher dimensional local class field theory*, preprint (1995).

[V1] S.V. VOSTOKOV. — *Explicit form of the law of reciprocity*, Izv. AN SSSR. Ser. Mat. **42** (1978), 1288–1231; English transl. in Math. USSR Izv. **13**, (1979).

[V2] S.V. VOSTOKOV. — *Explicit construction in class field theory of a multidimensional local field*, Izv. AN SSSR. Ser. Mat. **49** (1985), n° 2, 238–308; English transl. in Math. USSR Izv. **26**, (1986).

[Zh] I.B. ZHUKOV. — *Structure theorems for complete fields*, Trudy St. Petersburg Mat. Obshchestva **3** (1994); English transl. in Proceed. St. Petersburg Math. Society, AMS Translation Series, 2 **66** (1995), 175–192.

Ivan B. FESENKO
Department of Mathematics
University of Nottingham
University Park
NG7 2RD Nottingham UK

Obstructions de Manin transcendantes

David Harari

Manin a introduit en 1970 ([11]) une obstruction au principe de Hasse qui a permis d'expliquer tous les contre-exemples explicites connus à ce jour à ce principe. L'idée de cette obstruction (dont on rappellera la définition en 1.3) consiste, quand on s'intéresse au principe de Hasse pour une variété X définie sur un corps de nombres k, à faire intervenir le groupe de Brauer $\operatorname{Br} X$ de X.

Supposons X propre, lisse, et géométriquement intègre sur k. Notant $\overline{X} = X \times_k \bar{k}$ (où $\bar{k}$ est une clôture algébrique fixée de k), on dispose du sous-groupe $\operatorname{Br}_1 X = \ker(\operatorname{Br} X \to \operatorname{Br} \overline{X})$ de $\operatorname{Br} X$. Rappelons que $\operatorname{Br} \overline{X}$ est toujours une extension d'un groupe fini par un groupe divisible ([7], corollaire 3.4). Le groupe $\operatorname{Br}_1 X / \operatorname{Br} k$ est isomorphe au groupe de cohomologie galoisienne $H^1(k, \operatorname{Pic} \overline{X})$ ce qui permet souvent en pratique de le calculer et d'obtenir au moyen de ses éléments des contre-exemples au principe de Hasse. On trouve en particulier de tels contre-exemples parmi les surfaces fibrées en coniques comme les surfaces de Châtelet (voir par exemple [2] et [6]). Ces dernières sont définies par des équations affines du type $y^2 - az^2 = f(x)$, où f est un polynôme de degré 4.

Les éléments transcendants de $\operatorname{Br} X$ (c'est-à-dire qui ne sont pas dans $\operatorname{Br}_1 X$) sont a priori plus difficiles à représenter simplement; les exemples d'obstruction de Manin obtenus jusqu'à présent ne les faisaient pas intervenir ce qui aurait pu faire penser que c'est seulement le groupe $\operatorname{Br}_1 X$ qui est important pour l'étude de cette obstruction. Le but de cet article est de montrer que ce n'est pas le cas.

Nous donnons en effet (théorème 1 et proposition 2) un exemple d'obstruction de Manin pour une k-variété X dont le groupe de Brauer "algébrique" $\operatorname{Br}_1 X$ est trivial (c'est-à-dire réduit à $\operatorname{Br} k$). En particulier, l'obstruction est définie à l'aide d'un élément transcendant de $\operatorname{Br} X$.

Vers 1976, Colliot-Thélène et Sansuc (dans [4]) ont formulé une obstruc-

tion de Manin pour l'approximation faible (propriété arithmétique intime-
ment liée au principe de Hasse). Nous donnons également (théorème 2) un
exemple d'une telle obstruction définie au moyen d'un élément transcendant du groupe de Brauer.

Notons que nos contre exemples s'inspirent des contre-exemples connus sur les surfaces de Châtelet ; ils sont obtenus à partir de fibrés en
coniques au-dessus du plan. On pourra trouver des résultats généraux sur
l'obstruction de Manin au principe de Hasse pour les familles de variétés
dans [9].

D'autre part, le groupe de Brauer géométrique $\operatorname{Br} \overline{X}$ des variétés considérées dans cet article est fini. La question de trouver un contre-exemple
au principe de Hasse ou à l'approximation faible s'appuyant sur la partie
divisible de $\operatorname{Br} \overline{X}$ reste donc ouverte.

1. — Rappels et notations

1.1. — Symboles de Hilbert

Soit F un corps de caractéristique différente de 2 et $\operatorname{Br} F$ son groupe de
Brauer. Pour tout couple (f, g) d'éléments de F^*, on note (f, g) le *symbole
de Hilbert* de f et g : c'est un élément du sous-groupe $_2\operatorname{Br} F = H^2(F, \mathbb{Z}/2)$
constitué des éléments de $\operatorname{Br} F$ tués par 2. Il s'obtient en faisant le cup-
produit des classes de f et g dans $F^*/F^{*^2} = H^1(F, \mathbb{Z}/2)$. Le symbole de
Hilbert (f, g) est bilinéaire en f et g ; il est trivial si et seulement si l'équation
$x^2 - fy^2 - gz^2 = 0$ a une solution non triviale (les inconnues x, y, z étant à
valeurs dans F). C'est le cas en particulier si f ou g est un carré dans F^*,
ou encore si $g = -f$. Quand f n'est pas un carré dans F^*, le symbole (f, g)
est trivial si et seulement si g est une norme de l'extension $F(\sqrt{f})/F$.

En particulier, quand F est un corps complet pour une valuation
discrète à corps résiduel fini ou quand $F = \mathbb{R}$, le groupe $_2\operatorname{Br} F$ est
isomorphe à $\mathbb{Z}/2$ ([12], 10.7 et 13.3). Lorsque $F = \mathbb{Q}_p$ avec p premier ou
$F = \mathbb{R}$, on identifiera ce groupe à $\{1, -1\}$. On rappelle les règles de calcul
suivantes ([13], théorème 1 page 39) :

PROPOSITION 1. — *Soit $f = p^\alpha u$ et $g = p^\beta v$ (où α, β sont dans $\mathbb{Z}$ et u, v
inversibles dans $\mathbb{Z}_p$) deux éléments de $\mathbb{Q}_p$. Pour tout élément inversible x de
$\mathbb{Q}_p$, posons $[x]_p = 1$ si x est un carré modulo p et $[x]_p = -1$ sinon. Pour
tout élément inversible y de $\mathbb{Q}_2$, notons $\varepsilon(y)$ et $\omega(y)$ les classes respectives
modulo 2 de $(y - 1)/2$ et $(y^2 - 1)/8$. Alors :*
- *Si $p \neq 2$, on a :*

$$(f, g) = (-1)^{\alpha\beta\varepsilon(p)} [u^\beta]_p [v^\alpha]_p$$

- Si $p = 2$, on a :
$$(f, g) = (-1)^{\varepsilon(u)\varepsilon(v)+\alpha\omega(v)+\beta\omega(u)}$$
- Si a et b sont deux éléments de $\mathbb{R}^*$, on a $(a, b) = 1$ sauf si a et b sont tous les deux strictement négatifs.

Notons en particulier que si h est un élément de $\mathbb{Q}_p$ de valuation strictement plus grande que celle de g, on a $(f, g) = (f, g + h)$ si $p \neq 2$ et le même résultat vaut pour $p = 2$ à condition de supposer que h/g est de valuation 2-adique au moins 3.

1.2. — Groupe de Brauer non ramifié

Soit maintenant k un corps de caractéristique zéro et V une k-variété géométriquement intègre de corps des fonctions $F = k(V)$. On note $\mathrm{Br}_{\mathrm{nr}}(F/k)$ le *groupe de Brauer non ramifié* de V, c'est-à-dire le sous-groupe de $\mathrm{Br}(k(V))$ constitué des éléments dont les résidus en tous les anneaux de valuation discrète de corps des fractions $k(V)$ et contenant k sont triviaux. D'après le théorème de pureté de Grothendieck ([8], corollaire 6.2), ce n'est autre que le groupe de Brauer $\mathrm{Br}\, W = H^2_{\mathrm{ét}}(W, \mathbb{G}_m)$ d'un modèle projectif lisse W de V ; c'est un invariant k-birationnel ([8], corollaire 7.3).

(Rappelons que quand R est un anneau de valuation discrète dont le corps résiduel κ est parfait, on dispose du résidu $\mathrm{Br}\, K \to H^1(\kappa, \mathbb{Q}/\mathbb{Z})$, où K est le corps des fractions de R. Pour une définition de cette flèche, on pourra se reporter au paragraphe 2 de [8]).

Fixons une clôture algébrique $\bar{k}$ de k et notons $\bar{k}(\overline{V})$ le corps des fonctions de la $\bar{k}$-variété $\overline{V} = V \times_k \bar{k}$. Un élément de $\mathrm{Br}(k(V))$ sera dit *transcendant* s'il n'est pas tué par la flèche $\mathrm{Br}(k(V)) \to \mathrm{Br}(\bar{k}(\overline{V}))$. Dans le cas contraire, il sera dit *algébrique*. On emploiera la même terminologie pour les éléments du sous-groupe $\mathrm{Br}_{\mathrm{nr}}(k(V)/k)$ de $\mathrm{Br}(k(V))$ (qui est aussi le sous-groupe $\mathrm{Br}\, V$ quand V est propre et lisse sur k).

1.3. — Obstruction de Manin

On fixe désormais un corps de nombres k dont on note Ω_k l'ensemble des places. Pour toute place v de k, on note $j_v : \mathrm{Br}\, k_v \hookrightarrow \mathbb{Q}/\mathbb{Z}$ l'invariant de la théorie du corps de classes local.

Soit W une k-variété géométriquement intègre, projective et lisse. On dit que W contredit *le principe de Hasse* si W a des points dans tous les complétés de k mais n'a pas de point rationnel. Quand $W(k) \neq \emptyset$, on dit que W vérifie *l'approximation faible* si $W(k)$ est dense dans $\prod_{v \in S} W(k_v)$ (muni de la topologie produit des topologies v-adiques) pour tout ensemble fini S de places de k. On dit qu'il y a *obstruction de Manin* (ou de Brauer-Manin) au principe de Hasse pour W si pour tout point (P_v) de $\prod\limits_{v \in \Omega_k} W(k_v)$, il existe un élément A de Br W tel que :

$$\sum_{v \in \Omega_k} j_v(A(P_v)) \neq 0 \text{ dans } \mathbb{Q}/\mathbb{Z}.$$

Il s'agit bien d'une obstruction au principe de Hasse d'après la loi de réciprocité du corps de classes global. Notons aussi que quand v est réelle ou non archimédienne, l'invariant j_v induit l'isomorphisme de $_2\mathrm{Br}\,k_v$ sur $\mathbb{Z}/2$ (lequel s'injecte dans $\mathbb{Q}/\mathbb{Z}$). La condition que W est propre (jointe à $A \in \mathrm{Br}\,W$) assure qu'en dehors d'un nombre fini de places v (les places réelles et les places de mauvaise réduction de W ou de A), on a $A(P_v) = 0$ pour tout point P_v de $W(k_v)$ ([5], 3).

On a de même (quand $W(k) \neq \emptyset$) une obstruction de Manin à l'approximation faible définie par la condition qu'il existe un point (P_v) de $\prod\limits_{v \in \Omega_k} W(k_v)$ et un élément A de Br W tels que $\sum\limits_{v \in \Omega_k} j_v(A(P_v)) \neq 0$ dans $\mathbb{Q}/\mathbb{Z}$. Plus précisément, on dit dans ce cas qu'il y a obstruction de Manin à l'approximation faible *associée à A*.

De même, quand un élément A de Br W vérifie $\sum\limits_{v \in \Omega_k} j_v(A(P_v)) \neq 0$ pour tout point $(P_v)_v$ de $\prod\limits_{v \in \Omega_k} W(k_v)$, on dira qu'il y a obstruction de Manin au principe de Hasse associée à A pour W. Bien entendu, un élément constant (c'est-à-dire provenant de Br k) de Br W ne peut fournir d'obstruction de Manin.

Si A est en outre un élément transcendant de Br W, nous parlerons d'obstructions de Manin "transcendantes".

Rappelons que si X est une k-variété lisse et U un ouvert de Zariski non vide de X, alors $U(k_v)$ est dense dans $X(k_v)$ pour la topologie v-adique ([2], lemme 3.1.2). On en déduit que si U est un ouvert lisse d'une k-variété V et $A \in \mathrm{Br}_{\mathrm{nr}}(k(V)/k) \subset \mathrm{Br}\,U$ est tel que $\sum\limits_{v \in \Omega_k} j_v(A(P_v)) \neq 0$ pour tout $(P_v)_v$ de $\prod\limits_{v \in \Omega_k} U(k_v)$, alors tout modèle projectif lisse de V est un contre-exemple au principe de Hasse (cette propriété est indépendante du modèle choisi d'après le lemme 3.1.1 de [2]). De même, si pour un certain élément $(P_v)_v$ de $\prod\limits_{v \in \Omega_k} U(k_v)$, on a $\sum\limits_{v \in \Omega_k} j_v(A(P_v)) \neq 0$, alors tout modèle projectif lisse de V est un contre-exemple à l'approximation faible. Pour cette dernière propriété, il faut bien noter que si on a seulement $A \in \mathrm{Br}\,U$

(ce qui n'empêche pas la somme $\sum_{v \in \Omega_k} j_v(A(P_v))$ d'être finie pour certains $(P_v)_v$), on ne peut rien conclure : il est essentiel que A appartienne au groupe de Brauer d'un modèle projectif lisse de V.

Par abus, nous dirons donc qu'une k-variété V vérifie le principe de Hasse si c'est le cas pour un modèle projectif lisse W de V. On parlera de même d'approximation faible, ou d'obstruction de Manin pour V.

2. — Un exemple d'obstruction de Manin transcendante au principe de Hasse

Notation : Soit $R = P/Q$ (avec P et Q polynômes non nuls premiers entre eux, en n variables à coefficients dans k) une fraction rationnelle. Par abus de langage, nous parlerons de la k-variété algébrique définie dans $\mathbb{A}_k^n$ par l'équation $R = 0$ pour désigner l'ouvert $Q \neq 0$ de l'hypersurface de $\mathbb{A}_k^n$ d'équation $P = 0$.

THÉORÈME 1. — *Soit V la $\mathbb{Q}$-hypersurface de $\mathbb{A}_{\mathbb{Q}}^4$ définie par l'équation :*

$$y^2 - g(t)z^2 = [f(x)^2 + \frac{2^7}{p}][1 + g(t)^2 - g(t)(f(x)^2 + \frac{2^7}{p} + 2)] \, ,$$

avec :
- *p nombre premier congru à -1 modulo 4 ;*
- *$f(x) = \dfrac{1}{(x^2 + 1)}$ et $g(t) = -\dfrac{2^6 p}{(t^2 + 1)} - 1$.*

Alors :
1. *L'élément $A = (g(t), (f(x)^2 + 2^7/p))$ de $\operatorname{Br} \mathbb{Q}(V)$ est un élément transcendant de $\operatorname{Br}_{\mathrm{nr}}(\mathbb{Q}(V)/\mathbb{Q})$.*
2. *Il y a obstruction de Manin au principe de Hasse associée à A pour V.*

Preuve : pour simplifier un peu les notations, posons $F(x) = f(x)^2 + 2^7/p$ et $G(x,t) = 1 + g(t)^2 - g(t)(f(x)^2 + 2^7/p + 2)$.

Soit U l'ouvert de V défini par $g(t)(y^2 - g(t)z^2)(t^2 + 1)(x^2 + 1) \neq 0$; on a $A \in \operatorname{Br} U$ (car $g(t)$ et $F(x)$ ne s'annulent pas sur U) et U est lisse.

Remarquons que l'élément $B = (g(t), G(x,t))$ de $\operatorname{Br} U$ est égal à A car le symbole de Hilbert $(g(t), y^2 - g(t)z^2)$ est trivial. Nous allons prouver les deux lemmes suivants :

LEMME 1. — *Soit l un nombre premier impair distinct de p, alors $U(\mathbb{Q}_l) \neq \emptyset$ et pour tout point l-adique M_l de $U(\mathbb{Q}_l)$, on a $A(M_l)$ trivial.*

LEMME 2. — On a :
1. L'ensemble $U(\mathbb{Q}_p)$ est non vide et pour tout point p-adique M_p de $U(\mathbb{Q}_p)$, on a $A(M_p)$ non trivial.
2. L'ensemble $U(\mathbb{Q}_2)$ est non vide et pour tout point 2-adique M_2 de $U(\mathbb{Q}_2)$, on a $A(M_2)$ trivial.

Preuve du lemme 1 : soit $M_l = (X, Y, Z, T)$ dans $U(\mathbb{Q}_l)$, notons v la valuation l-adique sur $\mathbb{Q}_l$. On a $A(M_l) = (g(T), F(X))$ et $v(1/p) = 0$.

Si $v(f(X)) < 0$, on a (d'après la proposition 1) $A(M_l) = (g(T), f(X)^2)$ donc $A(M_l)$ est trivial. Supposons donc $v(f(X)) \geq 0$. Alors, si $v(g(T)) > 0$, on a $B(M_l) = (g(T), 1)$ et si $v(g(T)) < 0$, on a $B(M_l) = (g(T), g(T)^2)$ donc dans ces deux cas $A(M_l)$ (qui est égal à $B(M_l)$) est trivial.

Si enfin $v(f(X)) \geq 0$ et $v(g(T)) = 0$, alors la seule possibilité pour que $A(M_l) = B(M_l)$ soit non trivial est qu'on ait $v(F(X))$ et $v(G(X,T))$ impaires et que $g(T)$ ne soit pas un carré modulo l ; mais ceci implique que $F(X)$ et $G(X,T)$ sont nuls modulo l (puisque $F(X)$ et $G(X,T)$ sont dans $\mathbb{Z}_l$) donc en particulier $(g(T) - 1)^2$ nul modulo l donc $g(T) = 1$ modulo l ce qui contredit le fait que $g(T)$ ne soit pas un carré modulo l. Ainsi, dans tous les cas on a bien $A(M_l) = 0$.

Pour montrer que $U(M_l)$ est non vide, il suffit de trouver T et X dans $\mathbb{Q}_l$ tels que $(g(T), H(X,T))$ soit trivial, où $H(X,T) = F(X)G(X,T)$ car on aura bien alors Y et Z dans $\mathbb{Q}_l$ tels que $Y^2 - g(T)Z^2 = H(X,T)$. Choisissons déjà T de valuation < 0, on obtient $g(T)$ congru à -1 modulo l. Si l ne divise pas $2^7 + 4p$, on choisit X de valuation < 0 ce qui donne $v(f(X)) > 0$ et donc $v(F(X)) = 0$, puis $G(X,T)$ congru à $4 + 2^7/p$ modulo l donc $v(G(X,T)) = 0$ et finalement $v(H(X,T)) = 0$ donc $(g(T), H(X,T))$ trivial. Supposons donc que l divise $2^7 + 4p$, alors si $l \neq 3$, on a $1 + 2^7/p$ inversible dans $\mathbb{Q}_l$ donc en prenant $X = 0$ (c'est-à-dire $f(X) = 1$), on obtient $v(F(X)) = 0$ et $G(X,T)$ congru à $5 + 2^7/p$ (donc à 1) modulo l et on a encore $v(H(X,T)) = 0$ puis $(g(T), H(X,T))$ trivial.

Enfin, si $l = 3$ avec $2^7 + 4p$ divisible par 3 (soit p congru à 1 modulo 3), on prend $T = 0$ ce qui donne $g(T)$ congru à 1 modulo 3 et ainsi $g(T)$ est un carré modulo 3 et on aura automatiquement $(g(T), H(X,T))$ trivial. Ceci achève la preuve du lemme 1.

Preuve du lemme 2 : soient X et T dans $\mathbb{Q}_p$, alors comme -1 n'est pas un carré modulo p, on a (en notant v la valuation p-adique) $v(f(X)) \geq 0$ et donc $v(F(X)) = -1$, tandis que $v(g(T)) = 0$ et $g(T)$ est égal à -1 modulo p ce qui implique (avec la proposition 1) que $(g(T), F(X)) = [-1]_p$ est non trivial donc pour tout point M_p de $U(\mathbb{Q}_p)$ on a $A(M_p)$ non trivial.

D'autre part on a bien $U(\mathbb{Q}_p)$ non vide car si l'on prend T et X quelconques tels que $f(X)$ et $g(T)$ soient non nuls, on aura $v(f(X)) \geq 0$

et $v(g(T)) = 0$; ainsi $v(F(X)) = -1$ et $v(G(X,T)) = v(2^7/p) = -1$ donc $v(H(X,T))$ sera paire et $(g(T), H(X,T))$ trivial et on pourra bien trouver Y et Z vérifiant $Y^2 - g(T)Z^2 = H(X,T)$. D'où le premier point.

Soit maintenant T dans $\mathbb{Q}_2$, on a (en notant maintenant v la valuation 2-adique) $v(T^2 + 1) \leq 1$ (car -1 n'est pas un carré modulo 4) donc $g(T) = -1 + 2^5 a$, avec a dans $\mathbb{Z}_2$. Si $v(f(X)) < 3$, on a $v(2^7/p) \geq v(f(X)^2) + 3$ donc d'après la proposition 1 on a $(g(T), F(X)) = (g(T), f(X)^2)$ qui est trivial. Si $v(f(X)) \geq 3$, on a $G(X,T)$ congru à 2^2 modulo 2^5 donc $(g(T), G(X,T)) = (g(T), 2^2)$ est encore trivial.

Enfin $U(\mathbb{Q}_2)$ est non vide car en prenant $X = T = 0$, on trouve (en appliquant la proposition 1) $(g(T), H(X,T)) = (-1, 5)$ qui est bien trivial.

Preuve du théorème 1 : d'après les lemmes 1 et 2, l'ouvert lisse U a des points dans tous les complétés de $\mathbb{Q}$ (l'existence de points réels est évidente). D'autre part, on a $A(M_l)$ trivial pour tout nombre premier l distinct de p et tout $\mathbb{Q}_l$-point M_l de U donc d'après le théorème $2.1.1^{[1]}$ de [9], l'élément A de $\mathrm{Br}\, U$ est en fait dans $\mathrm{Br}_{\mathrm{nr}}\,(\mathbb{Q}(V)/\mathbb{Q})$. Or $A(M_\infty)$ est trivial pour tout point réel M_∞ de U (parce que pour tout X de $\mathbb{R}$ on a $F(X) > 0$) et le lemme 2 dit que $A(M_p)$ est non trivial pour tout $\mathbb{Q}_p$-point M_p de U. Ainsi, pour tout point (M_v) de $\prod_{v \in \Omega} U(\mathbb{Q}_v)$, (on note Ω l'ensemble des places de $\mathbb{Q}$), on a $\sum_{v \in \Omega} j_v(A(M_v)) \neq 0$; de ce fait (d'après les remarques à la fin de 1.3), il y a obstruction de Manin associée à A au principe de Hasse pour tout modèle projectif lisse W de V (on a bien $A \in \mathrm{Br}\, W$ puisque $A \in \mathrm{Br}_{\mathrm{nr}}\,(\mathbb{Q}(V)/\mathbb{Q})$).

Il reste enfin à prouver que A est un élément transcendant de $\mathrm{Br}_{\mathrm{nr}}\,(\mathbb{Q}(V)/\mathbb{Q})$. Notons $K = \overline{\mathbb{Q}}(x,t)$ le corps des fonctions rationnelles en deux variables sur $\overline{\mathbb{Q}}$. La variété $\overline{V}$ est fibrée en coniques (via (x,t)) au-dessus de $\mathbf{A}^2_{\overline{\mathbb{Q}}}$; la conique générique C a pour corps des fonctions $K(C) = \overline{\mathbb{Q}}(V)$. Comme C est une conique, le noyau de la flèche $\mathrm{Br}\, K \to \mathrm{Br}\,(K(C))$ est engendré par la classe de l'algèbre de quaternions associée à C ([3], proposition 1.5), c'est-à-dire par l'élément $(g(t), F(x)G(x,t))$ de $\mathrm{Br}\, K$. Ainsi, si A n'était pas un élément transcendant de $\mathrm{Br}_{\mathrm{nr}}\,(\mathbb{Q}(V)/\mathbb{Q})$ (c'est-à-dire si A s'annulait dans $\mathrm{Br}\,(\overline{\mathbb{Q}}(V))$), l'élément $(g(t), F(x))$ de $\mathrm{Br}\, K$ (dont l'image dans $\mathrm{Br}\,(K(C))$ est précisément l'image de A dans $\mathrm{Br}\,(\overline{\mathbb{Q}}(V))$) serait nul ou égal à $(g(t), F(x)G(x,t))$. Ainsi, il nous suffit de prouver que les éléments $(g(t), F(x))$ et $(g(t), G(x,t))$ de $\mathrm{Br}\, K$ sont non triviaux.

---

[1] Le recours à ce résultat n'est en fait pas indispensable : on peut montrer que A est non ramifié par des calculs algébriques de résidus tout à fait similaires aux calculs arithmétiques du lemme 1.

La fonction $g(t)$ n'est pas un carré dans $\overline{\mathbb{Q}}(t)$, notons L le corps $K(\sqrt{g(t)})$. Il est clair que $F(x)$ n'est pas une norme de l'extension L/K (vu que $F(x)$ ne dépend que de x et n'est pas un carré dans K) donc on a déjà le résultat pour $(g(t), F(x))$.

Posons $\gamma = 2^7/p + 2$ et $u(t) = g(t)/(1 + g(t)^2 - \gamma g(t))$; notons également $P(x,t) = (x^2 + 1)^2 - u(t)$. On a $G(x,t)/P(x,t) = (1 + g(t)^2 - \gamma g(t))/(x^2 + 1)^2$ donc, comme $(g(t), 1 + g(t)^2 - \gamma g(t))$ est trivial dans Br K (vu que la fonction $1 + g(t)^2 - \gamma g(t) = (1 + g(t))^2 - g(t)(\gamma + 2)$ est une norme de l'extension L/K), on a $(g(t), G(x,t)) = (g(t), P(x,t))$ dans Br K.

Notons Z la courbe d'équation $P(x,t) = 0$, c'est-à-dire (moyennant l'abus de notation du début du paragraphe 2) la $\overline{\mathbb{Q}}$-variété définie dans le plan affine par l'équation :

$$(1) \qquad\qquad (x^2 + 1)^2 = u(t).$$

On peut voir cette équation comme une équation du quatrième degré en x à coefficients dans $\overline{\mathbb{Q}}(t)$, ce qui définit un plongement de $\overline{\mathbb{Q}}(t)$ dans le corps des fonctions M de Z, qui est le corps de rupture de cette équation (il est donc de degré 4 sur $\overline{\mathbb{Q}}(t)$). Ainsi M est engendré sur $\overline{\mathbb{Q}}(t)$ par $\sqrt{-1 - \sqrt{u(t)}}$ donc la seule extension quadratique de $\overline{\mathbb{Q}}(t)$ que contient M est obtenue en adjoignant une racine carrée de $u(t)$ à $\overline{\mathbb{Q}}(t)$ (sinon M serait biquadratique sur $\overline{\mathbb{Q}}(t)$).

Pour prouver que $P(x,t)$ (qui n'est pas un carré dans K) n'est pas une norme de l'extension L/K, il suffit de prouver que l'élément $g(t)$ de $\overline{\mathbb{Q}}(t)$ n'est pas un carré dans M car M est le corps résiduel du plan affine au point générique de la courbe Z d'équation $P(x,t) = 0$ (l'idée sous-jacente à tout ce calcul est de montrer que l'élément $(g(t), P(x,t))$ de Br K a un résidu non nul au point générique de la courbe Z). Il s'agit donc de voir que $g(t)/u(t) = 1 + g^2(t) - \gamma g(t)$ n'est pas un carré dans $\overline{\mathbb{Q}}(t)$, ce qui résulte aisément de ce que γ n'est pas égal à 2 ou -2. Ceci achève la preuve du théorème 1.

3. — Un exemple d'obstruction de Manin transcendante à l'approximation faible

Il est plus simple d'obtenir des obstructions à l'approximation faible (voir la fin du paragraphe 6 de [9] pour une remarque générale à ce sujet); l'exemple ci-dessous a servi de point de départ pour la construction de l'exemple du théorème 1 :

THÉORÈME 2. — *Soit V la $\mathbb{Q}$-variété de $\mathbb{A}_{\mathbb{Q}}^4$ définie par l'équation :*

$$y^2 - tz^2 = (x^2 + \frac{1}{p})(1 + t^2 - t(x^2 + \frac{1}{p} + 2)),$$

où p est un nombre premier impair. Alors :
1. *L'élément $A = (t, x^2 + 1/p)$ de $\mathrm{Br}\,(\mathbb{Q}(V))$ est un élément transcendant de $\mathrm{Br}_{\mathrm{nr}}\,(\mathbb{Q}(V)/\mathbb{Q})$.*
2. *La variété V possède un k-point lisse mais il y a obstruction de Manin à l'approximation faible associée à A pour V.*

Preuve : notons encore U l'ouvert $t(y^2 - tz^2) \neq 0$ de V. Un calcul similaire à celui du lemme 1 montre que pour tout nombre premier impair l distinct de p et tout $\mathbb{Q}_l$-point M_l de U, on a $A(M_l)$ trivial donc on a déjà $A \in \mathrm{Br}_{\mathrm{nr}}\,(\mathbb{Q}(V)/\mathbb{Q})$ (d'après le théorème 2.1.1 de [9]; on pourrait aussi avoir recours à un calcul algébrique de résidus). La transcendance de A se prouve par le même argument que dans le théorème 1, en utilisant le fait que $x^2 + 1/p$ et $1 + t^2 - (1/p + 2)t$ ne sont pas des carrés dans $\overline{\mathbb{Q}}(x, t)$, ce qui permet de voir que les éléments $(t, x^2 + 1/p)$ et $(t, 1 + t^2 - t(x^2 + 1/p + 2))$ de $\mathrm{Br}\,(\overline{\mathbb{Q}}(x, t))$ ne sont pas triviaux (pour le deuxième, on voit en effet que t n'est pas un carré dans le corps des fonctions de la courbe $1 + t^2 - t(x^2 + 1/p + 2)) = 0$).

Choisissons t dans $\mathbb{Z}$ non divisible par p et qui n'est pas un carré modulo p, puis x dans $\mathbb{Z}$, alors les valuations p-adiques de $x^2 + 1/p$ et $1 + t^2 - t(x^2 + 1/p + 2)$ sont toutes deux -1 et on a donc $(t, x^2 + 1/p)$ non trivial mais par contre le symbole $(t, (x^2 + 1/p)(1 + t^2 - t(x^2 + 1/p + 2))$ est trivial ce qui fournit un $\mathbb{Q}_p$-point M_p de U tel que $A(M_p)$ soit non trivial.

Enfin, en prenant $t = 1$, on trouve un point rationnel M de U (c'est évident en faisant le changement de variables $y' = y + z$ et $z' = y - z$, la fibre en $t = 1$ étant même $\mathbb{Q}$-rationnelle) et on a $A(M)$ trivial. On obtient un point (M_v) de $\prod_{v \in \Omega} U(\mathbb{Q}_v)$ en prenant pour M_v le point M aux places v autres que p et le point M_p à la place p et on a bien $\sum_{v \in \Omega} j_v(A(M_v)) \neq 0$ d'où le théorème 2.

Remarque : on peut en outre voir que dans cet exemple, l'obstruction de Manin à l'approximation faible est la seule pour un modèle projectif lisse X de V (cela signifie que tout point (P_v) de $\prod_{v \in \Omega} X(\mathbb{Q}_v)$ vérifiant $\sum_{v \in \Omega} j_v(\alpha(P_v)) = 0$ pour tout α de $\mathrm{Br}\,X$ est dans l'adhérence de $X(\mathbb{Q})$ pour la topologie produit des topologies v-adiques). On obtient ce résultat en fibrant V (via t) au-dessus de la droite affine et en appliquant le théorème 4.2.1 de [9], vu que toutes les fibres sont géométriquement intègres et que l'obstruction de Manin au principe de Hasse et à l'approximation faible est la seule pour ces fibres ([6], théorème 8.11). Nous verrons du reste au paragraphe suivant que $\mathrm{Br}\,X$ est engendré par A modulo les constantes, ce qui fait qu'un point (P_v) de $\prod_{v \in \Omega} X(\mathbb{Q}_v)$ est dans l'adhérence de $X(\mathbb{Q})$ si et seulement s'il vérifie $\sum_{v \in \Omega} j_v(A(P_v)) = 0$.

4. — Compléments

4.1. — Calcul du groupe de Brauer

Nous allons montrer que dans les deux exemples que nous avons considérés, le groupe de Brauer non ramifié de V est en fait engendré par A (modulo les constantes) ce qui fait qu'*aucun élément algébrique du groupe de Brauer ne peut induire d'obstruction de Manin.*

PROPOSITION 2. — *Soit V la $\mathbb{Q}$-variété du théorème 1 (resp. du théorème 2). Alors, le groupe* $\mathrm{Br}_{\mathrm{nr}}(\mathbb{Q}(V)/\mathbb{Q})$ *est engendré par A modulo les constantes.*

Preuve : nous faisons la preuve pour la variété du théorème 1 (l'autre preuve est analogue) ; gardons les notations de la preuve de ce théorème et considérons V comme fibrée (par t) au-dessus de la droite affine. D'après la proposition 5.1 de [2] (voir aussi [10], proposition 2.1.1), le groupe de Brauer non ramifié de la fibre générique est engendré (modulo les éléments qui viennent de $\mathrm{Br}(\mathbb{Q}(t))$) par la classe de A car les polynômes $F(x)$ et $G(x,t)$ sont irréductibles sur $\mathbb{Q}(t)$; ainsi, un élément A' de $\mathrm{Br}_{\mathrm{nr}}(\mathbb{Q}(V)/\mathbb{Q})$ (qui doit a fortiori être non ramifié sur la fibre générique) est égal à A modulo un élément A_0 de $\mathrm{Br}_{\mathrm{nr}}(\mathbb{Q}(V)/\mathbb{Q})$ qui provient d'un élément $\mathcal{A}$ de $\mathrm{Br}(\mathbb{Q}(t))$. Mais pour tout point m de $\mathbb{A}_{\mathbb{Q}}^1$, l'élément A_0 doit être non ramifié au point générique de la fibre en m (notée V_m), laquelle est géométriquement intègre ; de ce fait l'élément $\mathcal{A}$ de $\mathrm{Br}(\mathbb{Q}(t))$ est non ramifié en m car si K_m est le corps des fonctions de V_m et k_m le corps résiduel de $m \in \mathbb{A}_{\mathbb{Q}}^1$, la flèche $H^1(k_m, \mathbb{Q}/\mathbb{Z}) \to H^1(K_m, \mathbb{Q}/\mathbb{Z})$ (induite par l'inclusion $k_m \subset K_m$) est injective puisque k_m est algébriquement clos dans K_m ; or le résidu de A_0 au point générique de V_m n'est autre que l'image dans $H^1(K_m, \mathbb{Q}/\mathbb{Z})$ du résidu de $\mathcal{A}$ en m (par fonctorialité du résidu, cf [3], corollaire à la proposition 1.1).

On en conclut que $\mathcal{A}$ (qui n'est ramifié en aucun point de la droite affine) est dans $\mathrm{Br}\,\mathbb{Q}$ d'après la suite exacte de Faddeev (cf [9], preuve du lemme 4.1.1) et A_0 est constant. Ainsi $\mathrm{Br}_{\mathrm{nr}}(\mathbb{Q}(V)/\mathbb{Q})$ est réduit à A modulo les constantes. Ceci achève la preuve de la proposition 2.

4.2. — Remarques géométriques

Pour conclure, disons quelques mots sur la géométrie de la variété du théorème 2 ; notons déjà que cette variété est un revêtement double de $\mathbb{A}_{\mathbb{Q}}^3$ qui est ramifiée le long d'une surface quartique.

PROPOSITION 3. — *Soit V la $\mathbb{Q}$-hypersurface de $\mathbb{A}_{\mathbb{Q}}^4$ d'équation :*

$$y^2 - tz^2 = (x^2 + \frac{1}{p})(1 + t^2 - t(x^2 + \frac{1}{p} + 2))$$

(où p est un nombre premier impair).

Alors la variété V est $\mathbb{Q}$-unirationnelle mais $\overline{V} = V \times_{\mathbb{Q}} \overline{\mathbb{Q}}$ n'est pas rationnelle. Le groupe de Brauer non ramifié de la $\overline{\mathbb{Q}}$-variété $\overline{V}$ est engendré par l'élément $\overline{A} = (t, x^2 + 1/p)$.

Preuve : posons $t = u^2$, nous obtenons une nouvelle hypersurface V' de $\mathbb{A}_{\mathbb{Q}}^4$ qui domine V, et qui est rationnelle sur $\mathbb{Q}$: en effet en faisant les changements de variables $y' = y - uz$ et $z' = y + uz$ on voit que V' est $\mathbb{Q}$-birationnelle au produit d'une droite et d'une hypersurface de $\mathbb{A}_{\mathbb{Q}}^3$ d'équation $y'z' = H(x, t)$ où H est un polynôme non nul. D'autre part, on a vu (théorème 2) que le groupe de Brauer non ramifié de $\overline{V}$ était non nul (il contient l'élément non nul $\overline{A}$) ce qui implique la non rationalité de $\overline{V}$ (à cause de la nullité du groupe de Brauer non ramifé d'une extension transcendante pure de $\overline{\mathbb{Q}}$). Pour montrer que $\mathrm{Br}_{\mathrm{nr}}(\overline{\mathbb{Q}}(\overline{V})/\overline{\mathbb{Q}})$ est engendré par $\overline{A}$, on considère encore $\overline{V}$ comme fibrée par t au-dessus de la droite affine. La fibre générique est une surface de Châtelet sur le corps $\overline{\mathbb{Q}}(t)$ (qui est C_1 d'après le théorème de Tsen donc $\mathrm{Br}(\overline{\mathbb{Q}}(t)) = 0$) ; comme le polynôme $G(x, t) = 1 + t^2 - t(x^2 + 1/p + 2)$ est irréductible sur $\overline{\mathbb{Q}}(t)$, le groupe de Brauer non ramifié de cette fibre générique est engendré par $(t, x^2 + 1/p)$ (ceci résulte encore de la proposition 5.1 de [2]). Comme $\mathrm{Br}_{\mathrm{nr}}(\overline{\mathbb{Q}}(\overline{V})/\overline{\mathbb{Q}})$ en est un sous-groupe et qu'il contient bien $\overline{A}$, le groupe $\mathrm{Br}_{\mathrm{nr}}(\overline{\mathbb{Q}}(\overline{V})/\overline{\mathbb{Q}})$ est engendré par $\overline{A}$ (ainsi $\mathrm{Br}_{\mathrm{nr}}(\overline{\mathbb{Q}}(\overline{V})/\overline{\mathbb{Q}})$ est isomorphe à $\mathbb{Z}/2$).

Remarque : la variété $\overline{V}$ est fibrée en coniques au-dessus du plan affine ; ceci est à rapprocher du célèbre contre-exemple d'Artin et Mumford ([1]). Le groupe de Brauer y était utilisé (via la torsion du groupe $H^3(X, \mathbb{Z})$) pour donner un exemple de $\mathbb{C}$-variété X projective et lisse, unirationnelle mais non rationnelle. Dans [3], on trouvera des variations sur cet exemple, avec le point de vue du groupe de Brauer non ramifié (qui apparaît dans les travaux de D. Saltman), ainsi que des exemples utilisant des invariants cohomologiques de degré supérieur à 2.

Manuscrit reçu le 22 mars 1994

BIBLIOGRAPHIE

[1] M. Artin, D. Mumford. — *Some elementary examples of unirational varieties which are not rational,* Proc. Lond. Math. Soc. **25**, 75-95 (1972).

[2] J.-L. Colliot-Thélène, D. Coray, J.-J. Sansuc. — *Descente et principe de Hasse pour certaines variétés rationnelles,* J. reine angew. Math. **320**, 150-191 (1980).

[3] J.-L. Colliot-Thélène, M. Ojanguren. — *Variétés unirationnelles non rationnelles : au-delà de l'exemple d'Artin et Mumford,* Invent. Math. **97**, 141-158 (1989).

[4] J.-L. Colliot-Thélène, J.-J. Sansuc. — Trois notes, C.R. Acad. Sci. Paris **282**, 1113-1116 (1976); **284**, 967-970 (1977); **284**, 1215-1218 (1977).

[5] J.-L. Colliot-Thélène, J.-J. Sansuc. — *La descente sur les variétés rationnelles* II, Duke Math. J. **54**, 375-492 (1987).

[6] J.-L. Colliot-Thélène, J.-J. Sansuc, Sir Peter Swinnerton Dyer. — *Intersection of two quadrics and Châtelet surfaces,* J. reine angew. Math. **373** (1987); **374** (1987).

[7] A. Grothendieck. — *Le groupe de Brauer,* II, dans *Dix exposés sur la cohomologie des schémas,* Masson-North-Holland, Amsterdam 1968.

[8] A. Grothendieck. — *Le groupe de Brauer,* III : exemples et compléments, dans *Dix exposés sur la cohomologie des schémas,* Masson-North-Holland, Amsterdam 1968.

[9] D. Harari. — *Méthode des fibrations et obstruction de Manin,* Duke Math. J. **75**, 221-260 (1994).

[10] D. Harari. — *Principe de Hasse et approximation faible sur certaines hypersurfaces,* à paraître aux Annales de la Faculté des Sciences de Toulouse.

[11] Yu. I. Manin. — *Le groupe de Brauer-Grothendieck en géométrie diophantienne,* dans *Actes du Congrès Intern. Math. (Nice 1970),* Tome **1**, 401-411, Gauthiers-Villars, Paris 1971.

[12] J.–P. SERRE. — *Corps locaux*, Hermann, Paris 1968.

[13] J.–P. SERRE. — *Cours d'arithmétique*, PUF, Paris 1970.

David HARARI
(e-mail : harari@dmi.ens.fr)
E. N. S.
Département de mathématiques
et informatique
45 rue d'Ulm
75005 Paris
FRANCE

On Selmer Groups of Adjoint Modular Galois Representations

Haruzo Hida*

0. — Introduction

Let p be an odd prime. Starting from a modular Galois representation φ into $GL_2(\mathbb{I})$ for an irreducible component $\mathrm{Spec}(\mathbb{I})$ of the spectrum of the universal ordinary Hecke algebra of prime-to-p level N, we study the Selmer group $\mathrm{Sel}(\mathrm{Ad}(\varphi) \otimes \nu^{-1})_{/\mathbb{Q}}$ of Greenberg [G] for the adjoint representation of $\mathrm{Ad}(\varphi)$ on the trace zero subspace $V(\mathrm{Ad}(\varphi))$ of $M_2(\mathbb{I})$ and the universal character ν unramified outside p deforming the trivial character of $\mathcal{G}_{\mathbb{Q}} = \mathrm{Gal}(\overline{\mathbb{Q}}/\mathbb{Q})$. The Pontryagin dual of $\mathrm{Sel}(\mathrm{Ad}(\varphi))_{/\mathbb{Q}}$ is basically known to be a torsion $\mathbb{I}$-module of finite type by a result of Flach [F] and Wiles [W] under a suitable assumption on φ. The key point of the proof is to show for an arithmetic height 1 prime P, the subgroup $\mathrm{Sel}(\mathrm{Ad}(\varphi))[P]$ killed by P is finite. Our Selmer group $\mathrm{Sel}(\mathrm{Ad}(\varphi) \otimes \nu^{-1})$ is naturally a module over $\mathbb{I}[[\Gamma]]$ for $\Gamma = \mathrm{Im}(\nu)$ ($\cong \mathbb{Z}_p$). However, it is well known that for the augmentation ideal P of $\mathbb{I}[[\Gamma]]$, $\mathrm{Sel}(\mathrm{Ad}(\varphi) \otimes \nu^{-1})[P]$ has non-trivial $\mathbb{I}$-divisible subgroup, and hence the co-torsionness of $\mathrm{Sel}(\mathrm{Ad}(\varphi) \otimes \nu^{-1})$ over $\mathbb{I}[[\Gamma]]$ does not follow from the co-torsionness of $\mathrm{Sel}(\mathrm{Ad}(\varphi))_{/\mathbb{Q}}$ over $\mathbb{I}$. In this paper, under a suitable assumption, we prove a control theorem giving the following exact sequence :

$$0 \to \mathrm{Sel}(\mathrm{Ad}(\varphi))_{/\mathbb{Q}} \to \mathrm{Sel}(\mathrm{Ad}(\varphi) \otimes \nu^{-1})^{\Gamma} \to \mathbb{I}^* \to 0 \ ,$$

where $\mathbb{I}^*$ is the Pontryagin dual module of $\mathbb{I}$ on which Γ acts trivially. Actually this assertion is valid for more general 2-dimensional representations φ not necessarily modular (theorems 2.2 and 3.2) and also for $\mathrm{Sel}(\mathrm{Ad}(\varphi))_{/F}$ for a general number field F. Although the above exact sequence does not directly yield the co-torsionness of $\mathrm{Sel}(\mathrm{Ad}(\varphi) \otimes \nu^{-1})$, when φ is modular, we can deduce it from the co-torsionness of $\mathrm{Sel}(\mathrm{Ad}(\varphi))_{/\mathbb{Q}}$ using the fact

---

* The author is partially supported by a grant from NSF

that the p-th Hecke operator $T(p)$ is transcendental over $\mathbb{Z}_p$ in the universal ordinary Hecke algebra (see Theorem 3.3). For these, we consider the universal ordinary deformation ring R_F of φ restricted to $\mathrm{Gal}(\overline{\mathbb{Q}}/F)$ as in [W]. Then we have natural projection $\pi_F : R_F \to \mathbb{I}$, and we can identify $\mathrm{Sel}(\mathrm{Ad}(\varphi))_{/F}$ with the module (called a Mazur module) of 1-differentials of R_F as in [MT] which gives a tool of proving the above control theorem. To help the reader to understand the formal but subtle argument dealing with various deformation rings, we added to the main text a lengthy Appendix which describes a general theory of controlling deformation rings.

1. — Control of differential modules

In this section, we describe how a group action on a ring induces a group action on its differential modules.

1.1. Functoriality of differential modules. We start with a noetherian integral domain A with quotient field K. Let H be an A-algebra, and $\lambda : H \to B$ be an A-algebra homomorphism. The differential module is then defined by

$$
\begin{aligned}
C_1(\lambda; B) = \mathrm{Tor}_1^H(\mathrm{Im}(\lambda), B) &\cong \mathrm{Ker}(\lambda) \otimes_{H,\lambda} B \\
&\cong (\mathrm{Ker}(\lambda)/\mathrm{Ker}(\lambda)^2) \otimes_{H,\lambda} B.
\end{aligned}
$$

(C$_1$)

See [H2] Section 6 and [H3] Section 1 for a general theory of these modules including above isomorphisms. Suppose that we have two surjective A-algebra homomorphisms : $H \xrightarrow{\theta} T \xrightarrow{\mu} B$ with $\lambda = \mu \circ \theta$. Anyway, these modules are torsion modules over A if B is of finite type as an A-module. Then we recall Theorem 6.6 in [H2] :

PROPOSITION 1.1. — *Suppose the surjectivity of θ and μ. Then we have the following canonical exact sequences of H-modules :*

$$
\mathrm{Tor}_1^T(B, \mathrm{Ker}(\mu)) \to C_1(\theta; T) \otimes_T B \to C_1(\lambda; B) \to C_1(\mu; B) \to 0 ;
$$

Proof : we have an exact sequence of H-modules :

$$
0 \to \mathrm{Ker}(\theta) \to \mathrm{Ker}(\lambda) \xrightarrow{\theta} \mathrm{Ker}(\mu) \to 0 .
$$

Tensoring B over H with the above sequence, we obtain the desired result.

We now suppose that a finite group G acts on H through A-algebra automorphisms. Thus the finite group G acts on $\mathrm{Spec}(H)$. We consider the following condition :

(Nt) $\mathrm{Spec}(T)$ is the fixed point subscheme of G in $\mathrm{Spec}(H)$.

Let $\mathfrak{a}$ be the augmentation ideal of $\mathbb{Z}[G]$. Then the condition (Nt) is equivalent to

(Nt') $\mathrm{Ker}(\theta)$ is generated over H by $g(x) - x$ for $x \in H$ and $g \in G$, that is, $\mathrm{Ker}(\theta) = H\mathfrak{a}H$.

Let $\sigma \in G$. Then, under (Nt), the action of $\sigma - 1$ induces an A-linear map : $\mathrm{Ker}(\lambda) \to \mathrm{Ker}(\theta)$. If $x, y \in \mathrm{Ker}(\lambda)$, then

$$\sigma(xy) - xy = (\sigma(x) - x)(\sigma(y) - y) + x(\sigma(y) - y) + y(\sigma(x) - x)$$
$$\equiv (\sigma(x) - x)(\sigma(y) - y) \equiv 0 \bmod \mathrm{Ker}(\lambda)\,\mathrm{Ker}(\theta) .$$

Note that $C_1(\theta; T) \otimes_{T,\mu} B \cong (\mathrm{Ker}(\theta)/\mathrm{Ker}(\lambda)\,\mathrm{Ker}(\theta)) \otimes_{H,\lambda} B$. Thus the A-linear map induces a B-linear map $[\sigma - 1] : C_1(\lambda; B) \to C_1(\theta; T) \otimes_{T,\mu} B$. Under (Nt), $\sigma(x) - x$ for $x \in H$ and $\sigma \in G$ generates $\mathrm{Ker}(\theta)$ over H. Now assume that

(Sec) λ has a section $\iota : B \to H$ of $A[G]$ – modules.

Then for each $y \in H$, we can write $y = x \oplus \iota\lambda(y)$ for $x = y - \iota\lambda(y) \in \mathrm{Ker}(\lambda)$, and hence $\sigma(y) - y = \sigma(x) - x \in (\sigma - 1)\,\mathrm{Ker}(\lambda)$. Thus $[\sigma - 1](x)$ for $\sigma \in G$ and $x \in \mathrm{Ker}(\lambda)$ generates $C_1(\theta; T) \otimes_{T,\mu} B$ over B, and

$$\oplus_{\sigma \in G}[\sigma - 1] : \oplus_{\sigma \in G} C_1(\lambda; B) \to C_1(\theta; T) \otimes_{T,\mu} B \text{ is surjective.}$$

This shows that the image of $C_1(\theta; T) \otimes_{T,\mu} B$ in $C_1(\lambda; B)$ is equal to $\mathfrak{a}C_1(\lambda; A)$, and we have

COROLLARY 1.1. — *Suppose* (Nt) *and* (Sec). *Then we have*

$$C_1(\mu; B) \cong C_1(\lambda; B)/\mathfrak{a}C_1(\lambda; B) = H_0(G, C_1(\lambda; B)) ,$$

where $\mathfrak{a}$ *is the augmentation ideal of* $\mathbb{Z}[G]$.

We now put ourselves in a bit more general setting where μ is not necessarily surjective. We write B_0 for $\mathrm{Im}(\mu)$ and consider the following three algebra homomorphisms :

$$H \otimes_A B \xrightarrow{\theta \otimes \mathrm{id}} T \otimes_A B \xrightarrow{\mu \otimes \mathrm{id}} B_0 \otimes_A B \xrightarrow{m} B ,$$

where $m(a \otimes b) = ab$. Since $\mathrm{Ker}(\mu \otimes \mathrm{id})^j$ is a surjective image of $\mathrm{Ker}(\mu)^j \otimes_A B$, the natural map : $C_1(\lambda; B) \otimes_A B \to C_1(\lambda \otimes \mathrm{id}; B \otimes_A B)$ is surjective. When B is flat over A, the map is an isomorphism of $B \otimes_A B$-modules. Similarly, the natural maps

$$C_1(\lambda; B_0) \otimes_A B \to C_1(\lambda \otimes \mathrm{id}; B_0 \otimes_A B) ,$$

(Ext1) $$C_1(\mu; B_0) \otimes_A B \to C_1(\mu \otimes \mathrm{id}; B_0 \otimes_A B) \text{ and}$$

$$C_1(\theta; T) \otimes_A B \to C_1(\theta; T \otimes_A B)$$

are all surjective and are isomorphisms if B is flat over A. By Proposition 1.1, we get an exact sequence, writing B' for $B_0 \otimes_A B$,

(Ext2) $$\mathrm{Tor}_1^{B'}(\mathrm{Ker}(m), B) \to C_1(\mu \otimes \mathrm{id}; B') \otimes_{B'} B \to$$
$$C_1(m \circ (\mu \otimes \mathrm{id}); B) \to C_1(m; B) \to 0 .$$

We get from the short exact sequence : $0 \to \mathrm{Ker}(\mu) \to T \to B_0 \to 0$, an exact sequence : $\mathrm{Tor}_1^A(B_0, B_0) \to \mathrm{Ker}(\mu) \otimes_A B_0 \to T \otimes_A B_0 \to B_0 \otimes_A B_0 \to 0$, and as a part of it, we know the exactness of the following sequence :

$$\mathrm{Tor}_1^A(B_0, B_0) \to \mathrm{Ker}(\mu) \otimes_A B_0 \to \mathrm{Ker}(\mu \otimes \mathrm{id}) \to 0 .$$

Applying $\otimes_{T'} B$ to the last sequence, writing T' for $T \otimes_A B$, we have another exact sequence :

(Ext2′) $$\mathrm{Tor}_1^A(B_0, B_0) \otimes_{B'} B \to (\mathrm{Ker}(\mu) \otimes_A B_0) \otimes_{T'} B \to$$
$$C_1(\mu \otimes \mathrm{id}; B') \otimes_{B'} B \to 0$$

and

$$(\mathrm{Ker}(\mu) \otimes_A B_0) \otimes_{T'} B = (\mathrm{Ker}(\mu)/\mathrm{Ker}(\mu)^2) \otimes_A B_0 \otimes_{B'} B$$

$$= ((\mathrm{Ker}(\mu)/\mathrm{Ker}(\mu)^2) \otimes_{B_0} B_0) \otimes_A B \otimes_{B'} B$$

$$= (\mathrm{Ker}(\mu)/\mathrm{Ker}(\mu)^2) \otimes_{B_0} B = C_1(\mu; B_0) \otimes_{B_0} B .$$

This combined with (Ext2) shows the exactness of

(Ext3) $$C_1(\mu; B_0) \otimes_{B_0} B \to C_1(m \circ (\mu \otimes \mathrm{id}); B) \to C_1(m; B) \to 0 .$$

If B_0 is A-flat, then $\mathrm{Tor}_1^A(B_0, B_0) = 0$. When $B = B_0$, the above sequence is nothing but the well known exact sequence for the closed immersion μ

of $X = \mathrm{Spec}(B_0)$ into $Y = \mathrm{Spec}(T)$ over $S = \mathrm{Spec}(A) : \mathrm{Ker}(\mu)/\mathrm{Ker}(\mu)^2 \rightarrow \mu^* \Omega_{Y/S} \rightarrow \Omega_{X/S} \rightarrow 0.$

2. — Control Theorems of universal ordinary deformation rings

We fix a prime $p \geq 3$. For a number field X in $\overline{\mathbb{Q}}$, we write $\mathcal{G}_X = \mathrm{Gal}(\overline{\mathbb{Q}}/X)$ for the absolute Galois group over X. Let $\mathfrak{O}$ be a valuation ring finite flat over $\mathbb{Z}_p$ with residue field $\mathbb{F}$. We consider a p-ordinary deformation problem $\mathcal{D} = \mathcal{D}_X$ defined on the category CNL$_{\mathfrak{O}}$ of complete noetherian local $\mathfrak{O}$-algebras with residue field $\mathbb{F}$. Morphisms of CNL$_{\mathfrak{O}}$ are assumed to be local $\mathfrak{O}$-algebra homomorphisms. See [T] and Appendix for a general theory of such deformation problems. Let (R_X, ρ_X) be the universal couple of the deformation problem $\mathcal{D}_X$ of representations of $\mathcal{G}_X$. We study how the Galois action controls R_X.

2.1. Deformation problems. Let $\overline{\rho}$ be a continuous representation of $\mathcal{G}_E$ into $\mathrm{GL}_2(\mathbb{F})$ for a number field E. We consider the following condition for an algebraic extension F/E :

(AI_F) $\qquad$ $\overline{\rho}$ restricted to $\mathcal{G}_F$ is absolutely irreducible .

We assume (AI_F). For each prime ideal $\mathfrak{l}$, we write $F_{\mathfrak{l}}$ for the $\mathfrak{l}$-adic completion of F and $\mathcal{G}_{F_{\mathfrak{l}}}$ for the absolute Galois group over $F_{\mathfrak{l}}$. Let C be an integral ideal of E prime to p and write $Cl_E(Cp)$ for the strict ray class group of E modulo Cp. We also pick a character $\chi : Cl_E(Cp) \rightarrow \mathfrak{O}^\times$ such that the order of χ is prime to p. We write $C(\chi)$ for the conductor of χ and assume that $C|C(\chi)$. By class field theory, we may regard χ as a character of $\mathcal{G}_E$. We write χ_q for the restriction of χ to the decomposition subgroup $\mathcal{G}_{E_q}$ at each prime q. Let $\mathcal{M}$ be a finite set of primes outside p. We assume that $\mathcal{M}$ contains all prime factors of $C(\chi)$ outside p. We write $\mathcal{M}(\chi)$ for the set of primes in $\mathcal{M}$ dividing $C(\chi)$ and put $\mathcal{M}' = \mathcal{M} - \mathcal{M}(\chi)$. We write Σ for the union of $\mathcal{M}$ and the set of all prime factors of p. Then for q $\in \Sigma$, write I_q (resp. $\mathcal{N}_q$) for the inertia subgroup of $\mathcal{G}_{E_q}$ (resp. the p-adic cyclotomic character $\mathcal{N}$ restricted to $\mathcal{G}_{E_q}$). Here we normalize cyclotomic characters so that they take the *geometric* Frobenius at each unramified prime ideal $\mathfrak{l}$ to the norm of the ideal $\mathfrak{l}$. Under this convention, we consider a deformation problem of $\overline{\rho}$ on CNL$_{\mathfrak{O}}$. A deformation $\rho : \mathcal{G}_E \rightarrow \mathrm{GL}_2(A)$ of $\overline{\rho}$ is called of type $\mathcal{D} = \mathcal{D}_E$ if ρ satisfies the following five conditions (UNR), (χ_p), (Reg_p) for each prime $\mathfrak{p}|p$, (χ_q) for each prime q $\in \mathcal{M}(\chi)$ and $(\mathcal{N}_q)$ for each prime q $\in \mathcal{M}'$:

(UNR) $\qquad\qquad\qquad$ π is unramified outside Σ ;

$(\chi_{\mathfrak{p}})$ We have an exact sequence of $\mathcal{G}_{E_{\mathfrak{p}}}$-modules :

$$0 \to V(\rho_{1;\mathfrak{p}}) \to V(\rho) \to V(\rho_{2,\mathfrak{p}}) \to 0 \text{ with } V(\rho_{2,\mathfrak{p}}) \text{ } A\text{-free of rank } 1,$$

$\rho_{1,\mathfrak{p}}$ unramified and $\rho_{2,\mathfrak{p}}$ mod $m = \chi_{\mathfrak{p}}$ mod m on the inertia subgroup $I_{\mathfrak{p}}$.

Writing $\overline{\rho}_i$ for ρ_i mod m, we assume

$(\mathrm{Reg}_{\mathfrak{p}})$ $\overline{\rho}_{1,\mathfrak{p}} \ne \overline{\rho}_{2,\mathfrak{p}}$.

We assume the following conditions for $\mathfrak{q} \in \mathcal{M}$:

$(\chi_{\mathfrak{q}})$ As $I_{\mathfrak{q}}$ − modules, $V(\rho) \cong V(\mathrm{id}) \oplus V(\chi_{\mathfrak{q}})$ with $V(\chi_{\mathfrak{q}})$ A-free of rank 1 for $\mathfrak{q}|C(\chi)$,

$(\mathcal{N}_{\mathfrak{q}})$ For $\mathfrak{q} \in \mathcal{M}'$, we have an exact sequence, of $\mathcal{G}_{E_{\mathfrak{q}}}$-modules, non-split over $I_{\mathfrak{q}}$.

$$0 \to V(\rho_{1,\mathfrak{q}}) \to V(\rho) \to V(\rho_{2,\mathfrak{q}}) \to 0$$

where $V(\rho_{2,\mathfrak{q}})$ is A-free of rank 1, $\rho_{i,\mathfrak{q}}$ $(i = 1,2)$ is unramified and $\rho_{1,\mathfrak{q}}\rho_{2,\mathfrak{q}}^{-1} = \mathcal{N}_{\mathfrak{q}}$.

Since the order of χ is prime to p, $\chi_{\mathfrak{q}}$ for $\mathfrak{q} \in \mathcal{M}(\chi)$ is non-trivial. To make our deformation problem $\mathcal{D}_E$ non-empty, we assume that $\overline{\rho}$ satisfies the above five conditions. The contragredient of this deformation problem is studied in [W] and is denoted by $D = (\mathrm{Ord}, \Sigma, \mathfrak{O}, \mathcal{M})$ (for $E = \mathbb{Q}$) there. As shown in [W], the problem D is representable, and hence $\mathcal{D}_E$ is also representable. See [T] and Appendix for the proof in more general case. To apply the argument in [T] and Appendix to our situation here, we note the following facts : for the maximal extension F_Σ of F unramified outside Σ, the Galois group $G = \mathrm{Gal}(F_\Sigma/E)$ satisfies the condition (pF) in Appendix; any deformation of type $\mathcal{D}_E$ factors through G; the group $D \in S$ (resp. its subgroup I) is given by a choice of decomposition subgroups (resp. its inertia subgroup) at each $\mathfrak{p}^\cdot$, and the condition $(\mathrm{Reg}_{\mathfrak{p}})$ is the same as (RG_D) in Section A.2.2 for the decomposition subgroup D of $\mathfrak{p}$ in G. We write (R_E, ρ_E) for the universal couple for $\mathcal{D}_E$. Thus for each deformation $\rho : \mathcal{G}_E \to \mathrm{GL}_2(A)$ of type $\mathcal{D}_E$, there exists a unique local $\mathfrak{O}$-algebra homomorphism $\varphi : R_E \to A$ such that ρ is strictly equivalent to $\varphi\rho_E$. Here we say ρ is strictly equivalent to ρ' if $\rho(\tau) = x\rho'(\tau)x^{-1}$ for $x \in \widehat{\mathrm{GL}}_2(A) = 1 + m_A M_2(A)$ independent of τ. We write $\rho \approx \rho'$ if ρ is strictly equivalent to ρ'.

Let F be a finite extension of E. Write $\overline{\rho}_F$ (resp. χ_F) for the restriction of $\overline{\rho}$ (resp. χ) to $\mathcal{G}_F$. We consider the deformation problem $\mathcal{D}_F$ of $\overline{\rho}_F$ on $\mathrm{CNL}_{\mathfrak{O}}$

given as follows. Let $\mathcal{M}_F(\chi)$ (resp. $\mathcal{M}'_F$) be the set of prime ideals dividing $C(\chi_F)$ and prime to p (resp. the set of primes dividing primes in $\mathcal{M}'$). We write Σ_F for the union of $\mathcal{M}_F(\chi)$, $\mathcal{M}'_F$ and the set of all prime factors of p in F. A deformation ρ of $\overline{\rho}_F$ is a continuous representation $\rho : \mathcal{G}_F \to \mathrm{GL}_2(A)$ with $\rho \bmod m_A = \overline{\rho}_F$ for an object A in $\mathrm{CNL}_{\mathfrak{O}}$. A deformation ρ of $\overline{\rho}_F$ is of type $\mathcal{D}_F$ if ρ satisfies the following five conditions (UNR_F), $(\chi_{\mathcal{P},F})$, $(\mathrm{Reg}_{\mathcal{P},F})$ for each prime $\mathcal{P}$ of F dividing p, $(\chi_{\mathcal{Q},F})$ for $\mathcal{Q} \in \mathcal{M}_F(\chi)$ and $(\mathcal{N}_{\mathcal{Q}})$ for $\mathcal{Q} \in \mathcal{M}'_F$:

(UNR_F) $\qquad\qquad\qquad$ ρ is unramified outside Σ_F ;

$(\chi_{\mathcal{P},F})$ We have an exact sequence of $\mathcal{G}_{F_{\mathcal{P}}}$-modules for each prime ideal $\mathcal{P}|p$:

$$0 \to V(\rho_{1,\mathcal{P}}) \to V(\rho) \to V(\rho_{2,\mathcal{P}}) \to 0 \text{ with } V(\rho_{2,\mathcal{P}}) \text{ } A\text{-free of rank 1},$$

$$\rho_{1,\mathcal{P}} \text{ unramified and } \rho_{2,\mathcal{P}} \bmod m = \chi_{\mathcal{P}} \bmod m \text{ on } I_{\mathcal{P}} \text{ ;}$$

$(\mathrm{Reg}_{\mathcal{P},F})$ $\qquad\qquad$ $\overline{\rho}_{1,\mathcal{P}} \neq \overline{\rho}_{2,\mathcal{P}}$ for each prime ideal $\mathcal{P}|p$.

where $\overline{\rho}_{i,\mathcal{P}} = \rho_{i,\mathcal{P}} \bmod m$. Writing $\chi_{\mathcal{Q}}$ for the restriction of χ to $\mathcal{G}_{F_{\mathcal{Q}}}$, we assume :

$(\chi_{\mathcal{Q},F})$ As $I_{\mathcal{Q}}$−modules, $V(\rho) \cong V(\mathrm{id}) \oplus V(\chi_{\mathcal{Q}})$ with A-free of rank 1 for $\mathcal{Q} \in \mathcal{M}_F$,

$(\mathcal{N}_{\mathcal{Q}})$ For $\mathcal{Q} \in \mathcal{M}'_F$ we have an exact sequence, of $\mathcal{G}_{F_{\mathcal{Q}}}$-modules, non-split over $I_{\mathcal{Q}}$,

$$0 \to V(\rho_{1,\mathcal{Q}}) \to V(\rho) \to V(\rho_{2,\mathcal{Q}}) \to 0$$

where $V(\rho_{2,\mathcal{Q}})$ is A-free of rank 1, $\rho_{i,\mathcal{Q}}$ $(i = 1,2)$ is unramified and $\rho_{1,\mathcal{Q}}\rho_{2,\mathcal{Q}}^{-1} = \mathcal{N}_{\mathcal{Q}}$.

Then $\mathcal{D}_F$ is representable under (AI_F). We write $H = \mathrm{Gal}(F_{\Sigma}/F)$. Then H is a normal subgroup of G with $G/H = \Delta = \mathrm{Gal}(F/E)$. We take $S_G = \{D_{\mathfrak{p}}\}_{\mathfrak{p}|p}$ and $S_H = \{D_{\mathcal{P}}\}_{\mathcal{P}|p}$ for the decomposition subgroups $D_?$ for each prime "?". Then to the quadruple (G, H, S_G, S_H) the theory described in Appendix (Sections A.2.1-3) applies, and it is easy to deduce the representability from the argument in Section A.2.3. Hereafter, assuming (AI_F), we write (R_F, ρ_F) for the universal couple representing the problem $\mathcal{D}_F$.

2.2. Controlling universal deformation rings and Mazur modules. We now suppose that F/E is cyclic of degree d. Since ρ_E restricted to $H = \mathrm{Gal}(F_{\Sigma}/F)$ is a solution of the deformation problem $\mathcal{D}_F$, we have a non-trivial algebra homomorphism $\alpha : R_F \to R_E$ such that $\alpha\rho_F \approx \rho_E$.

If ρ is a deformation of type $\mathcal{D}_E$, then we have a unique $\varphi : R_E \to A$ such that $\varphi \rho_E \approx \rho$. Then $\rho|_H \approx \varphi \rho_E|_H \approx \varphi \alpha \rho_F$. Let us write $\mathcal{F}_F$ (resp. $\mathcal{F}$) for the deformation functor of the problem $\mathcal{D}_F$ (resp. $\mathcal{D}_E$). Then the Galois group $\Delta = \mathrm{Gal}(F/E)$ naturally acts on $\mathcal{F}_F$ as follows : taking $\sigma \in G$ and define $\rho^\sigma(g) = \rho(\sigma g \sigma^{-1})$ for $\rho \in \mathcal{F}_F(A)$ and $g \in H$. We take $c(\sigma) \in \mathrm{GL}_2(\mathcal{D})$ such that $c(\sigma) \equiv \overline{\rho}(\sigma) \bmod m_o$ and define $\rho^{[\sigma]} = c(\sigma)^{-1} \rho^\sigma c(\sigma) \in \mathcal{F}_F(A)$. The strict equivalence class of $\rho^{[\sigma]}$ is well defined depending only on the class of σ in Δ, and in this way Δ acts on $\mathcal{F}_F$ and R_F through $\mathcal{D}$-algebra automorphisms. We define a new functor $\mathcal{F}_F^\Delta$ by $\mathcal{F}_F^\Delta(A) = \{\rho \in \mathcal{F}_F(A)|\rho^{[\sigma]} \approx \rho\}$. Since $\varphi \alpha$ is the unique homomorphism bringing ρ_F down to $\rho|_H$, the deformation subfunctor :

$$\mathcal{F}_{E,F}(A) = \{\rho|_H \in \mathcal{F}_F(A) \mid \rho \in \mathcal{F}(B) \text{ for a flat } A\text{-algebra } B \text{ in } \mathrm{CNL}_\mathcal{D}\}$$

is representable by $(\mathrm{Im}(\alpha), \alpha \rho_F)$ under (AI_F), as long as $\alpha \rho_F$ can be extended to an element of $\mathcal{F}(B)$ for an algebra B flat over $\mathrm{Im}(\alpha)$ in $\mathrm{CNL}_\mathcal{D}$. The argument proving this is the same as the proof of Theorem A.2.3. To check the extendibility of $\alpha \rho_F$, we introduce the following assumptions. Let F_0 be the maximal subfield of F such that $d' = [F_0 : E]$ is prime to p. Let $\mathbb{S}$ be the set of primes of E ramifying in F. For each prime $\mathfrak{q}$ of E, we write $I_\mathfrak{q}$ (resp. $I(\mathfrak{q})$) for the inertia group of $\mathfrak{q}$ in G (resp. Δ). We also write $D(\mathfrak{q})$ for the decomposition subgroup of Δ at $\mathfrak{q}$. We assume for $\mathfrak{q}$ outside p

$(\mathrm{TR}_\mathfrak{q})$ $|I(\mathfrak{q})|$ is prime to p .

For $\mathfrak{p}|p$, we assume either $(\mathrm{TR}_\mathfrak{p})$ or

$(\mathrm{Ex}_\mathfrak{p})$ Every character of $I(\mathfrak{p}) \cap \mathrm{Gal}(F/F_0)$ with values in $A^\times$ can be

                   extended to a character of Δ having values in

                   $B^\times$ for a flat extension B of A such that it is

                   unramified outside $\mathfrak{p}$.

These conditions correspond the conditions (TR_D) and (Ex_D) in Section A.2.3. If $I(\mathfrak{p}) \cap I(\mathfrak{q}) \cap \mathrm{Gal}(F/F_0) = \{1\}$ for any two primes $\mathfrak{p}$ dividing p and an arbitrary $\mathfrak{q}$, then $(\mathrm{Ex}_\mathfrak{p})$ is satisfied. In particular, if F is a subfield of $\mathbb{Q}_\infty$, $(\mathrm{Ex}_\mathfrak{p})$ is satisfied, where $\mathbb{Q}_\infty$ is the unique $\mathbb{Z}_p$-extension of $\mathbb{Q}$.

   Take $\pi \in \mathcal{F}_F^\Delta(A)$. By Corollary A.1.2 combined with the argument in A.2.1, there exists a faithfully flat A-algebra B in $\mathrm{CNL}_\mathcal{D}$ such that π extends to a representation $\pi_E : G \to \mathrm{GL}_2(B)$ with $\pi_E \equiv \overline{\rho} \bmod m_\mathcal{D}$. By $(\mathrm{TR}_\mathfrak{q})$, the unramifiedness at $q \nmid p$ in $\mathbb{S} - \mathcal{M}$ of $\overline{\rho}$ implies the unramifiedness of π_E, where $\mathbb{S}$ is the set of primes ramifying in F/E. We now look at the restriction of π_E to the inertia group $I = I_\mathfrak{q}$ for primes $\mathfrak{q}$ in $\mathcal{M}$ or dividing

p. By $(\mathrm{TR_q})$ and the fact that the decomposition group at q acts through conjugation on the maximal tame quotient of the inertia subgroup by the cyclotomic character $\mathcal{N}$, the conditions (χ_Q) (resp. $(\mathcal{N}_Q)$) for $\overline{\rho}$ implies (χ_q) (resp. $(\mathcal{N}_q)$) for π_E at $q = Q \cap E$. We look at the restriction of π_E to $I_\mathfrak{p}$. Since the characteristic polynomial of $\overline{\rho}(\sigma)$ for an element σ of $I_\mathfrak{p}$ has two distinct roots in $\mathbb{F}$ by $(\mathrm{Reg_p})$ and $(\chi_\mathfrak{p})$, that of $\pi_E(\sigma)$ again has two distinct roots a and b in B by Hensel's lemma. Then writing V for $\mathrm{Ker}(\pi_E(\sigma) - a \cdot id)$, $V(\pi_E)/V$ is B-free of rank 1, and on V, $D_\mathfrak{p}$ acts by a character $\eta_\mathfrak{p}$. Replacing a by b if necessary, we may assume that $\overline{\eta}_\mathfrak{p}$ is trivial on $I(\mathfrak{p})$. If $\mathfrak{p}$ satisfies $(\mathrm{TR_p})$, the argument is the same as q above. Now suppose $(\mathrm{Ex_p})$. Since $\overline{\eta}_\mathfrak{p}$ is trivial on $I(\mathfrak{p})$, $\eta_\mathfrak{p}$ factors through the p-primary quotient of $I(\mathfrak{p})$. Thus by $(\mathrm{Ex_p})$, we can lift $\eta_\mathfrak{p}$ to a p-power order character ξ of Δ unramified outside $\mathfrak{p}$. Replacing π_E by $\pi_E \otimes \xi^{-1}$, we may assume that π_E satisfies $(\chi_\mathfrak{p})$ because $\xi \equiv 1 \bmod m$. Thus π_E is a deformation of type $\mathcal{D}_E$, and we get

PROPOSITION 2.1. — Assume (AI_F), $(\mathrm{TR_q})$ for q outside p and one of $(\mathrm{Ex_p})$ or $(\mathrm{TR_p})$ for $\mathfrak{p}|p$. Then $\pi \in \mathcal{F}_F^\Delta(A)$ can be extended to an element π_E of $\mathcal{F}(B)$ for a faithfully flat A-algebra B in $\mathrm{CNL}_{\mathfrak{D}}$. Moreover the functor $\mathcal{F}_{E,F}$ is represented by $(\mathrm{Im}(\alpha), \alpha\rho_F)$.

For each integral ideal C of a number field X, we write $\mathrm{Cl}_X(Cp^e)$ for the strict ray class group modulo Cp^e. We allow $e = \infty$, and $\mathrm{Cl}_X(Cp^\infty) = \varprojlim_e \mathrm{Cl}_X(Cp^e)$. Then by class field theory, there exists an abelian extension X_∞/X unramified outside Cp such that $\mathrm{Gal}(X_\infty/X) \cong \mathrm{Cl}_X(Cp^\infty)$. We consider the character $\det(\rho_F) : \mathcal{G}_F \to R_F^\times$. By (χ_Q), the restriction of this character to the inertia subgroup I_Q factors through a finite quotient. Thus there exists an integral ideal C prime to p of F such that $\det(\rho_F)$ factors through $Z_F = \mathrm{Cl}_F(Cp^\infty)$, and hence there exists an algebra homomorphism : $\mathfrak{D}[[Z_F]] \to R_F$ taking $z \in Z_F$ to $\det(\rho_E(z))$. We take C maximal among the ideals satisfying the above condition. We write Λ_F for the image of $\mathfrak{D}[[Z_F]]$ in R_F which is an object in $\mathrm{CNL}_{\mathfrak{D}}$.

We now modify the deformation problem $\mathcal{D}_E$ on $\mathrm{CNL}_{\mathfrak{D}}$ and create a new one $\mathcal{D}_\Lambda$ defined over the category CNL_Λ of complete noetherian local Λ_E-algebras with residue field $\mathbb{F}$ by adding the following condition to the conditions of $\mathcal{D}_E$:

(det) $\det(\rho)$ for each deformation $\rho : \mathcal{G}_E \to \mathrm{GL}_2(A)$ of type $\mathcal{D}_E$

coincides with $\det(\rho_E)$ composed with the inclusion $i : \Lambda_E \to A$.

For any deformation $\rho : \mathcal{G}_E \to \mathrm{GL}_2(A)$ of type $\mathcal{D}_\Lambda$, it is automatically a deformation of type $\mathcal{D}_E$. Thus we have a unique $\mathfrak{D}$-algebra homomorphism

$\phi : R_E \to A$ such that $\phi \rho_E \approx \rho$ and $\phi(\det(\rho_E)) = \det(\rho)$, which implies that ϕ is actually a Λ_E-algebra homomorphism. Therefore (R_E, ρ_E) represents the new problem $\mathcal{D}_\Lambda$. We consider another deformation functor $\mathcal{F}_{\Lambda,E,F}$ defined on CNL_Λ by

$$\mathcal{F}_{\Lambda,E,F}(A) = \{\rho|_H \in \mathcal{F}_{\Lambda,F}(A) | \rho \in \mathcal{F}_\Lambda(B) \text{ for a flat } A\text{-algebra } B \text{ in } \mathrm{CNL}_\Lambda\} .$$

By Lemma A.2.1 in Appendix, actually $\mathcal{F}_{\Lambda,E,F}(A) = \{\rho|_H \in \mathcal{F}_{\Lambda,F}(A) | \rho \in \mathcal{F}_\Lambda(A)\}$. By the argument proving Theorem A.2.1, we can conclude that this functor is represented by $(\mathrm{Im}(\alpha)\Lambda_E, \rho_E|_H)$, where we write $\mathrm{Im}(\alpha)\Lambda_E$ for the image of $\mathrm{Im}(\alpha)\widehat{\otimes}_{\mathcal{D}}\Lambda$ in R_E. Here is the argument : let ρ and ρ' be two deformations of $\overline{\rho}$ over E of type $\mathcal{D}_\Lambda$ with values in $\mathrm{GL}_2(A)$. Suppose that $\rho \approx \rho'$ on H. Under (AI_F), $\rho \cong \rho' \otimes \xi$ for some character ξ of $\mathrm{Gal}(F/E)$ (see Corollary A.2.1). Since in our deformation problem, the determinant is fixed, we have $\xi^2 = 1$. Since $\overline{\rho} \cong \overline{\rho} \otimes \xi$ and ξ is quadratic, if $\xi \neq 1$, $\xi \bmod m$ is non-trivial, because p is odd. By [DHI] Proposition 4.1, $\overline{\rho}$ is an induced representation of a character of $\mathrm{Ker}(\xi)$, which violates (AI_F). Thus ξ is trivial. The algebra $\mathrm{Im}(\alpha)$ may not be a Λ-algebra for $\Lambda = \Lambda_E$. We thus find

$$\mathrm{Hom}_{\Lambda-\mathrm{alg}}(\mathrm{Im}(\alpha)\Lambda_E, A) \cong \{\pi|_H \mid \pi \in \mathcal{F}(A),\ \det(\pi) = \det(\rho_E)\}/ \approx$$
$$\cong \mathcal{F}_\Lambda(A) \cong \mathrm{Hom}_{\Lambda-\mathrm{alg}}(R_E, A) \text{ for } \Lambda-\text{algebras } A.$$

Thus under (AI_F), we have $\mathrm{Im}(\alpha)\Lambda_E = R_E$.

THEOREM 2.1. — *Assume* (AI_F) *and one of* $(\mathrm{Ex_p})$ *or* $(\mathrm{TR_p})$ *for* $\mathfrak{p}|p$, $(\mathrm{TR_q})$ *for* $\mathfrak{q}$ *outside* p. *Then we have* $\mathrm{Im}(\alpha)\Lambda_E = R_E$. *If a prime factor* $\mathfrak{p}$ *of* p *ramifies totally in the maximal* p-*extension of* E *in* F, *then* $\mathrm{Im}(\alpha) = R_E$.

Proof : we only need to prove the last assertion. Here we give a short argument restricting ourselves to our special case. See Appendix for the treatment in more general cases. We argue similarly as above by replacing the deformation problem $\mathcal{D}_\Lambda$ by $\mathcal{D}_E$. Thus we pick two deformations ρ and ρ' of $\overline{\rho}$ over E of type $\mathcal{D}_E$ with values in $\mathrm{GL}_2(A)$. Suppose that $\rho \approx \rho'$ on H. Under (AI_F), $\rho \cong \rho' \otimes \xi$ for some character ξ of $\mathrm{Gal}(F/E)$. If $\xi \neq 1$, by our assumptions and $(\chi_\mathfrak{p})$, $\xi \rho_{1,\mathfrak{p}} = \rho'_{2,\mathfrak{p}}$ and $\xi \rho_{2,\mathfrak{p}} = \rho'_{1,\mathfrak{p}}$. Since $\xi \bmod m$ is trivial (because ξ is of p-power order), this contradicts to $(\mathrm{Reg_p})$. Thus $\xi = 1$ by the total ramification of $\mathfrak{p}$, and we find

$$\mathrm{Hom}_{\mathcal{D}-\mathrm{alg}}(\mathrm{Im}(\alpha), A) \cong \mathcal{F}_{E,F}(A) \supset \mathrm{Hom}_{\mathcal{D}-\mathrm{alg}}(R_E, A) \text{ for } \mathcal{D}-\text{algebras } A ,$$

which shows the result, because $\mathrm{Im}(\alpha) \subset R_E$.

Let $\mathfrak{a}$ be the ideal of R_F generated by $[\sigma](x) - x$ for all $x \in R_F$ and a generator $\sigma \in \Delta$. Then $\pi = \rho_F \bmod \mathfrak{a} : H \to \mathrm{GL}_2(A)$ for $A = R_F/\mathfrak{a}$ satisfies $\pi^{[\sigma]} \approx \pi$ and hence, by proposition 2.1, π extends to a representation π_E in $\mathcal{F}(B)$ for a faithfully flat A-algebra B in $\mathrm{CNL}_\mathfrak{O}$. On the other hand, we have the Galois representation ρ_E attached to R_E with $\alpha \circ \rho_F \approx \rho_E$ on H. By the universality of (R_E, ρ_E), we have an $\mathfrak{O}$-algebra homomorphism $\theta : R_E \to R$ such that $\theta \circ \rho_E \approx \pi_E$. We conclude from $\theta \alpha \rho_F \approx \theta \rho_E \approx \pi_E$ that $\theta \alpha$ coincides with the inclusion map of A into B. Thus α is injective on $A = R/\mathfrak{a}$. This shows (Nt) in 1.1 for $(R_F, \mathrm{Im}(\alpha))$ in place of (H, T) there. See Theorem A.2.3 for a more general result of this type.

By definition, the ideal C defining Λ_F is invariant under Δ, because $\overline{\rho}$ is Δ-invariant. Thus Δ naturally acts on Λ_F. Identifying $Z_F = \mathrm{Gal}(F_\infty/F)$, we have the restriction map $\mathrm{res} : Z_F \to Z_E \ (= \mathrm{Gal}(E_\infty/E))$. This induces a Δ-equivariant algebra homomorphism $\mathrm{res} : \Lambda_F \to \Lambda_E$. We now take a closed $\mathfrak{O}$-subalgebra Λ_F' in R_F which is stable under the action of Δ. We can take Λ_F' to be Λ_F, but some other choice is also possible. The map $\alpha : \Lambda_F' \to \alpha(\Lambda_F')$ coincides with res on the image of Λ_F and is equivariant under the Galois action. Let $\mathbb{I}$ be a normal integral Λ_F'-algebra which is a member of $\mathrm{CNL}_\mathfrak{O}$. Let $\pi : R_E \to \mathbb{I}$ be a Λ_F'-algebra homomorphism. We put

$$\lambda_F' = m(\pi\alpha \otimes_{\Lambda_F'} \mathrm{id}) : R_F \widehat{\otimes}_{\Lambda_F'} \mathbb{I} \to \mathbb{I} \text{ and } \mu_F' = m(\pi \otimes_{\Lambda_F'} \mathrm{id}) : R_E \widehat{\otimes}_{\Lambda_F'} \mathbb{I} \to \mathbb{I}$$

for the multiplication $m : \mathbb{I} \widehat{\otimes}_{\Lambda_F'} \mathbb{I} \to \mathbb{I}$, where $\mathbb{I}_0 = \mathrm{Im}(\pi)$. Here "$\widehat{\otimes}$" indicates the completion of the algebraic tensor product under the adic topology of the maximal ideal of the algebraic tensor product. Since the condition (Nt) is insensitive to tensor product, $\pi\alpha \otimes \mathrm{id} : R_F \widehat{\otimes}_{\lambda_F'} \mathbb{I} \to R_E \widehat{\otimes}_{\lambda_F'} \mathbb{I}$ again satisfies (Nt) if α is surjective. Note that $R_F \widehat{\otimes}_{\lambda_F'} \mathbb{I}$ is an $\mathbb{I}$- algebra and hence λ_F' has a trivial section of $\mathbb{I}$-modules. Thus we get from Corollary 1.1.

THEOREM 2.2. — *Assume* (AI_F) *and one of* $(\mathrm{Ex})_\mathfrak{p}$ *or* $(\mathrm{TR}_\mathfrak{p})$ *for* $\mathfrak{p}|p$, $(\mathrm{TR}_\mathfrak{q})$ *for* $\mathfrak{q}$ *outside* p. *Let* $\mathbb{I}$ *be a normal integral* Λ_F'-*algebra in* $\mathrm{CNL}_\mathfrak{O}$ *and* $\pi : R_E \to \mathbb{I}$ *be a* Λ_F'-*algebra homomorphism which is a morphism in* CNL. *Then we have, for* $\Delta = \mathrm{Gal}(F/E)$,

(i) $\mathrm{Spec}(\mathrm{Im}(\alpha))$ *is the maximal subscheme of* $\mathrm{Spec}(R_F)$ *fixed under the action of* Δ;

(ii) *If* α *is surjective, then* $C_1(\mu_F'; \mathbb{I}) \cong C_1(\lambda_F'; \mathbb{I})/(\sigma - 1)C_1(\lambda_F'; \mathbb{I})$, *where* σ *is a generator of* Δ.

Suppose that α is surjective and $\mathbb{I}$ is a Λ_F'-module of finite type. We now take a subalgebra Λ_E' of R_E containing $\alpha(\Lambda_F')$. Note here that $C_1(\mu_F'; \mathbb{I})$ can be different from $C_1(\mu'_E; \mathbb{I})$ for $\mu'_E = m(\pi \otimes_{\Lambda'_E} \mathrm{id}) : R_E \otimes_{\Lambda'_E} \mathbb{I} \to \mathbb{I}$. We now compute the difference using the following diagram :

$$
\begin{array}{ccc}
\Lambda'_E \otimes_{\alpha(\Lambda'_F)} \mathbb{I} & \xrightarrow{\ \theta\ } & \Lambda'_E \otimes_{\Lambda'_E} \mathbb{I} \to \mathbb{I} \\
\downarrow & & \downarrow \\
R_E \otimes_{\alpha(\Lambda'_F)} \mathbb{I} & \xrightarrow{\ \mathrm{id}\,\otimes\theta\ } & R_E \otimes_{\Lambda'_E} \mathbb{I} \to \mathbb{I} .
\end{array}
$$

We have an exact sequence, writing $T = R_E \otimes_{\Lambda'_E} \mathbb{I}$,

$$
\mathrm{Tor}_1^T(\mathbb{I}, \mathrm{Ker}(\mu'_E)) \to C_1(\mathrm{id}\otimes\theta; T) \otimes_T \mathbb{I} \to C_1(\mu'_F; \mathbb{I}) \to C_1(\mu'_E; \mathbb{I}) \to 0 .
$$

By the diagram, since the multiplication map has a section of Λ'_E-modules, we have

$$
C_1(\mathrm{id}\otimes\theta; T) \cong C_1(\theta; \Lambda'_E \otimes_{\Lambda'_E} \mathbb{I}) \otimes_{\Lambda'_E} R_E \ \text{and}
$$

$$
C_1(\theta; \Lambda'_E \otimes_{\Lambda'_E} \mathbb{I}) \cong C_1(\mathrm{id}\otimes\theta; T) \otimes_T \mathbb{I} .
$$

Since θ is a scalar extension to $\mathbb{I}$ of the multiplication map : $\Lambda'_E \otimes_{\alpha(\Lambda'_F)} \Lambda'_E \to \Lambda'_E$. Thus we see that $C_1(\theta; \Lambda'_E \otimes_{\Lambda'_E} \mathbb{I}) \cong \Omega_{\Lambda'_E/\alpha(\Lambda'_F)} \otimes_{\Lambda'_E} \mathbb{I}$, and we have an exact sequence :

(Ext4) $\mathrm{Tor}_1^T(\mathbb{I}, \mathrm{Ker}(\mu'_E)) \to \Omega_{\Lambda'_E/\alpha(\Lambda'_F)} \otimes_{\Lambda'_E} \mathbb{I} \to$

$$
C_1(\lambda'_F; \mathbb{I})/(\sigma - 1)C_1(\lambda'_F; \mathbb{I}) \to C_1(\mu'_E; \mathbb{I}) \to 0 .
$$

Then as seen in [H3] Lemma 1.11, if $\Lambda'_F \cong \mathfrak{O}[[W_F]]$ and $\Lambda'_E \cong \mathfrak{O}[[W_F]]$ for a p-profinite subgroup W_X of Z_X, we have $\Omega_{\Lambda'_E/\Lambda'_F} \cong \Lambda'_E \otimes_Z W_E/\mathrm{res}(W_F)$, and hence $C_1(\mathrm{id}\otimes\theta; T) \otimes_T \mathbb{I} \cong \mathbb{I} \otimes_Z W_E/\mathrm{res}(W_F)$. We write λ_X and μ_X for λ'_X and μ'_X when $\Lambda'_X = \Lambda_X$. Assume that Λ'_F contains Λ_F. By a similar argument using the following diagram with exact rows :

$$
\begin{array}{ccc}
\Lambda'_F \widehat{\otimes}_{\Lambda_F} \mathbb{I} & \xrightarrow{\ \theta\ } & \Lambda'_F \otimes_{\Lambda'_F} \mathbb{I} \to \mathbb{I} \\
\downarrow & & \downarrow \\
R_F \widehat{\otimes}_{\Lambda_F} \mathbb{I} & \xrightarrow{\ \mathrm{id}\,\otimes\theta\ } & R_F \otimes_{\Lambda'_F} \mathbb{I} \to \mathbb{I} ,
\end{array}
$$

we get the following exact sequence :

(Ext'4) $\Omega_{\Lambda'_F/\Lambda_F} \widehat{\otimes}_{\Lambda_F} \mathbb{I} \to C_1(\lambda_F; \mathbb{I}) \to C_1(\lambda'_F; \mathbb{I}) \to 0 .$

Here $\mathbb{I}$ may not be a Λ_F-module of finite type, but we assume it is Λ'_F-module of finite type. The exactness of the above sequence follows from the compactness of these modules and [EGA] IV, 0.20.7.18.

3. — Selmer groups

Keeping notation introduced in the previous section, under a suitable assumption, we deduce the control theorem for the Selmer group $\text{Sel}(\text{Ad}(\varphi) \otimes \nu^{-1})_{/\mathbb{Q}}$ from the control theorem of the deformation rings (Theorem 2.2). Then assuming the transcendence of $\varphi_{1,p}(\phi_p)$ over $\mathbb{Z}_p$ for the geometric Frobenius ϕ_p at p and the co-torsionness of $\text{Sel}(\text{Ad}(\varphi))_{/\mathbb{Q}}$ over $\mathbb{I}$, we prove the co-torsionness of $\text{Sel}(\text{Ad}(\varphi) \otimes \nu^{-1})_{/\mathbb{Q}}$ over $\mathbb{I}[[\Gamma]]$.

3.1. Selmer groups. Here we suppose that $E = \mathbb{Q}$; so, instead of writing primes in E as q, we use the Roman character q for that. Let $\mathbb{I}$ be an integral normal local domain complete under $m_{\mathbb{I}}$-adic topology for the maximal ideal $m_{\mathbb{I}}$. We assume that $\mathbb{I}$ is an algebra over $\mathfrak{O}$ and $\mathbb{I}/m_{\mathbb{I}} = \mathbb{F}$. Let $\varphi : \mathcal{G}_{\mathbb{Q}} \to \text{GL}_n(\mathbb{I})$ be a continuous Galois representation. Thus $E = \mathbb{Q}$ in the notation of the previous section. Let $V(\varphi) = \mathbb{I}^n$ be the representation space of φ and $W(\varphi)$ be a subspace of $V(\varphi)$ stable under $\mathcal{G}_{\mathbb{Q}_p}$. We define two Galois modules $V(\varphi)^* = V \otimes_{\mathbb{I}} \mathbb{I}^*$ and $W(\varphi)^* = W(\varphi) \otimes_{\mathbb{I}} \mathbb{I}^*$, where $\mathbb{I}^*$ is the Pontryagin dual module of $\mathbb{I}$. Let $\mathbb{Q}_\infty/\mathbb{Q}$ be the unique $\mathbb{Z}_p$-extension. We identify $\text{Gal}(\mathbb{Q}_\infty/\mathbb{Q})$ with $\Gamma = 1 + p\mathbb{Z}_p$ by the cyclotomic character $\mathcal{N}$. Here $\mathcal{N}$ satisfies $\mathcal{N}(\phi_q) = q$ for the geometric Frobenius ϕ_q at q. We write $\nu = \nu_\infty$ for the inclusion of $\text{Gal}(\mathbb{Q}_\infty/\mathbb{Q})$ into $\mathfrak{O}[[\Gamma]]$. Let Γ_m be the subgroup of Γ of index p^m, and write $\mathbb{Q}_m$ for the fixed field of Γ_m. We write the projection $\pi_m : \mathfrak{O}[[\Gamma]] \to \mathfrak{O}[[\Gamma/\Gamma_m]]$ and put $\nu_m = \pi_m \circ \nu$. We put for $m = 1, 2, \ldots, \infty$,

$$V(\varphi \otimes \nu_m^{-1}) = V(\varphi) \widehat{\otimes}_{\mathfrak{O}} V(\nu_m^{-1}), \ W(\varphi \otimes \nu_m^{-1}) = W(\varphi) \widehat{\otimes}_{\mathfrak{O}} V(\nu_m^{-1}) ,$$

where $V(\nu_m^{-1}) = \mathfrak{O}[\Gamma/\Gamma_m]$ for finite m.

LEMMA 3.1. — *We have the following isomorphisms of* $\mathbb{I}[[\Gamma]]$-*modules for* $m = 1, 2, \ldots, \infty$,

$$H^1(\mathcal{G}_{\mathbb{Q}_m}, V(\varphi)^*) \cong H^1(\mathcal{G}_{\mathbb{Q}}, V(\varphi \otimes \nu_m^{-1})^*) ,$$

$$H^1(I_p \cap \mathcal{G}_{\mathbb{Q}_m}, V(\varphi)^*/W(\varphi)^*) \cong H^1(I_p, V(\varphi \otimes \nu_m^{-1})^*/W(\varphi \otimes \nu_m^{-1})^*) ,$$

$$\Pi_{\gamma \in \Gamma/\Gamma_m} H^1(\gamma I_q \gamma^{-1}, V(\varphi)^*/W(\varphi)^*) \cong H^1(I_q, V(\varphi \otimes \nu_m^{-1})^*/W(\varphi \otimes \nu_m^{-1})^*)$$
for $q \neq p$,

where I_q *is the inertia subgroup at* q *of* $\mathcal{G}_{\mathbb{Q}}$, $\mathbb{I}[[\Gamma]]$ *acts through coefficients on the right-hand side, and on the left-hand side* $\mathbb{I}$ *acts through coefficients but* Γ *acts through the group* $\mathcal{G}_{\mathbb{Q}_\infty}$ *by conjugation.*

Proof : note that $V(\varphi \otimes \nu_m^{-1})^*$ is the induced representation of (injective type) of $V(\varphi)^*$ restricted to $\mathcal{G}_{\mathbb{Q}_m}$ to $\mathcal{G}_{\mathbb{Q}}$, that is, $V(\varphi)\widehat{\otimes}_{\mathcal{D}}V(\nu_m^{-1})$ is isomorphic to the space of continuous functions $\mathrm{Cont}(\Gamma/\Gamma_m, V(\varphi)^*)$ on Γ with values in $V(\varphi)^*$, on which $\mathcal{G}_{\mathbb{Q}}$ acts by $g\phi(\gamma) = \varphi(g)\phi(g^{-1}\gamma)$ for $\phi \in \mathrm{Cont}(\Gamma/\Gamma_m, V(\varphi)^*)$. Thus by Shapiro's lemma, we have the first isomorphism in the lemma. The second isomorphism can be proven in the same manner because $I_p/I_p \cap \mathcal{G}_{\mathbb{Q}_m} \cong \Gamma/\Gamma_m$. Since $I_q \subset \mathcal{G}_{\mathbb{Q}_\infty}$, the third isomorphism is obvious and induced by the first.

We now fix a representation $\bar{\rho}$ satisfying the conditions defining $\mathcal{D}$. Let $(R_{\mathbb{Q}}, \rho_{\mathbb{Q}})$ be the universal couple representing the problem $\mathcal{D}$ for $\bar{\rho}$. We assume that $\mathrm{Spec}(\mathbb{I})$ gives a closed subscheme of the normalization of the reduced part of $\mathrm{Spec}(R_{\mathbb{Q}})$. Let $\pi : R_{\mathbb{Q}} \to \mathbb{I}$ be the projection and $\varphi = \pi\rho_{\mathbb{Q}}$ be the representation of $\mathcal{G}_{\mathbb{Q}}$ into $\mathrm{GL}_2(\mathbb{I})$. Since p is odd, we can decompose $V(\varphi \otimes_{\mathbb{I}} \varphi^\vee) = \mathrm{End}_{\mathbb{I}}(V(\varphi))$ into the sum of trace zero space $\mathfrak{sl}_2(V(\varphi))$ and the center $Z(\varphi)$. We write the representation on $\mathfrak{sl}_2(V(\varphi))$ as $\mathrm{Ad}(\varphi)$, that is, $V(\mathrm{Ad}(\varphi)) = \mathfrak{sl}_2(V(\varphi))$. We have a filtration (χ_p) and $(\mathcal{N}_q)$ of the Galois representation of $\mathcal{G}_{\mathbb{Q}_q}$ for $q \in \mathcal{M}' \cup \{p\}$: $V(\rho_{\mathbb{Q},1,q}) \subset V(\rho_{\mathbb{Q}})$. This induces the filtration on $\mathfrak{sl}_2(V(\varphi))$: $0 \subset V_q^+(\mathrm{Ad}(\varphi)) \subset V_q^-(\mathrm{Ad}(\varphi)) \subset V(\mathrm{Ad}(\varphi))$ given as follows (see [H6]) :

$$V_q^+(\mathrm{Ad}(\varphi)) = \{\phi \in \mathfrak{sl}_2(V(\varphi)) | \phi(V(\rho_{\mathbb{Q},1,q})) = 0\} \text{ and}$$
$$V_q^-(\mathrm{Ad}(\varphi)) = \{\phi \in \mathfrak{sl}_2(V(\varphi)) | \phi(V(\rho_{\mathbb{Q},1,q})) \subset V(\rho_{\mathbb{Q},1,q})\} .$$

Let F be an algebraic extension of $\mathbb{Q}$. Suppose (AI_F) and that $\bar{\rho}$ satisfies the condition defining $\mathcal{D}_F$. Then the above filtration stable under $\mathcal{G}_{F_{\mathbb{Q}}}$ induces a filtration for each prime $\mathcal{Q}$ in $\mathcal{M}'_F \cup \{\mathcal{P}|p\}$: $0 \subset V_{\mathcal{Q}}^+(\mathrm{Ad}(\varphi)) \subset V_{\mathcal{Q}}^-(\mathrm{Ad}(\varphi)) \subset V(\mathrm{Ad}(\varphi))$. In other words, if $\sigma\mathcal{G}_{\mathbb{Q}_q}\sigma^{-1} \supseteq \mathcal{G}_{F_{\mathbb{Q}}}$, then $V_{\mathcal{Q}}^\pm(\mathrm{Ad}(\varphi)) = \sigma V_q^\pm(\mathrm{Ad}(\varphi))$. For each $\mathbb{I}$-direct summand W of $V(\mathrm{Ad}(\varphi))$ or $V(\mathrm{Ad}(\varphi) \otimes \nu_m^{-1})$ for $0 \leq m < \infty$, we define $W^* = W \otimes_{\mathbb{I}} \mathbb{I}^*$ for the Pontryagin dual $\mathbb{I}^*$ of $\mathbb{I}$. When $m = \infty$, for each $\mathbb{I}[[\Gamma]]$-submodule W of $V(\mathrm{Ad}(\varphi) \otimes \nu^{-1}) = V(\mathrm{Ad}(\varphi))\widehat{\otimes}_{\mathcal{D}}V(\nu^{-1})$, we define $W^* = W \otimes_{\mathbb{I}[[\Gamma]]} \mathbb{I}[[\Gamma]]^*$ for the Pontryagin dual $\mathbb{I}[[\Gamma]]^*$ of $\mathbb{I}[[\Gamma]]$. We also put

$$V_{\mathcal{Q}}^\pm(\mathrm{Ad}(\varphi)) \otimes \nu_m^{-1}) = V_{\mathcal{Q}}^\pm(\mathrm{Ad}(\varphi))\widehat{\otimes}_{\mathcal{D}}V(\nu_m^{-1}) \text{ for } m = 0, 1, \ldots, \infty .$$

We let the Galois group act on W^* through W and write Φ for one of $\mathrm{Ad}(\varphi)$ and $\mathrm{Ad}(\varphi) \otimes \nu_m^{-1}$ for $m = 0, 1, \ldots, \infty$. For each prime ideal $\mathcal{Q}$ of F and a subset $\mathcal{L}$ of $\mathcal{M}'_F$, we consider the following subgroup $L_{\mathcal{Q}}$ of $H^1(\mathcal{G}_{F_{\mathcal{Q}}}, V(\Phi)^*)$:

$$L_{\mathcal{Q}} = \ker(H^1(\mathcal{G}_{F_{\mathcal{Q}}}, V(\Phi)^*) \to H^1(I_{\mathcal{Q}}, V(\Phi)^*/V_{\mathcal{Q}}^+(\Phi)^*)) \text{ for } \mathcal{Q} \in \mathcal{L} \cup \{\mathcal{P}|p\} ,$$
$$L_{\mathcal{Q}} = \ker(H^1(\mathcal{G}_{F_{\mathcal{Q}}}, V(\Phi)^*) \to H^1(I_{\mathcal{Q}}, V(\Phi)^*)) \text{ for } \mathcal{Q} \text{ outside } \mathcal{L} \cup \{\mathcal{P}|p\} .$$

Then associated Selmer group of Φ over F is defined by

(Sel) $\mathrm{Sel}_{\mathcal{L}}(\Phi)_{/F} = \cap_Q \ker(H^1(\mathcal{G}_F, V(\Phi)^*) \to H^1(\mathcal{G}_{F_Q}, V(\Phi)^*)/L_Q))$.

The Selmer group defined in [G] Section 4 is equal to $\mathrm{Sel}_\emptyset(\mathrm{Ad}(\varphi))_{/F}$, which we write simply as $\mathrm{Sel}(\mathrm{Ad}(\varphi))_{/F}$. By a general theory due to Greenberg [G] p. 217, the Pontryagin dual $\mathrm{Sel}^*(\mathrm{Ad}(\varphi))_{/F}$ of $\mathrm{Sel}(\mathrm{Ad}(\varphi))_{/F}$ is of finite type over $\mathbb{I}$. In this case, starting from a modular Galois representation $\overline{\rho}$, it is basically known by Flach [F] and Wiles [W] that $\mathrm{Sel}^*(\mathrm{Ad}(\varphi))_{/\mathbb{Q}}$ is an $\mathbb{I}$-torsion module of finite type if $\mathrm{Spec}(\mathbb{I})$ gives an irreducible component of $R_{\mathbb{Q}}$. It has been conjectured by Greenberg that $\mathrm{Sel}^*(\mathrm{Ad}(\varphi))_{/F}$ (resp. $\mathrm{Sel}^*(\mathrm{Ad}(\varphi)) \otimes \nu^{-1})_{/F})$ is an $\mathbb{I}$-torsion (resp. $\mathbb{I}[[\Gamma]]$-torsion) module of finite type if $\mathbb{I}$ is sufficiently large.

If $F/\mathbb{Q}$ is a Galois extension, by definition, $\mathrm{Gal}(F/\mathbb{Q})$ naturally acts on $\mathrm{Sel}^*(\mathrm{Ad}(\varphi))_{/F}$, and the restriction map of cohomology takes $\mathrm{Sel}(\mathrm{Ad}(\varphi))_{/E}$ into $\mathrm{Sel}(\mathrm{Ad}(\varphi))_{/F}$. By Lemma 3.1, for the subfield $\mathbb{Q}_m$ of $\mathbb{Q}_\infty$, we have the following commutative diagram for $n > m$:

$$
\begin{array}{ccc}
\mathrm{Sel}_{\mathcal{L}}(\mathrm{Ad}(\varphi))_{/\mathbb{Q}_m} & \cong & \mathrm{Sel}_{\mathcal{L}}(\mathrm{Ad}(\varphi) \otimes V_m^{-1})_{/\mathbb{Q}} \\
\downarrow \text{res} & & \downarrow i_{m,n} \\
\mathrm{Sel}_{\mathcal{L}}(\mathrm{Ad}(\varphi))_{/\mathbb{Q}_n} & \cong & \mathrm{Sel}_{\mathcal{L}}(\mathrm{Ad}(\varphi) \otimes V_n^{-1})_{/\mathbb{Q}}
\end{array}
$$

where $i_{m,n}$ is induced by the natural inclusion $V(\mathrm{Ad}(\varphi) \otimes \nu_m^{-1})^* \subset V(\mathrm{Ad}(\varphi) \otimes \nu_n^{-1})^*$ induced by the dual map of the projection $\mathfrak{O}[\Gamma/\Gamma_n] \to \mathfrak{O}[\Gamma/\Gamma_m]$. Since the formation of Galois cohomology commutes with injective limit of coefficients, we get the following version of a result of Greenberg [G] Proposition 3.2 :

(Sel1) $\varinjlim_m \mathrm{Sel}_{\mathcal{L}}(\mathrm{Ad}(\varphi))_{/\mathbb{Q}_m} = \varinjlim_m \mathrm{Sel}_{\mathcal{L}}(\mathrm{Ad}(\varphi) \otimes \nu_m^{-1})_{/\mathbb{Q}}$

$$
= \mathrm{Sel}_{\mathcal{L}}(\mathrm{Ad}(\varphi) \otimes \nu^{-1})_{/\mathbb{Q}} = \mathrm{Sel}_{\mathcal{L}}(\mathrm{Ad}(\varphi))_{/\mathbb{Q}_\infty} \;.
$$

3.2. Mazur modules and Selmer groups. We return to the situation in 2.1. We assume (AI$_F$) and the conditions defining $\mathcal{D}_F$ for $\overline{\rho}$. Take a Λ-algebra homomorphism $\pi : R_F \to \mathbb{I}$ and suppose that $\mathbb{I}$ is a Λ_F-module of finite type. Let $\varphi = \pi\rho_F$. We consider the scalar extension $R = R_F \otimes_{\Lambda_F} \mathbb{I}$, which is naturally an $\mathbb{I}$-algebra. Then we consider the module $\Omega_{R/\mathbb{I}}$ of m_R-adically continuous 1-differentials over $\mathbb{I}$. Then

$$
\mathrm{Hom}_R(\Omega_{R/\mathbb{I}}, M) = \mathrm{Hom}_{R_F}(\Omega_{R_F/\Lambda_F}, M) = \mathrm{Der}_{\Lambda_F}(R_F, M)
$$

for each topological $\mathbb{I}$-module M of finite type or an injective limit of such modules. Here every homomorphism and derivation as above is supposed to be continuous under the $m_{\mathbb{I}}$-adic topology. We consider the ring $R_F[M] = R_F \oplus M$ with $M^2 = 0$. Then for $\delta \in \mathrm{Der}_{\Lambda_F}(R_F, M)$, we have an Λ_F-algebra homomorphism $\iota(\delta) : R_F \to R_F[M]$ given by $r \mapsto r \oplus \delta(r)$. Any Λ_F-algebra homomorphism, inducing the identity modulo M, is of the form $\iota(\delta) : R_F \to R_F[M]$ for a derivation δ.

We consider the deformation $\rho : \mathcal{G}_F \to \mathrm{GL}_2(R_F[M])$ of ρ_F. Then we can write down $\rho = \rho_F \oplus u'$ for $u' : \mathcal{G}_F \to \mathrm{End}_{\mathbb{I}}(M \oplus M)$. Define $u(\sigma) = u'(\sigma)\rho_F(\sigma)^{-1}$. Then

$$u(\sigma\tau) = u'(\sigma\tau)\rho_F(\sigma\tau)^{-1} = (\rho_F(\sigma)u'(\tau) + u'(\sigma)\rho_F(\tau))\rho_F(\sigma\tau)^{-1}$$
$$= \mathrm{Ad}(\rho_F)(\sigma)u(\tau) + u(\sigma) .$$

Note that $\det(1 \oplus u) = \mathrm{Tr}(u)$ for $u \in \mathrm{End}_{\mathbb{I}}(M \oplus M)$. Thus by (det) in 2.2, u is a 1-cocycle, under the adjoint action $\mathrm{Ad}(\rho_F)$ on $\mathrm{Ad}(M) = V(\mathrm{Ad}(\varphi)) \otimes_{\mathbb{I}} M$ in $\mathrm{End}_{\mathbb{I}}(M \oplus M)$, having values in $\mathrm{Ad}(M)$. One can check that the map : $\rho \mapsto$ the cohomology class of u from the set of deformations of ρ_F of type $\mathcal{D}_F$ is an injection. For primes $\mathcal{Q}$ in $\mathcal{M}_F(\chi)$, by $(\chi_{\mathcal{Q}})$, the order of $\rho(I_{\mathcal{Q}})$ is prime to p, and thus (a) $u(I_{\mathcal{Q}}) = 0$. For $\mathcal{Q} \in \mathcal{M}'_F$ or $\mathcal{Q}|p$, we have (b) $u(\mathcal{G}_{F_{\mathcal{Q}}}) \subset V_{\mathcal{Q}}^-(\mathrm{Ad}(\varphi)) \otimes_{\mathbb{I}} M$ and (c) $u(I_{\mathcal{Q}}) \subset V_{\mathcal{Q}}^+(\mathrm{Ad}(\varphi)) \otimes_{\mathbb{I}} M$. Since $\mathcal{G}_{F_{\mathcal{Q}}}$ normalizes $I_{\mathcal{Q}}$, the non-splitting of the exact sequence in $(\mathcal{N}_{\mathcal{Q}})$ or $(\mathrm{Reg}_{\mathcal{P}})$ shows that (c) implies (b). It is obvious that if we are given a 1-cocycle u satisfying (a) and (c), $\rho = \rho_F \oplus u\rho_F$ is a deformation of type $\mathcal{D}_{\Lambda}$. Thus we get a version of the results in [MT] Proposition 25 and [HT] Proposition 2.3.10 :

THEOREM 3.1. — *Suppose that $\mathbb{I}$ is a Λ_F-module of finite type. We have*

$$\mathrm{Hom}_R(\Omega_{R/\mathbb{I}}, \mathbb{I}^*) = C_1(\lambda; \mathbb{I})^* \cong \mathrm{Sel}_{\mathcal{M}'}(\mathrm{Ad}(\varphi))_{/F} ,$$

where λ is given by $m \circ (\pi \otimes \mathrm{id})$ for the multiplication $m : \mathbb{I}_0 \otimes_{\Lambda_F} \mathbb{I} \to \mathbb{I}$ with $\mathbb{I}_0 = \mathrm{Im}(\pi)$ and "" indicates the Pontryagin dual module.*

3.3. Control theorem of Selmer groups of $\mathrm{Ad}(\varphi)$. We now assume that $E = \mathbb{Q}$ and $F = \mathbb{Q}_j \subset \mathbb{Q}_\infty$. Then it is easy to check that if $\overline{\rho}$ satisfies the condition of $\mathcal{D}_E$, then $\overline{\rho}$ restricted to $\mathcal{G}_F$ satisfies the conditions of $\mathcal{D}_F$. We assume $(\mathrm{AI}_{\mathbb{Q}_\infty})$. Then we write α_j (resp. Λ_j, m_j) for the morphism $\alpha : R_j = R_F \to R_{\mathbb{Q}}$ (resp. $\mathcal{O}[[\Gamma_j]]$) in Λ_F, the multiplication $m : \mathbb{I}_0 \otimes_{\Lambda_j} \mathbb{I} \to \mathbb{I}$). We write ρ_j for the universal representation realized on R_j. Now let $\mathrm{Spec}(\mathbb{I})$ be an irreducible component of the normalization of $\mathrm{Spec}(R_{\mathbb{Q}})$ and write

$\pi : R_\mathbb{Q} \to \mathbb{I}$ for the projection map. We assume that $\mathbb{I}$ is a *torsion-free* Λ_0-module of finite type. We apply the above theorem to $\pi_j = \pi \alpha_j : R_j \to \mathbb{I}$. We write $\lambda_j = m_j \circ (\pi_j \otimes \mathrm{id})$. Under $(\mathrm{AI}_{\mathbb{Q}_\infty})$, by Theorem 2.1, $\mathrm{Im}(\pi_j) = \mathrm{Im}(\pi) = \mathbb{I}_0$ independent of j. Then similarly to α_j, we can construct an algebra homomorphism $\alpha_{j,k} : R_k \to R_j$ so that $\alpha_{j,k} \circ \rho_k \approx \rho_j$. Then $\alpha_{j,k}$ induces a projection map

$$C_1(\pi_k; \mathbb{I}_0) \to C_1(\pi_j; \mathbb{I}_0), C_1(m_k; \mathbb{I}) \to C_1(m_j; \mathbb{I}) \text{ and } C_1(\lambda_k; \mathbb{I}) \to C_1(\lambda_j; \mathbb{I}) .$$

Since R_F is topologically of finite type over Λ_j, all these modules are made of compact modules. Note that projective limit is an exact functor on the category of compact modules. Then we take the projective limit of these modules, and write them as $C_1(\lambda_\infty; \mathbb{I})$, $C_1(\pi_\infty; \mathbb{I})$ and $C_1(m_\infty; \mathbb{I})$. Then we have an exact sequence from (Ext4) in 2.2, for the generator γ_j of $\Gamma_j = \mathrm{Gal}(\mathbb{Q}_\infty/\mathbb{Q}_j)$, $k \geq j$ and $T_j = R_j \otimes_{\Lambda_j} \mathbb{I}$

(Ext5) $\mathrm{Tor}_1^{T_j}(\mathbb{I}, \mathrm{Ker}(\mu_{\mathbb{Q}_j})) \to \mathbb{I} \otimes_\mathbb{Z} \Gamma_j/\Gamma_k \to C_1(\lambda_k; \mathbb{I})/(\gamma_j - 1)C_1(\lambda_k; \mathbb{I}) \to$
$\quad C_1(\lambda_j; \mathbb{I}) \to 0 .$

Since these are compact modules, after taking projective limit with respect to k, we still have the following exact sequence :

(Ext6) $\mathrm{Tor}_1^{T_j}(\mathbb{I}, \mathrm{Ker}(\mu_{\mathbb{Q}_j})) \to \mathbb{I} \otimes_{\mathbb{Z}_p} \Gamma_j(\cong \mathbb{I}) \to C_1(\lambda_\infty; \mathbb{I})/(\gamma_j - 1)C_1(\lambda_\infty; \mathbb{I}) \to$
$\quad C_1(\lambda_j; \mathbb{I}) \to 0 .$

Suppose $R_\mathbb{Q}$ is reduced. Then if $R_\mathbb{Q}$ is a $\Lambda_\mathbb{Q}$–module of finite type, $C_1(\lambda_0; \mathbb{I})$ is a torsion $\mathbb{I}$-module of finite type. We now show that $C_1(\lambda_\infty; \mathbb{I})/(\gamma - 1)C_1(\lambda_\infty; \mathbb{I})$ contains actually a copy of $\mathbb{I}$. For that, we look into the following exact sequence obtained from (Ext3) in 1.1 :

$$C_1(\pi_j; \mathbb{I}_0) \otimes_{\mathbb{I}_0} \mathbb{I} \xrightarrow{\iota_j} C_1(\lambda_j; \mathbb{I}) \to C_1(m_j; \mathbb{I}) \to 0 \text{ for } j = 0, 1, \ldots, \infty .$$

We study $\mathrm{Ker}(\iota_j)$. By definition, Λ_j is isomorphic to $\mathfrak{O}[[\Gamma_j]]$. We write Λ for $\mathfrak{O}[[\Gamma_j]]$. By (Ext2-3) in 1.1, this module is a surjective image of $\mathrm{Tor}_1^{\mathbb{I}^\circ}(\mathrm{Ker}(m_j), \mathbb{I})$ if $\mathbb{I}_0$ is flat over Λ, where we write $\mathbb{I}^\circ$ for $\mathbb{I}_0 \otimes_{\Lambda_j} \mathbb{I}$. Suppose for the moment that $\mathbb{I}_0$ is Λ-flat. We write Λ° for $\Lambda \otimes_{\Lambda_j} \Lambda$. We first compute $\mathrm{Tor}_1^{\Lambda^\circ}(\mathrm{Ker}(m_j), \Lambda)$ when $\mathbb{I} = \Lambda$. Identifying Λ with $\mathfrak{O}[[T]]$, we see easily that $\Lambda^\circ \cong \mathfrak{O}[[T]][[S]]/((1+S)^q - (1+T)^q)$ for $q = p^j$. Then $\mathrm{Ker}(m_j)$ is a principal ideal generated by $S - T$, and decomposing $(1+S)^q - (1+T)^q = (S-T)f(S,T)$, we have an isomorphism $\mathrm{Ker}(m_j) \cong \Lambda^\circ/(f(S,T))$. Thus we have an exact sequence :

$$0 \to (S - T)\Lambda^\circ \to \Lambda^\circ \xrightarrow{f(S,T)} \Lambda^\circ \to \mathrm{Ker}(m_j) \to 0 .$$

Tensoring Λ, we get $\mathrm{Tor}_1^{\Lambda^\circ}(\mathrm{Ker}(m_j),\Lambda) = 0$ and $\varprojlim_j \mathrm{Tor}_1^{\Lambda^\circ}(\mathrm{Ker}(m_j),\Lambda)$
$= 0$. Since $\mathbb{I}$ is a Λ-module of finite type, $\mathrm{Ker}(\iota_j)$ is a torsion $\mathbb{I}$-module in general. In particular, if $C_1(\lambda_0,\mathbb{I})$ is of $\mathbb{I}$-torsion, $C_1(\pi_0,b_0) \otimes_{\mathbb{I}_0} \mathbb{I}$ is of $\mathbb{I}$-torsion.

Writing $\mathbb{I}_j$ for $\mathbb{I}_0 \widehat{\otimes}_{\Lambda_j} \mathbb{I}$, we have projective systems of surjective homomorphisms $\mathbb{I}_j \to \mathbb{I}_{j-1}$ and $\mathrm{Ker}(m_j) \to \mathrm{Ker}(m_{j-1})$. Writing $\mathbb{I}_\infty$ for $\varprojlim_j \mathbb{I}_j$, we know that $\mathrm{Ker}(m_\infty) = \varprojlim_j \mathrm{Ker}(m_j)$ and that $\mathrm{Tor}_1^{\mathbb{I}_\infty}(\mathrm{Ker}(m_\infty),\mathbb{I}) = \varprojlim_j \mathrm{Tor}_1^{\mathbb{I}_j}(\mathrm{Ker}(m_j),\mathbb{I})$. Note that $\mathbb{I}_\infty = \mathbb{I}_0 \widehat{\otimes}_{\mathfrak{O}} \mathbb{I}$. It is obvious that $\mathrm{Ker}(m_\infty)$ is the ideal of the diagonal Δ in $\mathrm{Spec}(\mathbb{I}_0) \times_{\mathrm{Spec}(\mathfrak{O})} \mathrm{Spec}(\mathbb{I})$. Obviously Δ is irreducible and spanned by $S - T$. Thus $\mathrm{Ker}(m_\infty)$ is $\mathbb{I}_\infty$-free. This implies that $\mathrm{Tor}_1^{\mathbb{I}_\infty}(\mathrm{Ker}(m_\infty),\mathbb{I}) = 0$. Thus we have an exact sequence of $\mathbb{I}[[\Gamma]]$-modules :

$$0 \to C_1(\pi_\infty;\mathbb{I}_0) \otimes_{\mathbb{I}_0} \mathbb{I} \xrightarrow{\iota_\infty} C_1(\lambda_\infty;\mathbb{I}) \to C_1(m_\infty;\mathbb{I}) \to 0 .$$

When $\mathbb{I}_0$ is not flat over Λ but Λ-torsion free, then $\mathbb{I}_0$ can be embedded into a Λ-free subalgebra $\mathbb{I}'$ of $\mathbb{I}$ such that $\mathbb{I}'/\mathbb{I}_0$ is pseudo-null. Thus we can repeat the above argument in the category of $\mathbb{I}$-modules with pseudo-morphisms. We then get the above exact sequence with pseudo-null $\mathrm{Ker}(\iota_\infty)$.

We now study $C_1(m_\infty;\mathbb{I}_0)$. Then, by definition, we have $C_1(m_j;\mathbb{I}_0) = \Omega_{\mathbb{I}_0/\Lambda_j}$, where $\mathrm{Hom}_{\mathbb{I}_0}(\Omega_{\mathbb{I}_0/\Lambda_j},M)$ is naturally isomorphic to the module $\mathrm{Der}_{\Lambda_j}(\mathbb{I}_0,M)$ of continuous derivations over Λ_j for all compact $\mathbb{I}_0$-modules M. As we have seen (see also [H3] Lemma 1.11), we have $\mathrm{Ker}(m_\infty) = (S - T)\mathbb{I}_0 \widehat{\otimes}_{\mathfrak{O}} \mathbb{I}_0$. This shows that $C_1(m_\infty;\mathbb{I}_0) \cong \mathbb{I}_0$ and that $C_1(m_\infty;\mathbb{I}_0)$ is a torsion $\mathbb{I}[[\Gamma]]$-module of finite type.

Suppose either that $R_\mathbb{Q}$ is reduced and is a Λ-module of finite type or $R_\mathbb{Q}$ is reduced and $\mathrm{Spec}(\mathbb{I}_0)$ is an irreducible component of $\mathrm{Spec}(R_\mathbb{Q})$. Then $C_1(\lambda_0;\mathbb{I})$ is of torsion and cannot contain a submodule isomorphic to $\mathbb{I}$. Thus the inclusion of $\mathbb{I}$ into $C_1(\lambda_\infty;\mathbb{I})$ composed with the projection from $C_1(\lambda_j;\mathbb{I})$ onto $\mathbb{I}$ is a non-zero $\mathbb{I}$-linear map, and it is injective. Writing M for $\mathrm{Sel}^*_{\mathcal{M}'}(\mathrm{Ad}(\varphi) \otimes \nu^{-1})_{/\mathbb{Q}}$, we have a commutative diagram with exact rows :

$$
\begin{array}{ccccccc}
\mathbb{I} \otimes_{\mathbb{Z}_p} \Gamma_j & \xrightarrow{\varepsilon_j} & M/(\gamma^{p^j}-1)M & \to & C_1(\lambda_j;\mathbb{I}) & \to & 0 \\
\cap & & \downarrow & & \downarrow & & \\
0 \to \mathbb{I} \otimes_{\mathbb{Z}_p} \Gamma & \xrightarrow{\varepsilon_0} & M/(\gamma-1)M & \to & C_1(\lambda_0;\mathbb{I}) & \to & 0 ,
\end{array}
$$

where the last two vertical arrows are surjective and the first vertical map is injective induced by the *inclusion* : $\Gamma_j \subset \Gamma$. Thus ε_j is injective.

As is remarked in Section A.1 (see (AI) after Corollary A.1.3), the condition $(\mathrm{AI}_{\mathbb{Q}_\infty})$ is equivalent to $(\mathrm{AI}_{\mathbb{Q}})$. We record what we have proven.

THEOREM 3.2. — *Suppose* $(\mathrm{AI}_{\mathbb{Q}})$, *the conditions of $\mathcal{D}$ for $\bar{\rho}$ and that $\mathbb{I}$ is a torsion-free Λ_0-module of finite type giving the normalization of an irreducible component of* $\mathrm{Spec}(R_{\mathbb{Q}})$. *Let* $\mathrm{Sel}_{\mathcal{L}}^*(\mathrm{Ad}(\varphi) \otimes \nu^{-1})_{/\mathbb{Q}}$ *be the Pontryagin dual module of the Selmer group* $\mathrm{Sel}_{\mathcal{L}}(\mathrm{Ad}(\varphi) \otimes \nu^{-1})_{/\mathbb{Q}}$. *We have the following two exact sequences of $\mathbb{I}$-modules :*

$$\mathbb{I} \otimes_{\mathbb{Z}_p} \Gamma_j \xrightarrow{\varepsilon_j} \mathrm{Sel}_{\mathcal{M}'}^*(\mathrm{Ad}(\varphi) \otimes \nu^{-1})_{/\mathbb{Q}}/(\gamma^{p^j}-1)\,\mathrm{Sel}_{\mathcal{M}'}^*(\mathrm{Ad}(\varphi) \otimes \nu^{-1})_{/\mathbb{Q}} \to C_1(\lambda_j; \mathbb{I}) \to 0$$

$$C_1(\pi_\infty; \mathbb{I}_0) \otimes_{\mathbb{I}_0} \mathbb{I} \xrightarrow{\iota_\infty} \mathrm{Sel}_{\mathcal{M}'}^*(\mathrm{Ad}(\varphi) \otimes \nu^{-1})_{/\mathbb{Q}} \to \mathbb{I} \to 0$$

with pseudo-null kernel $\mathrm{Ker}(\iota_\infty)$, *which vanishes when $\mathbb{I}_0$ is flat over Λ. Moreover suppose that $R_{\mathbb{Q}}$ is reduced and either that $R_{\mathbb{Q}}$ is a Λ-module of finite type or that $\mathrm{Spec}(\mathbb{I}_0)$ is an irreducible component of* $\mathrm{Spec}(R_{\mathbb{Q}})$. *Then ε_j is injective.*

3.4. Cotorsionness of the Selmer group over $\mathbb{I}[[\Gamma]]$**.** We write Λ_j' for the subalgebra of R_0 topologically generated over $\mathfrak{O}$ by $\alpha(\rho_{F,1,\mathcal{P}}(\phi_{\mathcal{P}}))$ for $F = \mathbb{Q}_j$, where $\phi_{\mathcal{P}}$ is the geometric Frobenius element in $D_{\mathcal{P}}/I_{\mathcal{P}}$. Taking a unit u in $\mathfrak{O}$ such that $\rho_{\mathbb{Q},1,p}(\phi_p) \equiv u \bmod m_{\mathbb{Q}}$, we assume that, with the notation of $(\chi_{\mathcal{P},F})$,

(Ind) *the subalgebra of $\mathbb{I}_0$ topologically generated over $\mathfrak{O}$ by $\varphi_{1,p}(\phi_p)$ is isomorphic to the one variable power series ring $\mathfrak{O}[[X]]$ via $\varphi_{1,p}(\phi_p) - u \mapsto X$.*

Since π takes $\rho_{\mathbb{Q},1,p}(\phi_p)$ to $\varphi_{1,p}(\phi_p)$, $\rho_{\mathbb{Q},1,p}(\phi_p) - u$ is analytically independent over $\mathfrak{O}$. Since p totally ramified in $\mathbb{Q}_\infty$, α takes $\rho_{F,1,\mathcal{P}}(\phi_{\mathcal{P}})$ to $\rho_{\mathbb{Q},1,p}(\phi_p)$. Thus (Ind) implies

(Ind$_j$) *the subalgebra of R_j topologically generated over $\mathfrak{O}$ by $\rho_{F,1,\mathcal{P}}(\phi_{\mathcal{P}})$ is isomorphic to $\mathfrak{O}[[X]]$ via $\rho_{F,1,\mathcal{P}}(\phi_{\mathcal{P}}) - u \mapsto X$ and is stable under* $\mathrm{Gal}(\mathbb{Q}_j/\mathbb{Q})$.

Thus $\Lambda' = \Lambda_j'$ is independent of j and R_j is naturally an algebra over Λ'. We also suppose that $\mathbb{I}_0$ is a Λ_0-torsion free module of finite type. Since $\mathbb{I}_0$ is an integral domain, $\mathbb{I}_0$ is a Λ'-torsion free module. We write λ_j' for the composite of $\pi_j \otimes \mathrm{id} : R_j \widehat{\otimes}_{\Lambda'} \mathbb{I} \to \mathbb{I}_0 \widehat{\otimes}_{\Lambda'} \mathbb{I}$ and the multiplication : $\mathbb{I}_0 \widehat{\otimes}_{\Lambda'} \mathbb{I} \to \mathbb{I}$, where "$\widehat{\otimes}$" indicates the completion under the adic topology of the maximal ideal of the algebraic tensor product. We prove the following theorem in this section :

THEOREM 3.3. — *Suppose* $(AI_\mathbb{Q})$, (Ind), *that* $\mathbb{I}$ *is a torsion-free* Λ-*module of finite type giving the normalization of an irreducible component of* $\mathrm{Spec}(R_\mathbb{Q})$ *and that* $\mathrm{Sel}^*_{\mathcal{M}'}(\mathrm{Ad}(\varphi))_{/\mathbb{Q}}$ *is a torsion* $\mathbb{I}$-*module. Then we have*

(i) $\mathrm{Sel}^*_{\mathcal{M}'}(\mathrm{Ad}(\varphi)) \otimes \nu^{-1})_{/\mathbb{Q}}$ *is a torsion* $\mathbb{I}[[\Gamma]]$-*module of finite type;*

(ii) *There is a pseudo-isomorphism of* $\mathrm{Sel}^*_{\mathcal{M}'}(\mathrm{Ad}(\varphi)) \otimes \nu^{-1})_{/\mathbb{Q}}$ *into* $M \times \mathbb{I}$ *for a torsion* $\mathbb{I}[[\Gamma]]$-*module* $M = C_1(\lambda'_\infty; \mathbb{I})$ *such that* $M/(\gamma - 1)M$ *is a torsion* $\mathbb{I}$-*module and* M *is pseudo-isomorphic to* $C_1(\pi_\infty; \mathbb{I}_0) \otimes_{\mathbb{I}_0} \mathbb{I}$;

(iii) *If* $\mathrm{Sel}^*_{\mathcal{M}'}(\mathrm{Ad}(\varphi))_{/\mathbb{Q}}$ *is a pseudo-null* $\mathbb{I}$-*module and* $\Lambda' = \mathbb{I}$, *then* $\mathrm{Sel}^*_{\mathcal{M}}(\mathrm{Ad}(\varphi)) \otimes \nu^{-1})_{/\mathbb{Q}}$ *is pseudo isomorphic to* $\mathbb{I}$, *on which* Γ *acts trivially;*

(iv) *If* $\mathbb{I}_0$ *is formally smooth over* $\mathfrak{O}$, *then we have the following exact sequence of* $\mathbb{I}[[\Gamma]]$-*modules :*

$$0 \to C_1(\pi_\infty; \mathbb{I}) \to C_1(\lambda'_\infty; \mathbb{I}) \to \widehat{\Omega}_{\mathbb{I}/\Lambda'} \to 0,$$

where $\widehat{\Omega}_{\mathbb{I}/\Lambda'}$ *is the module of continuous 1-differentials or equivalently is the* $m_\mathbb{I}$-*adic completion of* $\Omega_{\mathbb{I}/\Lambda'}$ *(which is a torsion* $\mathbb{I}$-*module of finite type by* (Ind).

By the theorem, $\mathrm{Sel}^*_{\mathcal{L}}(\mathrm{Ad}(\varphi)) \otimes \nu^{-1})_{/\mathbb{Q}}$ is a torsion $\mathbb{I}[[\Gamma]]$-module of finite type for any subset $\mathcal{L}$ of $\mathcal{M}'$. The theorem combined with Theorem 3.2 reduces the study of $\mathrm{Sel}^*_{\mathcal{L}}(\mathrm{Ad}(\varphi)) \otimes \nu^{-1})_{/\mathbb{Q}}$ to the study of $C_1(\pi_\infty; \mathbb{I})$ if $\mathbb{I}_0$ is formally smooth over $\mathfrak{O}$.

Here is a concrete case where the theorem applies. For a positive integer N prime to p, let $h^{\mathrm{ord}}(N; \mathfrak{O})$ be the universal ordinary Hecke algebra for $\mathrm{GL}(2)_{/\mathbb{Q}}$. Then $h^{\mathrm{ord}}(N; \mathfrak{O})$ is an algebra over $\mathfrak{O}[[\Gamma \times (\mathbb{Z}/Np\mathbb{Z})^\times]]$. The algebra structure is given so that $h^{\mathrm{ord}}(N; \mathfrak{O})/P_\kappa h^{\mathrm{ord}}(N; \mathfrak{O})$ is isomorphic to the ordinary Hecke algebra of weight $\kappa + 1$ of level Np, where P_κ is the prime ideal of Λ generated by $\gamma - \mathcal{N}(\gamma)^\kappa$. Take a primitive character ψ modulo Np and suppose that ψ has order prime to p. We take the algebra direct summand $h(\psi)$ of $h^{\mathrm{ord}}(N; \mathfrak{O})$ on which $(\mathbb{Z}/Np\mathbb{Z})^\times$ acts by ψ. Take a maximal ideal m of $h(\psi)$ with residue field $\mathbb{F}$ and write H for the m-adic completion of the Hecke algebra $h(\psi)$. Then we have a unique isomorphism class of Galois representations $\rho : G \to \mathrm{GL}_2(H)$ as in [H1] (see also [DHI] Section 1) if $\overline{\rho} = \rho \bmod m$ is absolutely irreducible. In this case, $\overline{\rho}$ and ρ satisfy the requirement of the deformation problem $\mathcal{D}_\mathbb{Q}$. Since H is reduced and Λ-free of finite rank (see [H1] and [H] Chapter 7), the reducedness and the finiteness of $R_\mathbb{Q}$ over $\Lambda_\mathbb{Q}$ follows from Wiles' result [W] Theorem 3.3 asserting that $(R_\mathbb{Q}, \rho_\mathbb{Q}) \cong (H, \rho)$ under the assumption (AI_F) for $F = \mathbb{Q}(\sqrt{(-1)^{(p-1)/2}p})$. Thus under this condition, for each irreducible component $\mathrm{Spec}(\mathbb{I})$ of the normalization of $\mathrm{Spec}(H)$ and the projection $\pi : H \to \mathbb{I}$, $\mathrm{Sel}^*_{\mathcal{M}'}(\mathrm{Ad}(\varphi))_{/\mathbb{Q}}$ for $\varphi = \pi\rho$ is a torsion $\mathbb{I}$-module of finite type. This also follows from a result of Flach [F] in some special cases. The condition (Ind) can be verified in this case as follows. Note that $\rho_{\mathbb{Q},1,p}(\phi_p) = T(p)$ in H. Thus $\varphi_{1,p}(\phi_p)$ specializes to an algebraic integer a_κ in $\mathcal{O}$ modulo P_κ with $|a_\kappa| = p^{\kappa/2}$ for any archimedean absolute value $|\ |$ on $\mathbb{Q}(a_\kappa)$ for infinitely many κ. Thus $\varphi_{1,p}(\phi_p)$ is transcendental over $\mathcal{O}$ and hence (Ind) is satisfied. Thus in this case $\mathrm{Sel}^*_{\mathcal{M}'}(\mathrm{Ad}(\varphi) \otimes \nu^{-1})$ is a torsion $\mathbb{I}[[\Gamma]]$-module of finite type.

We give two proofs of the theorem. The first one is just a repetition of the argument in the previous section replacing Λ_j by Λ', which is easy but we need to assume an additional assumption that $\mathbb{I}_0$ is a Λ'-module of finite type. The other one works in general, but we need to use the theory of imperfection modules in [EGA] IV.0.20.6. Anyway we look into the following exact sequence obtained from (Ext3) in 1.1 for $j = 0, 1, \ldots, \infty$:

$$C_1(\pi_j; \mathbb{I}_0) \otimes_{\mathbb{I}_0} \mathbb{I} \xrightarrow{\iota'_j} C_1(\lambda'_j; \mathbb{I}) \to C_1(m'_j; \mathbb{I}) \to 0$$

where $m' : \mathbb{I}_0 \widehat{\otimes}_{\Lambda'} \mathbb{I} \to \mathbb{I}$ is the multiplication map and $\lambda'_j : R_j \widehat{\otimes}_{\Lambda'} \mathbb{I} \to \mathbb{I}$ is the composition $m' \circ (\pi\alpha_j \otimes \mathrm{id})$. Note here that the first term $C_1(\pi_j; \mathbb{I}_0) \otimes_{\mathbb{I}_0} \mathbb{I}$ is the same as the case studied in 3.3.

We study $\mathrm{Ker}(\iota'_j)$. As we remarked, here we assume that $\mathbb{I}_0$ is a Λ'-module of finite type and will deal with the general case later. Then $\mathrm{Ker}(\iota'_j)$ is a surjective image of $\mathrm{Tor}_1^{\mathbb{I}'}(\mathrm{Ker}(m'), \mathbb{I})$ if $\mathbb{I}_0$ is flat over Λ and is a surjective image up to pseudo-null error in general (see (Ext2) and (Ext2')), where we write $\mathbb{I}'$ for $\mathbb{I}_0 \otimes_{\Lambda'} \mathbb{I}$. Note that this fact holds independently of j. We have the long exact sequence for $M' = \mathrm{Tor}_1^{\mathbb{I}_0}(\mathrm{Ker}(m'), \mathbb{I}_0)$ and $\mathbb{I}'_0 = \mathbb{I}_0 \otimes_{\Lambda'} \mathbb{I}_0$:

$$0 \to M' \to I \otimes_{\mathbb{I}'} I \to I \to \Omega_{\mathbb{I}_0/\Lambda'} \to 0$$

obtained out of the following short exact sequence :

$$0 \to I \to \mathbb{I}'_0 \to \mathbb{I}_0 \to 0 .$$

Thus M' is an $\mathbb{I}_0$-module of finite type. Since $\Omega_{\mathbb{I}_0/\Lambda'}$ is a torsion $\mathbb{I}_0$-module, localizing at a prime P outside $\mathrm{Supp}(\Omega_{\mathbb{I}_0/\Lambda'})$, we have $I_P/I_P^2 = 0$, and by Nakayama's lemma, $I_P = 0$, which implies $M'_P = 0$. Thus $\mathrm{Supp}(M') \subset \mathrm{Supp}(\Omega_{\mathbb{I}_0/\Lambda'})$, which shows that M' is a torsion $\mathbb{I}_0$-module of finite type. Thus $C_1(\pi_\infty; \mathbb{I}_0) \otimes_{\mathbb{I}_0} \mathbb{I}$ is pseudo-isomorphic to $C_1(\lambda'_\infty; \mathbb{I})$ as $\mathbb{I}[[\Gamma]]$-modules. By Theorem 2.2, we have

$$C_1(\lambda'_\infty; \mathbb{I})/(\gamma - 1)C_1(\lambda'_\infty; \mathbb{I}) \cong C_1(\lambda'_0; \mathbb{I}) .$$

Thus if $C_1(\lambda_0; \mathbb{I}) \cong \mathrm{Sel}^*_{\mathcal{M}'}(\mathrm{Ad}(\varphi))_{/\mathbb{Q}}$ is of $\mathbb{I}$-torsion, $C_1(\pi_0; \mathbb{I}_0) \otimes_{\mathbb{I}_0} \mathbb{I}$ and hence $C_1(\lambda'_0; \mathbb{I})$ are $\mathbb{I}$-torsion modules of finite type. The Krull dimension of $C_1(\lambda'_0; \mathbb{I})$ over $\mathbb{I}$ satisfies $\dim_{\mathbb{I}}(C_1(\lambda'_0; \mathbb{I})) < \dim(\mathbb{I})$. By [EGA] IV.0.16.2.3.1, we have

$$\dim(\mathbb{I}[[\Gamma]]) = \dim(\mathbb{I}) + 1 > \dim_{\mathbb{I}[[\Gamma]]}(C_1(\lambda'_\infty; \mathbb{I})/(\gamma - 1)C_1(\lambda'_\infty; \mathbb{I})) + 1$$
$$\geq \dim_{\mathbb{I}[[\Gamma]]}(C_1(\lambda'_\infty; \mathbb{I})).$$

Thus $C_1(\lambda'_\infty; \mathbb{I})$ is a torsion $\mathbb{I}[[\Gamma]]$-module, and hence $C_1(\pi_\infty; \mathbb{I}_0) \otimes_{\mathbb{I}_0} \mathbb{I}$ is a torsion $\mathbb{I}[[\Gamma]]$-module of finite type. Then Theorem 3.2 tells us that $\mathrm{Sel}^*_{\mathcal{M}'}(\mathrm{Ad}(\varphi) \otimes \nu^{-1})_{/\mathbb{Q}}$ is a torsion $\mathbb{I}[[\Gamma]]$-module of finite type.

A principal ingredient of the second proof is the theory of imperfection modules in [EGA] IV §0.20.6; so, we recall the theory here and generalize it in the case of compact adic rings. Here $\mathfrak{D}$ is a valuation ring finite and flat over $\mathbb{Z}_p$ with residue field $\mathbb{F}$. Let Λ be a base ring which is an object of $\mathrm{CNL} = \mathrm{CNL}_{\mathfrak{D}}$. If $X \to Y$ is a morphism in CNL, we write $\widehat{\Omega}_{Y/X}$ for the m_X-adic completion of the module of one differentials of Y over X. Then $\widehat{\Omega}_{Y/X}$ is a Y-module of finite type and hence is compact. We consider local algebra homomorphisms in CNL : $\Lambda \to A \xrightarrow{u} B \xrightarrow{v} C$. Then by [EGA] 0.20.7.18 we have an exact sequence

$$\widehat{\Omega}_{B/A} \widehat{\otimes}_B C \xrightarrow{v_*} \widehat{\Omega}_{C/A} \xrightarrow{u_*} \widehat{\Omega}_{C/B} \to 0 ,$$

where "$\widehat{\otimes}$" indicates the m_C-adic completion. We define the imperfection module $\Upsilon_{C/B/A}$ following [EGA] 0.20.6.1.1 by

$$\Upsilon_{C/B/A} = \mathrm{Ker}(\widehat{\Omega}_{B/A} \widehat{\otimes}_B C \xrightarrow{v_*} \widehat{\Omega}_{C/A}) .$$

Then we have the following commutative diagram with exact rows (see [EGA] 0.20.6.16) :

$$0 \to \widehat{\Omega}_{A/\Lambda} \widehat{\otimes}_A C \xrightarrow{j_1} (\widehat{\Omega}_{A/\Lambda} \widehat{\otimes}_A C) \oplus (\widehat{\Omega}_{B/\Lambda} \widehat{\otimes}_B C) \xrightarrow{p_1} \widehat{\Omega}_{B/\Lambda} \widehat{\otimes}_B C \to 0$$
$$\downarrow u_* \otimes \mathrm{id} \qquad\qquad \downarrow (vu)_* \oplus \mathrm{id} \qquad\qquad \downarrow v_*$$
$$0 \to \widehat{\Omega}_{B/\Lambda} \widehat{\otimes}_B C \xrightarrow{j_0} (\widehat{\Omega}_{C/\Lambda} \oplus (\widehat{\Omega}_{B/\Lambda} \widehat{\otimes}_B C)) \xrightarrow{p_0} \widehat{\Omega}_{C/\Lambda} \to 0,$$

where, writing $f = u_* \otimes \mathrm{id}$ and $g = v_*$, $j_1(x) = x \oplus f(x)$, $p_1(y \oplus z) = z - f(y)$, $j_0(x) = g(x) \oplus x$ and $p_0(y \oplus z) = g(z) - y$. We put

$$\Upsilon^C_{B/A/\Lambda} = \mathrm{Ker}(\widehat{\Omega}_{A/\Lambda} \widehat{\otimes}_A C \to \widehat{\Omega}_{B/\Lambda} \widehat{\otimes}_B C) .$$

Then by the snake lemma, we have an exact sequence (cf. [EGA] 0.20.6.17) :

(E1)
$$0 \to \Upsilon_{B/A/\Lambda}^C \to \Upsilon_{C/A/\Lambda} \to \Upsilon_{C/B/\Lambda} \to \widehat{\Omega}_{B/A} \widehat{\otimes}_B C \to \widehat{\Omega}_{C/A} \to \widehat{\Omega}_{C/B} \to 0 \,.$$

Now suppose that v is surjective, and hence $\widehat{\Omega}_{C/B} = 0$. By [EGA] 0.20.7.20, we have another exact sequence :

(E1')
$$\operatorname{Ker}(v)/\operatorname{Ker}(v)^2 \to \widehat{\Omega}_{B/\Lambda} \widehat{\otimes}_B C \to \widehat{\Omega}_{C/\Lambda} \to 0 \,.$$

If $B/\operatorname{Ker}(v)^2 \to C$ has a section of Λ-algebras (for example, if C is formally smooth over Λ), then

$$0 \to \operatorname{Ker}(v)/\operatorname{Ker}(v)^2 \to \Omega_{B/\Lambda} \otimes_B C \to \Omega_{C/\Lambda} \to 0$$

is exact (cf. [EGA] 0.20.6.10). Since taking m_B-adic completion is a left exact functor, we have the exactness of

$$0 \to \operatorname{Ker}(v)/\operatorname{Ker}(v)^2 \to \widehat{\Omega}_{B/\Lambda} \widehat{\otimes}_B C \to \widehat{\Omega}_{C/\Lambda} \to 0 \,.$$

This shows that if $B/\operatorname{Ker}(v)^2 \to C$ has a section of Λ-algebras, then

$$\Upsilon_{C/B/\Lambda} \cong \operatorname{Ker}(v)/\operatorname{Ker}(v)^2$$

and the following sequence is exact :

(E2) $\quad 0 \to \Upsilon_{B/A/\Lambda}^C \to \Upsilon_{C/A/\Lambda} \to \operatorname{Ker}(v)/\operatorname{Ker}(v)^2 \to \widehat{\Omega}_{B/A} \widehat{\otimes}_B C \to \widehat{\Omega}_{C/A} \to 0 \,.$

Let $\mathbb{K}$ be a finite extension of the quotient field $\mathbb{L}$ of $A = \mathfrak{O}[[t]]$ for a variable t. Let $\mathbb{I}$ be an A-subalgebra of $\mathbb{K}$ integral over A with quotient field $\mathbb{K}$. Thus $\dim(\mathbb{I}) = 2$. Since A is a Japanese ring, $\mathbb{I}$ is an A-module of finite type. Thus we have a surjection $v : A[[X_1, \ldots, X_r]] \to \mathbb{I}$. This yields an exact sequence
$$\operatorname{Ker}(v)/\operatorname{Ker}(v)^2 \to \oplus_i \mathbb{I} dX_i \to \Omega_{\mathbb{I}/A} \to 0 \,.$$

Since $\Omega_{\mathbb{K}/\mathbb{L}} = 0$, $\Omega_{\mathbb{I}/A}$ is a torsion $\mathbb{I}$-module of finite type. We also have the following exact sequence :

$$0 \to \Upsilon_{\mathbb{I}/A/\mathfrak{O}} \to \widehat{\Omega}_{A/\mathfrak{O}} \otimes_A \mathbb{I} \to \widehat{\Omega}_{\mathbb{I}/\mathfrak{O}} \to \Omega_{\mathbb{I}/A} \to 0 \,.$$

Since any continuous derivation of A can be extended to $\mathbb{K}$, the image of $\widehat{\Omega}_{A/\mathfrak{O}} \otimes_A \mathbb{I} = \mathbb{I} dT \cong \mathbb{I}$ in $\widehat{\Omega}_{\mathbb{I}/\mathfrak{O}}$ is $\mathbb{I}$-free of rank 1. Since $\Omega_{\mathbb{I}/A}$ is a torsion $\mathbb{I}$-module, $\widehat{\Omega}_{A/\mathfrak{O}} \otimes_A \mathbb{I}$ has to inject into $\widehat{\Omega}_{\mathbb{I}/\mathfrak{O}}$. Thus $\Upsilon_{\mathbb{I}/A/\mathfrak{O}} = 0$. Let $s \in m_{\mathbb{I}}$, and suppose s is analytically independent over $\mathfrak{O}$. Let $\Lambda = \mathfrak{O}[[s]] \subset \mathbb{I}$. We consider $\widehat{\Omega}_{\mathbb{I}/\Lambda}$. Then we have, taking a surjective algebra homomorphism $v' : \Lambda[[X_1, \ldots, X_r]] \to \mathbb{I}$,

$$\operatorname{Ker}(v') / \operatorname{Ker}(v')^2 \to \oplus_i \mathbb{I} dX_i \to \Omega_{\mathbb{I}/A} \to 0 \, .$$

If t and s are analytically independent over $\mathfrak{O}$, then $\mathbb{I}$ becomes integral over the power series ring $\mathfrak{O}[[T, S]] \cong \mathfrak{O}[[t, s]]$, which is impossible because $\dim(\mathbb{I}) = 2$. Thus t and s are analytically dependent, and the evaluation map $v'' : \mathfrak{O}[[T, S]] \to \mathbb{I}$ at (t, s) has non-trivial kernel P. The prime P cannot have height 2 or more because $\dim(\operatorname{Im}(v'')) = 2$. Thus P is of height 1, and it is therefore generated by a single element $f(T, S)$ because $\mathfrak{O}[[T, S]]$ is a unique factorization domain. Then we have

$$\frac{\partial f}{\partial T}(t, s) dt + \frac{\partial f}{\partial S}(t, s) ds = 0 \text{ in } \widehat{\Omega}_{\mathbb{I}/\mathfrak{O}}, \text{ and}$$

$$\widehat{\Omega}_{\mathfrak{O}[[t,s]]/\mathfrak{O}} \cong (\mathfrak{O}[[t, s]] dt \oplus \mathfrak{O}[[t, s]] ds) / (\frac{\partial f}{\partial T}(t, s) dt + \frac{\partial f}{\partial S}(t, s) ds) \, .$$

Suppose $\frac{\partial f}{\partial S}(t, s) = 0$. Then $\frac{\partial f}{\partial S}(T, S)$ is divisible by $f(T, S)$, that is, $\frac{\partial f}{\partial S} = fg$ for $g \in \mathfrak{O}[[T, S]]$ and hence $\frac{\partial^2 f}{\partial S^2}(t, s) = (g\frac{\partial f}{\partial S} + f\frac{\partial g}{\partial S})(t, s) = 0$. Repeating this argument, we find $\frac{\partial^n f}{\partial S^n}(t, s) = 0$ for all n, and hence $f(T, S) \in \mathfrak{O}[[T]]$ because $\mathfrak{O}$ is of characteristic 0. This is in contradiction to the analytic independence of t. Thus $\frac{\partial f}{\partial S}(t, s) \neq 0$. Similarly we know that $\frac{\partial f}{\partial T}(t, s) \neq 0$. Since ds and dt has a linear relation, ds is $\mathbb{I}$-linearly independent. We thus conclude for an analytically independent s, $\widehat{\Omega}_{\mathbb{I}/\mathfrak{O}[[s]]}$ is a torsion $\mathbb{I}$-module. We now look at the following exact sequence :

$$0 \to \Upsilon_{\mathbb{I}/\mathfrak{O}[[s]]/\mathfrak{O}} \to \widehat{\Omega}_{\mathfrak{O}[[s]]/\mathfrak{O}} \widehat{\otimes}_{\mathfrak{O}[[s]]} \mathbb{I} \to \widehat{\Omega}_{\mathbb{I}/\mathfrak{O}} \to \widehat{\Omega}_{\mathbb{I}/\mathfrak{O}[[s]]} \to 0 \, .$$

Since s is analytically independent, $\widehat{\Omega}_{\mathfrak{O}[[s]]/\mathfrak{O}} \widehat{\otimes}_{\mathfrak{O}[[s]]} \mathbb{I} \cong \mathbb{I}$ via $ds \mapsto 1$. Since $\widehat{\Omega}_{\mathbb{I}/\mathfrak{O}[[s]]}$ is a torsion $\mathbb{I}$-module and $\widehat{\Omega}_{\mathbb{I}/\mathfrak{O}} \otimes_{\mathbb{I}} \mathbb{K}$ is of dimension 1, $\widehat{\Omega}_{\mathfrak{O}[[s]]/\mathfrak{O}} \widehat{\otimes}_{\mathfrak{O}[[s]]} \mathbb{I}$ has to inject into $\widehat{\Omega}_{\mathbb{I}/\mathfrak{O}}$. This shows that $\Upsilon_{\mathbb{I}/\mathfrak{O}[[s]]/\mathfrak{O}} = 0$.

We now consider the situation where we have a surjective $\mathfrak{O}[[s]]$-algebra homomorphism $\pi : R \to \mathbb{I}$ for an object R of $\mathrm{CNL}_{\mathfrak{O}}$. By (E1), we have the following exact sequence :

$$0 \to \Upsilon^{\mathbb{I}}_{R/\mathfrak{O}[[s]]/\mathfrak{O}} \to \Upsilon_{\mathbb{I}/\mathfrak{O}[[s]]/\mathfrak{O}} \to \Upsilon_{\mathbb{I}/R/\mathfrak{O}} \to \widehat{\Omega}_{R/\mathfrak{O}[[s]]} \widehat{\otimes}_R \mathbb{I} \to \widehat{\Omega}_{\mathbb{I}/\mathfrak{O}[[s]]} \to 0 \, ,$$

and $\Upsilon_{\mathbb{I}/R/\mathfrak{O}} \cong \operatorname{Ker}(\pi)/\operatorname{Ker}(\pi)^2 = C_1(\pi; \mathbb{I})$ if $\mathbb{I}$ is formally smooth over $\mathfrak{O}$. By the above result, this yields a short exact sequence :

$$0 \to \Upsilon_{\mathbb{I}/R/\mathfrak{O}} \to \widehat{\Omega}_{R/\mathfrak{O}[[s]]} \widehat{\otimes}_R \mathbb{I} \to \widehat{\Omega}_{\mathbb{I}/\mathfrak{O}[[s]]} \to 0 \, .$$

Now we study how large the difference of $\Upsilon_{\mathbb{I}/R/\mathfrak{D}}$ and $C_1(\pi, \mathbb{I}) = \text{Ker}(\pi)/\text{Ker}(\pi)^2$, when $\mathbb{I}$ is not formally smooth over $\mathfrak{D}$. The key point here is that $\Upsilon_{\mathbb{I}/R/\mathfrak{D}}$ is independent of the choice of s. We pick $t' \in R$ so that $\pi(t') = t$ as above. We regard R as an $\mathfrak{D}[[t]]$-algebra through the algebra homomorphism of $\mathfrak{D}[[t]]$ into R taking t to t'. Then we have again an exact sequence :

$$0 \to \Upsilon_{\mathbb{I}/R/\mathfrak{D}} \to \widehat{\Omega}_{R/\mathfrak{D}[[t]]} \widehat{\otimes}_R \mathbb{I} \to \widehat{\Omega}_{\mathbb{I}/\mathfrak{D}[[t]]} \to 0 .$$

We have a surjective $\mathbb{I}$-linear map $r : C_1(\pi, \mathbb{I}) \to \Upsilon_{\mathbb{I}/R/\mathfrak{D}}$ from (E1) and (E1'). Let $\mathbb{I}' = \mathbb{I} \otimes_{\mathfrak{D}[[t]]} \mathbb{I}$ and $m : \mathbb{I}' \to \mathbb{I}$ be the multiplication. In this case, if $\mathbb{I}$ is flat over $\mathfrak{D}[[t]]$, by (Ext2 and Ext2'), $\text{Ker}(r)$ is a surjective image of $\text{Tor}_1^{\mathbb{I}'}(\text{Ker}(m), \mathbb{I})$, which is a torsion $\mathbb{I}$-module of finite type, because $\mathbb{I}$ is an $\mathfrak{D}[[t]]$-module of finite type. Even if $\mathbb{I}$ is not flat over $\mathfrak{D}[[t]]$, one can embed $\mathbb{I}$ into an $\mathfrak{D}[[t]]$-flat module with pseudo-null cokernel. Thus $\text{Ker}(r)$ is a surjective image of $\text{Tor}_1^{\mathbb{I}'}(\text{Ker}(m), \mathbb{I})$ up to pseudo-null error. The error is annihilated by a power $m_{\mathbb{I}}^M$ for a positive M independently of R and the choice of t' with $\pi(t') = t$ (but depending on $\mathbb{I}$ and t). Thus without any assumption, $\text{Ker}(r)$ is a torsion $\mathbb{I}$-module of finite type killed by a non-trivial ideal $\mathfrak{a}$ of $\mathbb{I}$ independent of R.

We now give the second proof. Here we do not assume the integrality of $\mathbb{I}$ over Λ'. Since the result over $\mathbb{I}$ is just a scalar extension of the result over $\mathbb{I}_0$, we only need to prove the assertions (i)-(iv) replacing $\mathbb{I}$ by $\mathbb{I}_0$. Thus hereafter, we assume that $\text{Im}(\pi_0) = \mathbb{I}$ and discard the assumption that $\mathbb{I}$ is integrally closed. Thus hereafter, we write $\mathbb{I}$ instead of $\mathbb{I}_0$ for $\text{Im}(\pi_0)$. We pick $t \in \mathbb{I}$ so that $\mathbb{I}$ is integral over $\mathfrak{D}[[t]]$. We take $t_j \in R_j$ so that $\alpha_{j,k}(t_k) = t_j$ and $\pi_0(t_0) = t$. Then we apply the above theory to $R = R_j$, $s = \rho_{F,1,\mathcal{P}}(\phi_{\mathcal{P}}) \in R_j$ for $F = \mathbb{Q}_j$ and $t' = t_j$. Then we have an exact sequence of compact modules :

$$C_1(\pi_j, \mathbb{I}) \xrightarrow{r_j} \Upsilon_{\mathbb{I}/R_j/\mathfrak{D}} \to 0 ,$$

where $\text{Ker}(r_j)$ is the image of a torsion $\mathbb{I}$-module $X = \text{Tor}_1^{\mathbb{I}'}(\text{Ker}(m), \mathbb{I})$ of finite type (independent of j) up to a bounded $\mathbb{I}$-pseudo-null error. Taking the projective limit with respect to j, we find that $r_\infty : C_1(\pi_\infty; \mathbb{I}) \to \Upsilon_{\mathbb{I}/R_\infty/\mathfrak{D}} = \varprojlim_j \Upsilon_{\mathbb{I}/R_j/\mathfrak{D}}$ is surjective because of the compactness of these modules and that $\text{Ker}(r_\infty)$ is an $\mathbb{I}$-torsion module of finite type. Thus r_∞ is an $\mathbb{I}[[\Gamma]]$-pseudo-isomorphism. By our assumption : $\mathbb{I} = \text{Im}(\pi_0)$, we have $\widehat{\Omega}_{R_j/\Lambda'} \widehat{\otimes}_R \mathbb{I} \cong C_1(\lambda'_j; \mathbb{I})$. By taking the projective limit of the exact sequences : $0 \to \Upsilon_{\mathbb{I}/R_j/\mathfrak{D}} \to \widehat{\Omega}_{R_j/\Lambda'} \widehat{\otimes}_R \mathbb{I} (\cong C_1(\lambda'_j; \mathbb{I})) \to \widehat{\Omega}_{\mathbb{I}/\Lambda'} \to 0$ for

$\Lambda' = \mathfrak{O}[[s]]$, we get another exact sequence :

$$0 \to \Upsilon_{\mathbb{I}/R_\infty/\mathfrak{O}} \to C_1(\lambda'_\infty; \mathbb{I}) \to \widehat{\Omega}_{\mathbb{I}/\Lambda'} \to 0 \text{ for } R_\infty = \varprojlim_j R_j \ .$$

Since $\widehat{\Omega}_{\mathbb{I}/\Lambda'}$ is a torsion I-module of finite type, $\Upsilon_{\mathbb{I}/R_\infty/\mathfrak{O}}$ is pseudo-isomorphic to $C_1(\pi_\infty; \mathbb{I})$ as $\mathbb{I}[[\Gamma]]$-modules. Thus $C_1(\pi_\infty; \mathbb{I})$ is pseudo-isomorphic to $C_1(\lambda'_\infty; \mathbb{I})$ as $\mathbb{I}[[\Gamma]]$-modules. By Theorem 2.2, we have

$$C_1(\lambda'_\infty; \mathbb{I})/(\gamma - 1)C_1(\lambda'_\infty; \mathbb{I}) \cong C_1(\lambda'_0; \mathbb{I}) \ .$$

As in the first proof, the assumption that $\mathrm{Sel}^*_{\mathcal{M}'}(\mathrm{Ad}(\varphi))_{/\mathbb{Q}}$ is a torsion $\mathbb{I}$-module tells us that $C_1(\pi_0; \mathbb{I})$ is of $\mathbb{I}$-torsion. Again by the exact sequence :

$$\mathrm{Ker}(r_0) \to C_1(\pi_0; \mathbb{I}) \to C_1(\lambda'_0; \mathbb{I}) \to \widehat{\Omega}_{I/\Lambda'} \to 0 \ ,$$

the $\mathbb{I}$-torsionness of $\mathrm{Ker}(r_0)$, $C_1(\pi_0; \mathbb{I})$ and $\widehat{\Omega}_{\mathbb{I}/\Lambda'}$ tells us the same for $C_1(\lambda'_0; \mathbb{I})$. Then we conclude the assertions (i)-(iii) as in the first proof. If $\mathbb{I}$ is formally smooth over $\mathfrak{O}$, $\mathrm{Ker}(r_j) = 0$ for all j. The assertion (iv) follows from this immediately.

Appendix : control of universal deformation rings of representations

In this appendix, we give a general theory of controlling the deformation rings of representations of a normal subgroup under the action of the quotient finite group.

A.1. — Extending representations

Let G be a profinite group with a normal closed subgroup H of finite index. We put $\Delta = G/H$. In this section, we describe when we can extend a representation π of a profinite group H to G (keeping the dimension of π). The theory is a version of Schur's theory of projective representations [CR] Section 11E.

A.1.1. — Representations with invariant trace

Let $\mathcal{O}$ be a complete noetherian local ring over $\mathbb{Z}_p$ with residue field $\mathbb{F}$. We consider the category $CNL = CNL_{\mathcal{O}}$ of complete noetherian local $\mathcal{O}$-algebras with residue field $\mathbb{F}$. Any algebra A in this section will be assumed to be an object of CNL. For each continuous representation $\rho : H \to GL_n(A)$ and $\sigma \in G$, we define $\rho^\sigma(g) = \rho(\sigma g \sigma^{-1})$.

We take a representation $\pi : H \to GL_n(A)$ for an artinian local $\mathcal{O}$-algebra A with residue field $\mathbb{F}$. We assume one of the following conditions :

(AI_H) $\bar{\rho} = \pi \bmod m_A$ is absolutely irreducible for the maximal ideal m_A of A;

(Z_H) The centralizer of $\bar{\rho}(H)$ as an algebraic subgroup of $GL(n)_{/\mathbb{F}}$ is the center of $GL(n)$.

Of course the first condition implies the second. There are some other cases where the last condition is satisfied; for example, (Z_H) holds if the following condition is satisfied :

(Red_H) $\bar{\rho}$ is upper triangular with distinct n characters ρ_i at diagonal entries, and its image contains a unipotent subgroup U' such that $U'/(U', U') = U/(U, U)$ for the unipotent radical U.

LEMMA A.1.1. — Suppose (Z_H). Then the centralizer of π in $GL_n(A)$ is $A^\times$.

We assume the following condition :

(C) $\pi = c(\sigma)^{-1}\pi^\sigma c(\sigma)$ with some $c(\sigma) \in GL_n(A)$ for each $\sigma \in G$.

If we find another $c'(\sigma) \in GL_n(A)$ satisfying $\pi = c'(\sigma)^{-1}\pi^\sigma c'(\sigma)$, we have

$$\pi = c'(\sigma)^{-1}c(\sigma)\pi c(\sigma)^{-1}c'(\sigma),$$

and hence by Lemma 1.1, $c(\sigma)^{-1}c'(\sigma)$ is a scalar. In particular, for $\sigma, \tau \in G$,

$$c(\sigma\tau)^{-1}\pi^{\sigma\tau}c(\sigma\tau) = \pi = c(\tau)^{-1}\pi^\tau c(\tau) = c(\tau)^{-1}c(\sigma)^{-1}\pi^{\sigma\tau}c(\sigma)c(\tau),$$

and hence, $b(\sigma, \tau) = c(\sigma)c(\tau)c(\sigma\tau)^{-1} \in A^\times$. Thus $c(\sigma)c(\tau) = b(\sigma, \tau)c(\sigma\tau)$. This shows by the associativity of the matrix multiplication that

$$(c(\sigma)c(\tau))c(\rho) = b(\sigma, \tau)c(\sigma\tau)c(\rho) = b(\sigma, \tau)b(\sigma\tau, \rho)c(\sigma\tau\rho) \text{ and}$$
$$c(\sigma)(c(\tau)c(\rho)) = c(\sigma)b(\tau, \rho)c(\tau\rho) = b(\tau, \rho)b(\sigma, \tau\rho)c(\sigma\tau\rho),$$

and hence $b(\sigma, \tau)$ is a 2-cocycle of G. If $h \in H$, then

$$\pi(g) = c(h\tau)^{-1}\pi(h\tau g\tau^{-1}h^{-1})c(h\tau) =$$
$$c(h\tau)^{-1}\pi(h)c(\tau)\pi(g)c(\tau)^{-1}\pi(h)^{-1}c(h\tau).$$

Thus $c(h\tau)^{-1}\pi(h)c(\tau) \in A^\times$. Thus if we let $h \in G$ act on the space $C(G; M_n(A))$ of continuous functions $f : G \to M_n(A)$ by $f|h(g) = \pi(h)^{-1}f(hg)$, then c is an eigenfunction belonging to a character $\xi : H \to A^\times$. Now we take $\eta : G \to A^\times$ such that $\eta(h\tau) = \xi^{-1}(h)\eta(\tau)$ for all

$h \in H$. For example, writing $G = \bigsqcup_{\tau \in R} H\tau$ (disjoint), we may define $\eta(h\tau) = \xi^{-1}(h)$. We replace c by ηc. Then c satisfies that

(π) $c(h\tau) = \pi(h)c(\tau)$ for all $h \in H$.

Since $c(1)$ commutes with $\mathrm{Im}(\pi)$, $c(1)$ is scalar. Thus we may also assume

(Id) $c(1) = 1$.

Note that for $h, h' \in H$,

$$b(h\sigma, h'\tau) = c(h\sigma)c(h'\tau)c(h\sigma h'\tau)^{-1} =$$
$$\pi(h)c(\sigma)\pi(h')c(\tau)c(\sigma\tau)^{-1}\pi(h\sigma h'\sigma^{-1})^{-1}$$
$$= \pi(h)\pi^{\sigma}(h')b(\sigma,\tau)\pi(h\sigma h'\sigma^{-1})^{-1} = b(\sigma,\tau).$$

Thus b is a 2–cocycle factoring through Δ.

If $b(\sigma, \tau) = \zeta(\sigma)\zeta(\tau)\zeta(\sigma\tau)^{-1}$ is further a coboundary of $\zeta : \Delta \to A^{\times}$, we modify c by $\zeta^{-1}c$. Since ζ factors through Δ, this modification does not destroy (π). Then $c(\sigma\tau) = c(\sigma)c(\tau)$ and $c(h\tau)c(\tau)$ for $h \in H$. Thus c extends π to G. Let d be another extension of π. Then $\chi(\sigma) = c(\sigma)d(\sigma)^{-1} \in A^{\times}$ is a character of G. Thus $c = d \otimes \chi$.

We consider another condition

(Inv) $\mathrm{Tr}(\pi) = \mathrm{Tr}(\pi^{\sigma})$ for all $\sigma \in G$.

Under (AI_F), it has been proven by Carayol and Serre [C] that (Inv) is actually equivalent to (C). Thus we have

THEOREM A.1.1. — Let $\pi : H \to GL_n(A)$ be a continuous representation for a p–adic artinian local ring A. Suppose either (AI_H) and (Inv) or (Z_H) and (C). Then we can choose c satisfying (π). Then $b(\sigma, \tau) = c(\sigma)c(\tau)c(\sigma\tau)^{-1}$ is a 2–cocycle of Δ with values in $A^{\times}$, and if its cohomology class in $H^2(\Delta, A^{\times})$ vanishes, then there exists a continuous representation π_E of G into $GL_n(A)$ extending π. Moreover all other extensions of π are of the form $\pi_E \otimes \chi$ for a character χ of Δ with values in $A^{\times}$. In particular, if $H^2(\Delta, A^{\times}) = 0$, then any representation π satisfying either (AI_H) and (Inv) or (Z_H) and (C) can be extended to G.

COROLLARY A.1.1. — If Δ is a p–group, then any representation π with values in $GL_n(\mathbb{F})$ for a finite field of characteristic p satisfying either (AI_H) and (Inv) or (Z_H) and (C) can be extended to G.

This follows from the fact that $|\mathbb{F}^{\times}|$ is prime to p, and hence $H^2(\Delta, \mathbb{F}^{\times}) = 0$. When Δ is cyclic, then $H^2(\Delta, A^{\times}) \cong A^{\times}/(A^{\times})^d$ for $d = |\Delta|$. If for a generator σ of G, $\xi = c(\sigma^d)\pi(\sigma^d)^{-1} \in (A^{\times})^d$, then b is a coboundary of $\zeta(\sigma^j) = \xi^{j/d}$. By extending scalar to $B = A[X]/(X^d - \xi)$, in $H^2(G, B^{\times})$, the class of b vanishes. Thus we have

COROLLARY A.1.2. — *Suppose either (AI_F) and* (Inv) *or* (Z_F) *and* (C). *If Δ is a cyclic group of order d, then π can be extended to a representation of G into $GL_n(B)$ for a local A–algebra B which is A–free of rank at most d.*

Let $\bar{\rho} = \pi \bmod m_A$. We suppose that $\bar{\rho}$ can be extended to G. Then we may assume that the cohomology class of $b(\sigma, \tau) \bmod m_A$ vanishes in $H^2(G, \mathbb{F}^\times)$. Thus we can find $\zeta : G \to A^\times$ such that

$$a(\sigma, \tau) = b(\sigma, \tau)\zeta(\sigma)\zeta(\tau)\zeta(\sigma\tau)^{-1} \bmod m_A \equiv 1.$$

Then a has values in $\widehat{\mathbb{G}}_m(A) = 1 + m_A$. In particular, if the Sylow p-subgroup S of G is cyclic, we have $H^2(S, \widehat{\mathbb{G}}_m(A)) \cong \widehat{\mathbb{G}}_m(A)/\widehat{\mathbb{G}}_m(A)^{|S|}$. Write ξ for the element in $\widehat{\mathbb{G}}_m(A)$ corresponding to a. Then for $B = A[X]/(X^{|S|} - \xi)$, the cohomology class of a vanishes in $H^2(S, \widehat{\mathbb{G}}_m(B))$. This implies that in $H^2(S, B^\times)$, the cohomology class of b vanishes.

COROLLARY A.1.3. — *Suppose either (AI_H) and* (Inv) *or* (Z_H) *and* (C). *Suppose Δ has a cyclic Sylow p-subgroup of order g. If $\bar{\rho}$ can be extended to G, then π can be extended to a representation of G into $GL_n(B)$ for a local A–algebra B which is A–free of rank at most g.*

We now prove the following fact :

(AI) *When Δ is cyclic of odd order and $n = 2$, the condition (AI_H) is equivalent to (AI_G).*

We start a bit more generally. Let ρ be an absolutely irreducible representation of G into $GL_n(K)$ for a field K. For the moment, n is arbitrary. We assume that Δ is cyclic of order prime to n. We prove that ρ cannot contain a character of H as a representation of H, which shows the equivalence when $n = 2$. Suppose by absurdity that ρ restricted to H contains a character χ. If χ is invariant under the conjugate action of Δ, χ can be extended to a character of G, and it is easy to see in this case, ρ has to contain an extension of χ, and hence reducible. Thus χ is not invariant under Δ. If χ is invariant under a subgroup $H' \supset H$ of G, again by the same argument as above, ρ gets reducible on H' containing a character χ' of H' extending χ. Thus we may assume that conjugates of χ' under $\Delta' = G/H'$ are all distinct. By Mackey's theorem, the induced representation $\text{Ind}(\chi')$ for H' to G is irreducible. By Frobenius reciprocity or Shapiro's lemma, the induced representation $\text{Ind}(\chi')$ has a unique quotient isomorphic to ρ. Thus $|\Delta'| = n$, which contradicts to the assumption that the order of Δ is prime to n. Of course, one can generalize the above argument for more general Δ not necessarily cyclic.

A.2. — Deformation functors of group representations

We suppose that G satisfies the following condition (cf. [T]) :

(pF)   *All open subgroup of G has finite p-Frattini quotient .*

We fix a representation $\bar{\rho} : G \to GL_n(\mathbb{F})$ satisfying (Z_H). In this section, we study various deformation problems of $\bar{\rho}$ and relation among the universal rings.

A.2.1. — Full deformations

We consider a deformation functor $\mathcal{F}_H : CNL \to SETS$ given by

$$\mathcal{F}_H(A) = \{\rho : H \to GL_n(A) \mid \rho \equiv \bar{\rho} \bmod m_A\} / \approx$$

where "$\approx$" is the strict equivalence in $GL_n(A)$, this is, the conjugation by elements in $\widehat{GL}_n(A) = 1 + m_A M_n(A)$. The functor $\mathcal{F}_H$ is representable ([T] Theorem 3.3) under (Z_H). We write (R_H, ρ_H) for the universal couple. Since ρ_G restricted to H is an element in $\mathcal{F}_H(R_H)$, we have an $\mathcal{O}$-algebra homomorphism $\alpha : R_H \to R_G$ such that $\alpha \rho_H = \rho_G|_H$.

We like to determine $\mathrm{Ker}(\alpha)$ and $\mathrm{Im}(\alpha)$ in terms of Δ. By choosing a lift $c_0(\sigma) \in GL_n(\mathcal{O})$ for $\sigma \in G$ such that $c_0(\sigma) \equiv \bar{\rho}(\sigma) \bmod m_{\mathcal{O}}$, we can define for any $\rho \in \mathcal{F}_G(A)$, $\rho^\sigma(g) = \rho(\sigma g \sigma^{-1})$ and $\rho^{[\sigma]}(g) = c_0(\sigma)^{-1} \rho^\sigma(g) c_0(\sigma)$ in $\mathcal{F}_H(A)$. In this way, Δ acts via $\sigma \longmapsto [\sigma]$ on $\mathcal{F}_H$ and R_H. Then as seen in Section 1, we can attach a 2-cocycle b on Δ with values in $\widehat{\mathbb{G}}_m(A)$ to any representation $\rho \in \mathcal{F}_H(A)$ with $\rho^{[\sigma]} \approx \rho$ in the following way. First choose a lift $c(\sigma)$ of $\bar{\rho}(\sigma)$ in $GL_n(A)$ for each $\sigma \in G$ such that $\rho = c(\sigma)^{-1} \rho^\sigma c(\sigma)$ and $c(h\tau) = \rho(h)c(\tau)$ for $h \in H$ and $\tau \in G$. Then we know that $c(\sigma)c(\tau) = b(\sigma, \tau)c(\sigma\tau)$ for a 2-cocycle b of Δ with values in $\widehat{\mathbb{G}}_m(A)$. If we change c by c' such that $c'(\sigma) = c(\sigma)\zeta(\sigma)$ for $\zeta(\sigma) \in \widehat{\mathbb{G}}_m(A)$, we see from $c(\sigma)c(\tau) = b(\sigma, \tau)c(\sigma\tau)$ that $c'(\sigma)c'(\tau) = b(\sigma, \tau)\zeta(\sigma)\zeta(\tau)c'(\sigma\tau)\zeta(\sigma\tau)^{-1}$. Thus the cocycle b' attached to c' is cohomologous to b, and the cohomology class $[b] = [\rho] \in H^2(\Delta, \widehat{\mathbb{G}}_m(A))$ is uniquely determined by ρ. If $[\rho] = 0$, then $b(\sigma, \tau) = \zeta(\sigma)^{-1}\zeta(\tau)^{-1}\zeta(\sigma\tau)$ for a 1-cochain ζ. We then modify c by $c\zeta$ and by constant so that $c(1) = 1$. Then c extends the representation ρ to a representation π of G (Theorem A.1.1).

LEMMA A.2.1. — *Suppose (Z_H) and that n is prime to p and $\rho^{[\sigma]} \approx \rho$ for $\rho \in \mathcal{F}_H(A)$. If $\det(\rho)$ can be extended to a character of G having values in an A-algebra B containing A, then ρ can be extended uniquely to a representation $\pi : G \to GL_n(B)$ whose determinant coincides with the extension to G of $\det(\rho)$.*

Proof : by applying "det" to c and b, we know that $[\det(\rho)] = [\det(b)] = [\rho]^n$. If n is prime to p, the vanishing of $[\rho]^n$ in $H^2(\Delta, \widehat{\mathbb{G}}_m(B))$ is equivalent to the vanishing of $[\rho]$. Thus if $\det(\rho)$ extends to G (that is $[\rho]^n = 0$), then ρ extends to a representation π of G which has determinant equal to the extension of $\det(\rho)$ prearranged. We now show the uniqueness of π. We get, out of π, other extensions $\pi \otimes \chi \in \mathcal{F}_G(B)$ for $\chi \in H^1(\Delta, \widehat{\mathbb{G}}_m(B)) = \mathrm{Hom}(\Delta, \widehat{\mathbb{G}}_m(B))$. Conversely, if π and π' are two extensions of ρ in $\mathcal{F}_G(B)$, then for $h \in H$, $\pi'(\sigma)\rho(h)\pi'(\sigma)^{-1} = \pi(\sigma)\rho(h)\pi(\sigma)^{-1}$ and hence $\pi(\sigma)^{-1}\pi'(\sigma)$ commutes with ρ. Then by Lemma A.1.1, $\chi(\sigma) = \pi(\sigma)^{-1}\pi'(\sigma)$ is a scalar in $\widehat{\mathbb{G}}_m(B)$.

$$\chi(\sigma\tau) = \pi(\sigma\tau)^{-1}\pi'(\sigma\tau) = \pi(\tau)^{-1}\pi(\sigma)^{-1}\pi'(\sigma)\pi'(\tau)$$
$$= \pi(\tau)^{-1}\chi(\sigma)\pi'(\tau) = \chi(\sigma)\chi(\tau).$$

Thus χ is an element in $H^1(\Delta, \widehat{\mathbb{G}}_m(B))$ and $\pi' = \pi \otimes \chi$, which shows that $\det(\pi')$ is equal to $\det(\pi)\chi^n$. If $\det(\pi') = \det(\pi)$, then $\chi^n = 1$. Since χ is of p-power order, if n is prime to p, $\chi = 1$.

Here is a consequence of the proof of the lemma :

COROLLARY A.2.1. — *Let $\pi_0 \in \mathcal{F}_G(B)$ be an extension of $\rho \in \mathcal{F}_H(A)$ for an A-algebra B containing A. Then we have*

$$\{\pi_0 \otimes \chi \mid \chi \in \mathrm{Hom}(\Delta, \widehat{\mathbb{G}}_m(B))\} = \{\pi \in \mathcal{F}_G(B) \mid \pi_{|H} = \rho\}.$$

It is easy to see that if $H^2(\Delta, \mathbb{F}) = 0$, then $H^2(\Delta, \widehat{\mathbb{G}}_m(A)) = 0$ for all A in CNL. Therefore we see, if $H^2(\Delta, \mathbb{F}) = 0$,

(*) $\mathcal{F}_H^\Delta(A) = H^0(\Delta, \mathcal{F}_H(A)) \cong \mathcal{F}_G(A)/\widehat{\Delta}(A)$ for $\widehat{\Delta}(A) = \mathrm{Hom}(\Delta, \widehat{\mathbb{G}}_m(A))$.

Here we let $\chi \in \widehat{\Delta}(A)$ act on $\mathcal{F}_G(A)$ via $\pi \longmapsto \pi \otimes \chi$. Suppose that $\mathcal{F}_H^\Delta$ is represented by a universal couple $(R_{H,\Delta}, \rho_{H,\Delta})$ and $[\rho_{H,\Delta}] = 0$ in $H^2(\Delta, \widehat{\mathbb{G}}_m(R_{H,\Delta}))$. Then for each $\rho \in \mathcal{F}_H^\Delta(A)$, we have $\varphi : R_{H,\Delta} \to A$ such that $\varphi\rho_{H,\Delta} \approx \rho$. Then $\varphi_*[\rho_{H,\Delta}] = [\rho]$ and therefore, $[\rho] = 0$ in $H^2(\Delta, \widehat{\mathbb{G}}_m(A))$. This shows again (*).

Let us show that the functor $\mathcal{F}_H^\Delta$ is representable by applying the Schlessinger criterion (see [Sch] and [T] Proposition 2.5). For a Cartesian diagram in $\mathrm{CNL}_\mathcal{O}$:

$$
\begin{array}{ccc}
A_3 = A_1 \times_A A_2 & \xrightarrow{\;\pi_1\;} & A_1 \\
\Big\downarrow{\scriptstyle \pi_2} & & \Big\downarrow{\scriptstyle \alpha_1} \\
A_2 & \xrightarrow{\;\alpha_2\;} & A,
\end{array}
$$

we need to check the bijectivity of the natural map

$$\gamma_H^\Delta : \mathcal{F}_H^\Delta (A_1 \times_A A_2) \longrightarrow \mathcal{F}_H^\Delta (A_1) \times_{\mathcal{F}_H^\Delta (A)} \mathcal{F}_H^\Delta (A_2).$$

We already know from the representability of $\mathcal{F}_H$ that

$$\gamma_H : \mathcal{F}_H (A_1 \times_A A_2) \cong \mathcal{F}_H (A_1) \times_{\mathcal{F}_H (A)} \mathcal{F}_H (A_2).$$

Since $\mathcal{F}_H^\Delta$ is a subfunctor of $\mathcal{F}_H$, $\mathcal{F}_H^\Delta (A_1) \times_{\mathcal{F}_H^\Delta (A)} \mathcal{F}_H^\Delta (A_2)$ is a subset of $\mathcal{F}_H (A_1) \times_{\mathcal{F}_H (A)} \mathcal{F}_H (A_2)$, and hence γ_H^Δ is injective. Take an element (ρ_1, ρ_2) of $\mathcal{F}_H^\Delta (A_1) \times_{\mathcal{F}_H^\Delta (A)} \mathcal{F}_H^\Delta (A_2)$. Then $\alpha_1 \rho_1 \approx \alpha_2 \rho_2$, that is, there exists $x \in \widehat{GL}_n(A)$ such that $x\alpha_1\rho_1 x^{-1} = \alpha_2\rho_2$. We may assume that α_1 is surjective (cf. [T] Proposition 2.5). Then we can lift x to $x' \in \widehat{GL}_n(A_1)$. Then replacing ρ_1 by $x'\rho_1 x'^{-1}$, we may assume that $\alpha_1\rho_1 = \alpha_1\rho_2$. Thus $\rho = \rho_1 \times_A \rho_2$ has values in $GL_n(A_1 \times_A A_2)$. It is easy to see that ρ is invariant under Δ. Thus $\gamma_H^\Delta(\rho) = (\rho_1, \rho_2)$, and therefore γ_H^Δ is surjective. Then it is obvious that $\mathcal{F}_H^\Delta$ is represented by $R_{H,\Delta} = R_H / \Sigma_{\sigma \in \Delta} R_H([\sigma] - 1) R_H$.

PROPOSITION A.2.1. — *Suppose* (Z_H). *Then* $\mathcal{F}_H^\Delta$ *is represented by* $(R_{H,\Delta}, \rho_{H,\Delta})$ *for* $R_{H,\Delta} = R_H / \mathfrak{a}$ *with* $\mathfrak{a} = \Sigma_{\sigma \in \Delta} R_H([\sigma] - 1) R_H$ *and* $\rho_{H,\Delta} = \rho_H \bmod \mathfrak{a}$. *If either* $[\rho_{H,\Delta}] = 0$ *in* $H^2(\Delta, \widehat{\mathbb{G}}_m(R_{H,\Delta}))$ *or* $H^2(\Delta, \mathbb{F}) = 0$, *then we have* $\mathcal{F}_G / \widehat{\Delta} \cong \mathcal{F}_H^\Delta$ *via* $\pi \longmapsto \pi|_H$.

We now consider the following subfunctor $\mathcal{F}_{G,H}$ of $\mathcal{F}_H$ given by

$$\mathcal{F}_{G,H}(A) = \{\rho|_H \in \mathcal{F}_H(A) \mid \rho \in \mathcal{F}_G(B) \text{ for a flat } A\text{–algebra } B \text{ in } CNL_\mathcal{O}\}.$$

Here the algebra B may not be unique and depends on A. Let us check that $\mathcal{F}_{G,H}$ is really a functor. If $\varphi : A \to A'$ is a morphism in CNL and $\rho|_H \in \mathcal{F}_{G,H}(A)$ with $\rho \in \mathcal{F}_G(B)$, B being flat over A, then $A'\widehat{\otimes}_A B$ is a flat A'–algebra in CNL. Then $(\varphi \otimes id)\rho \in \mathcal{F}_G(A'\widehat{\otimes}_A B)$ such that $\varphi(\rho|_H) = ((\varphi \otimes id)\rho)|_H$. Thus $\mathcal{F}_H(\varphi)$ takes $\mathcal{F}_{G,H}(A)$ into $\mathcal{F}_{G,H}(A')$, which shows that $\mathcal{F}_{G,H}$ is a well defined functor. For each $\rho \in \mathcal{F}_{G,H}(A)$, we have an extension $\rho \in \mathcal{F}_G(B)$. By the universality of (R_G, ρ_G), we have $\varphi : R_G \to B$ such that $\varphi \rho_G = \rho$. Then $\rho|_H = (\varphi \rho_G)|_H = \varphi(\rho_G|_H) = \varphi \alpha \rho_H$. This shows that $\varphi \alpha$ is uniquely determined by $\rho|_H \in \mathcal{F}_{G,H}(A)$. Therefore φ restricted to $\text{Im}(\alpha)$ has values in A and is uniquely determined by $\rho|_H \in \mathcal{F}_{G,H}(A)$. Conversely, supposing that $[\alpha \rho_H] = 0$ in $H^2(\Delta, \widehat{\mathbb{G}}_m(B))$ for a flat extension B of $\text{Im}(\alpha)$ in CNL, for a given $\varphi : \text{Im}(\alpha) \to A$ which is a morphism in CNL, we shall show that $\rho = \varphi \alpha \rho_H$ is an element of $\mathcal{F}_{G,H}(A)$. Anyway $\alpha \rho_H$ can be extended to G as an element in $\mathcal{F}_G(B)$, and hence $\alpha \rho_H \in \mathcal{F}_{G,H}(\text{Im}(\alpha))$. We note that ρ can be extended to G because $[\varphi \alpha \rho_H] = \varphi_*[\alpha \rho_H]$ which vanishes in $H^2(\Delta, \widehat{\mathbb{G}}_m(B'))$ for $B' = B\widehat{\otimes}_{\text{Im}(\alpha),\varphi}A$. Thus $\rho \in \mathcal{F}_{G,H}(A)$, and $\mathcal{F}_{G,H}$ is represented by $(\text{Im}(\alpha), \alpha\rho_H)$ as long as $[\alpha\rho_H] = 0$ in $H^2(\Delta, \widehat{\mathbb{G}}_m(B))$ for a flat extension B of $\text{Im}(\alpha)$ in CNL.

We have the following inclusions of functors : $\mathcal{F}_H \supset \mathcal{F}_H^\Delta \supset \mathcal{F}_{G,H} \supset \mathcal{F}_G/\widehat{\Delta}$, the last inclusion being given by $\rho \longmapsto \rho|_H$. The functor $\mathcal{F}_H^\Delta$ is represented by $R_H/\mathfrak{a}$ for $\mathfrak{a} = \Sigma_{\sigma\in\Delta}R_H([\sigma] - 1)R_H$. Because of the above inclusion, if $[\alpha\rho_H] = 0$ in $H^2(\Delta, \widehat{\mathbb{G}}_m(B))$ for a flat extension B of $\text{Im}(\alpha)$ in CNL, the ring $\text{Im}(\alpha)$ is a surjective image of $R_H/\mathfrak{a} = R_{H,\Delta}$. If $[\rho_{H,\Delta}] = 0$ (for $\rho_{H,\Delta} = \rho \bmod \mathfrak{a}$) in $H^2(\Delta, \widehat{\mathbb{G}}_m(B'))$ for a flat extension B' of $R_{H,\Delta}$ in CLN, then $\rho_{H,\Delta} \in \mathcal{F}_{G,H}(R_{H,\Delta})$ and thus $\mathcal{F}_H^\Delta = \mathcal{F}_{G,H}$.

PROPOSITION A.2.2. — *Assume* (Z_H) *and that* $[\alpha\rho_H] = 0$ *in* $H^2(\Delta, \widehat{\mathbb{G}}_m(B))$ *for a flat extension* B *of* $\text{Im}(\alpha)$ *in* CNL. *Then* $\mathcal{F}_{G,H}$ *is represented by* $(\text{Im}(\alpha), \alpha\rho_H)$. *If further* $[\rho_{H,\Delta}] = 0$ *in* $H^2(\Delta, \widehat{\mathbb{G}}_m(B'))$ *for a flat extension* B' *of* $R_{H,\Delta}$, *then we have* $\mathcal{F}_{G,H} = \mathcal{F}_H^\Delta$.

The character $\det(\rho_H)$ induces an $\mathcal{O}$–algebra homomorphism : $\mathcal{O}[[H^{ab}]] \to R_H$ for the maximal continuous abelian quotient H^{ab} of H. We write its image as Λ_H and write simply Λ for Λ_G. Thus we have a character $\det(\rho_H) : H \to \Lambda_H^\times$. We consider the category CNL_{Λ_H} of complete noetherian local Λ_H–algebras with residue field $\mathbb{F}$. We consider the functor $\mathcal{F}_{\Lambda_H,H} : CNL_{\Lambda_H} \to SETS$ given by

$$\mathcal{F}_{\Lambda_H,H}(A) = \{\rho : H \to GL_n(A) \mid \rho \equiv \overline{\rho} \bmod m_A \text{ and } \det(\rho) = \det(\rho_H)\}/\approx .$$

Pick $\rho : H \to GL_n(A) \in \mathcal{F}_{\Lambda_H,H}(A)$. Then regarding A as an $\mathcal{O}$–algebra naturally, we know that $\rho \in \mathcal{F}_H(A)$. Thus there is a unique morphism $\varphi : R_H \to A$ such that $\varphi\rho_H \approx \rho$. Then $\varphi(\det(\rho_H)) = \det(\rho)$, and φ is a morphism in CNL_{Λ_H}. Therefore (R_H, ρ_H) represents $\mathcal{F}_{\Lambda_H}$. Similarly to $\mathcal{F}_{G,H}$, we consider another functor on CNL_Λ :

$$\mathcal{F}_{\Lambda,G,H}(A) = \{\rho|_H \mid \rho \in \mathcal{F}_{\Lambda,G}(B) \text{ for a flat } A\text{–algebra } B \text{ in } CNL_\Lambda\}/\approx .$$

Take $\rho \in \mathcal{F}_{\Lambda,G,H}(A)$ such that $\rho = \rho'|_H$ for $\rho' \in \mathcal{F}_{\Lambda,G}(B)$. Then there exists a unique $\varphi : R_G \to B$ with $\det(\rho') = \varphi(\det(\rho_G))$. Since the Λ–algebra structure of B is given by $\det(\rho')$, φ induces a Λ–algebra homomorphism of $\text{Im}(\alpha)\Lambda$ into B for the algebra $\text{Im}(\alpha)\Lambda$ generated by $\text{Im}(\alpha)$ and Λ. From $\rho = (\varphi\rho_G)|_H = \varphi(\rho_G|_H) = \varphi\alpha\rho_H$, we see that the Λ–algebra homomorphism φ restricted $\text{Im}(\alpha)\Lambda$ is uniquely determined by ρ. Supposing that $[\alpha\rho_H]$ vanishes in $H^2(\Delta, \widehat{\mathbb{G}}_m(B))$ for a flat extension B of $\text{Im}(\alpha)$, we know that $[\alpha\rho_H]$ vanishes in $H^2(\Delta, \widehat{\mathbb{G}}_m(\text{Im}(\alpha)\Lambda \otimes_{\text{Im}(\alpha)} B))$. For any morphism $\varphi : \text{Im}(\alpha)\Lambda \to A$ in CNL_Λ, $[\varphi\alpha\rho_H] = \varphi_*[\alpha\rho_H]$ vanishes in $H^2(\Delta, \widehat{\mathbb{G}}_m(B'))$ for

$B' = A \otimes_{\mathrm{Im}(\alpha)} B$ which is flat over A. Thus we have an extension π of ρ to G having values in B'. Suppose further that n is prime to p. In this case, as already remarked, we can always extend ρ without extending A and without assuming the vanishing of $[\alpha\rho_H]$, because $\det(\rho)$ can be extended to G by $\varphi \circ \det(\rho_G)$. Thus we know :

$$\mathcal{F}_{\Lambda,G,H}(A) = \{\rho|_H \mid \rho \in \mathcal{F}_{\Lambda,G}(A)\}/\approx .$$

Since $\det(\rho)$ can be extended to G without changing A, there is a unique extension of π with values in $GL_n(A)$ such that $\det(\pi) = \iota \circ (\det(\rho_G))$, which implies that $\pi \in \mathcal{F}_{\Lambda,G}(A)$ and hence $\pi|_H \in \mathcal{F}_{\Lambda,G,H}(A)$. Thus $\mathcal{F}_{\Lambda,G,H}$ is represented by $(\mathrm{Im}(\alpha)\Lambda, \alpha\rho_H)$ if n is prime to p. We consider the natural transformation : $\mathcal{F}_{\Lambda,G} \to \mathcal{F}_{\Lambda,G,H}$ sending π to $\pi|_H$. As we have already remarked, the extension of $\rho \in \mathcal{F}_{\Lambda,G,H}(A)$ to $\pi \in \mathcal{F}_\Lambda(A)$ is unique if n is prime to p. Thus in this case, the natural transformation is an isomorphism of functors. Therefore $(R_G, \rho_G) \cong (\mathrm{Im}(\alpha)\Lambda, \alpha\rho_H)$. Thus we get

THEOREM A.2.1. — *Suppose (Z_H) and that either n is prime to p or $[\alpha\rho_H]$ vanishes in $H^2(\Delta, \widehat{\mathbb{G}}_m(B))$ for a flat extension B of $\mathrm{Im}(\alpha)$. Then $\mathcal{F}_{\Lambda,G,H}$ is representable by $(\mathrm{Im}(\alpha)\Lambda_G, \alpha\rho_H)$. Moreover if n is prime to p, we have the equality $R_G = \mathrm{Im}(\alpha)\Lambda_G$.*

Since α restricted to Λ_H coincides with the algebra homomorphism induced by the inclusion $H \subset G$, $\alpha(\Lambda_H) \subset \Lambda$. We put $R' = \mathrm{Im}(\alpha) \otimes_{\Lambda_H} \Lambda$. By definition, the character $1 \otimes \det(\rho_G)$ of G coincides on H with $(\alpha \circ \det(\rho_H)) \otimes 1$ in R'. Thus $\alpha\rho_H$ can be extended uniquely to $\rho'_G : G \to GL_n(R')$ such that $\det(\rho'_G) = 1 \otimes \det(\rho_G)$ if n is prime to p. Thus we have a natural map $\iota : R_G \to R'$ such that $\iota\rho_G = \rho'_G$. Since R_G is an algebra over Λ and $\mathrm{Im}(\alpha)$, it is an algebra over R'. Thus we have the structural morphism $\iota' : R' \to R_G$. By Theorem A.2.1, ι' is surjective. By definition, $\iota\alpha\rho_H = \iota\rho_H|_H = \iota\rho'_G|_H = \alpha\rho_H \otimes 1$ and $\iota \det(\rho_G) = \det(\rho'_G) = 1 \otimes \det(\rho_G)$. Thus $\iota'\iota\alpha\rho_H = \iota'(\alpha\rho_H \otimes 1) = \alpha\rho_H$ and $\iota'\iota \det(\rho_G) = \iota'(1 \otimes \det(\rho_G)) = \det(\rho_G)$. Thus $\iota'\iota$ is identity on Λ and $\mathrm{Im}(\alpha)$, and hence $\iota'\iota = id$. Similarly, $\iota\iota'\rho'_g = \iota\rho_G = \rho'_G$. This shows that

$$\iota\iota'(\alpha\rho_H \otimes 1) = \iota(\alpha\rho_H) = (\alpha\rho_H \otimes 1) \text{ and}$$
$$\iota\iota'(1 \otimes \det(\rho_G)) = \iota(\det(\rho_G)) = 1 \otimes \det(\rho_G).$$

Thus $\iota\iota'$ is again identity on $\mathrm{Im}(\alpha) \otimes 1$ and $1 \otimes \Lambda$, and $\iota\iota' = id$. Let X_p (resp. $X^{(p)}$) indicate the maximal p-profinite (resp. prime-to-p profinite) quotient of each profinite group X. Write ω for the restriction of $\det(\rho_G)$ to $(G^{ab})^{(p)}$. Define $\kappa : G^{ab} \to \mathcal{O}[[G_p^{ab}]]^\times$ by $\kappa(g) = \omega(g)[g_p]$ for the projection g_p of g into G_p^{ab}, where $[x]$ denotes the group element of $x \in G_p^{ab}$ in the group algebra. Assuming that $\mathbb{F}$ is big enough to contain all g–th roots of unity for the order g of $\mathrm{Im}(\omega)$, we can perform the same argument replacing $(\Lambda_H, \Lambda_G, \det(\rho_G))$ by $(\mathcal{O}[[H_p^{ab}]], \mathcal{O}[[G_p^{ab}]], 1 \otimes \kappa)$. Thus we get

COROLLARY A.2.2. — *Suppose (Z_H) and that n is prime to p. Then we have*

$$(R_G, \rho_G) \cong (\mathrm{Im}(\alpha) \otimes_{\Lambda_H} \Lambda_G, \alpha\rho_H \otimes \det(\rho_G)) \cong$$
$$(\mathrm{Im}(\alpha) \otimes_{\mathcal{O}[[H_p^{ab}]]} \mathcal{O}[[G_p^{ab}]], \alpha\rho_H \otimes \kappa).$$

In particular, R_G is flat over $\mathrm{Im}(\alpha)$.

By Hochschild–Serre spectral sequence, we have an exact sequence

$$
\begin{array}{ccccccc}
H_2(\Delta, \mathbb{Z}_p) & \longrightarrow & H_0(\Delta, H_1(H, \mathbb{Z}_p)) & \longrightarrow & H_1(G, \mathbb{Z}_p) & \longrightarrow & H_1(\Delta, \mathbb{Z}_p) & \longrightarrow & 0 \\
 & & \|\wr & & \|\wr & & \|\wr & & \\
 & & H_0(\Delta, H^{ab})_p & \longrightarrow & G_p^{ab} & \longrightarrow & \Delta_p^{ab} & \longrightarrow & 1,
\end{array}
$$

where the subscript "p" indicates the maximal p–profinite quotient. Suppose that $\mathbb{F}$ is big enough to contain all d_0–th roots of unity for the prime-to-p part d_0 of the order d of Δ. Then the inclusion $H \subset G$ induces the following commutative diagram :

$$
\begin{array}{ccc}
\alpha' : \mathcal{O}[[H_p^{ab}]] & \longrightarrow & \mathcal{O}[[G_p^{ab}]] \\
\downarrow & & \downarrow \\
\alpha : \Lambda_H & \longrightarrow & \Lambda_G.
\end{array}
$$

As seen in Corollary 2.2, this diagram is Cartesian. Thus Λ_G is flat over Λ_H. If $H_2(\Delta, \mathbb{Z}_p) = 0$ ($\Leftrightarrow H^2(\Delta, \mathbb{Q}_p/\mathbb{Z}_p) = 0$), $\mathrm{Spec}(\mathrm{Im}(\alpha')) \cong \mathrm{Spec}(\mathcal{O}[[H_p^{ab}]])^\Delta$, and hence $\mathrm{Spec}(\alpha(\Lambda_H)) = \mathrm{Spec}(\Lambda_H)^\Delta$. From the exact sequence, if $\mathcal{O}$ contains a primitive g–th roots of unity for the order g of Δ^{ab}, we get

$$\mathrm{Im}(\alpha') = H^0(\widehat{\Delta}(\mathcal{O}), \mathcal{O}[[G_p^{ab}]]),$$

where $\chi \in \widehat{\Delta}(\mathcal{O})$ takes $\Sigma_{g \in G_p^{ab}} a(g)[g]$ to $\Sigma_{g \in G_p^{ab}} a(g)\chi(g)[g]$.

A.2.2. — Nearly ordinary deformations

Now we impose the following additional condition to our deformation problem : let $S = S_G$ be a finite set of closed subgroups of G. For each $D \in S$, let $S(D)$ be a complete representative set for H–conjugacy classes of $\{gDg^{-1} \cap H \mid g \in G\}$. In the main text (Section 2), the data S is given by a choice of decomposition subgroups of $G = \mathrm{Gal}(F_\Sigma/E)$ at primes dividing p. For simplicity, we assume that $D \cap H \in S(D)$ always. Then the disjoint union $S_H = \bigsqcup_{D \in S} S(D)$ is a finite set, because $|S(D)| = |H\backslash G/D|$. Let P_D be a proper parabolic subgroup of $GL(n)_{/\mathcal{O}}$ defined over $\mathcal{O}$ indexed by $D \in S$. For each $D' \in S(D)$ such that $D' = H \cap gDg^{-1}$, we define $P_{D'} = c(g)P_{D}c(g)^{-1}$ for a lift $c(g) \in GL_n(\mathcal{O})$ of $\bar{p}(g)$. We assume

(NO) $\bar{p}(D) \subset P_D(\mathbb{F})$ for each $D \in S_G$.

Then we consider the following condition :

(NO$_H$)   there exists $g_D \in \widehat{GL}_n(A)$ for each $D \in S_H$
                        such that $g_D\rho(D)g_D^{-1} \subset P_D(A)$,

where $\widehat{GL}_n(A) = 1 + m_A M_n(A)$. We define a subfunctor $\mathcal{F}_?^{n.o}$ of the functor $\mathcal{F}_?$, with various restriction "?" introduced in the previous section, by

$$\mathcal{F}_?^{n.o}(A) = \{\rho \in \mathcal{F}_?(A) \mid \rho \text{ satisfies } (NO_X)\},$$

where X denotes either G or H depending on the group concerned. Then by (NO), (NO_X) and our choice of P_D, $\mathcal{F}_?^{n.o}(\mathbb{F}) = \{\bar{p}|_X\} \neq \emptyset$. Let us write $\mathfrak{gl}$ (resp. $\mathcal{P}_D$) for the Lie algebra of $GL_n(\mathbb{F})$ (resp. $P_D(\mathbb{F})$). Note that D acts on $\mathfrak{gl}$ and $\mathcal{P}_D$ by conjugation. We can identify $\mathfrak{gl}$ with $V \otimes V^* = \mathrm{Hom}_\mathbb{F}(V, V)$ for the representation space V of $\bar{p}$, where V^* is the contragredient of V. Then $\mathcal{P}_D$ can be identified with

$$\{g \in \mathrm{Hom}_\mathbb{F}(V, V) \mid gF_D \subset F_D\},$$

for a filtration $F_D : \{0\} = F_0 \subset F_1 \subset \cdots \subset F_r = V$. Here $gF_D \subset F_D$ implies that $gF_i \subset F_i$ for all i. The filtration F_D naturally induces a double filtration F_{AD} on $\mathfrak{gl}$ and $\mathcal{P}_D$. Since this filtration is compatible with that of $\mathcal{P}_D$, it induces a filtration of $\mathfrak{gl}/\mathcal{P}_D$, which is stable under the adjoint action $Ad(\bar{p})$ of $\bar{p}$. As shown in [T] Proposition 6.2, under the following regularity condition for every $D \in S_X$,

(Reg$_D$) $H^0(D, \mathfrak{gl}/\mathcal{P}_D) = 0$,

$\mathcal{F}_X^{n.o}$ is representable for $X = H$ or G. We can think of a stronger condition :

(RG$_D$) $H^0(D, gr(\mathfrak{gl}/\mathcal{P}_D)) = 0$.

This condition is stronger than the condition (Reg$_D$), because on $gr(\mathfrak{gl}/\mathcal{P}_D)$, D acts through the Levi–quotient of P_D. Writing the representation of D on F_i/F_{i-1} as $\bar{\rho}_{D,i}$, (RG$_D$) is equivalent to

$$(RG'_D) \qquad\qquad \mathrm{Hom}_D(\bar{\rho}_{D,i}, \bar{\rho}_{D,j}) = 0 \quad \text{if } i > j.$$

In the same manner as in the previous section, we can check that Δ acts on $\mathcal{F}_H^{n.o}$. Take $D \in S$ and put $D' = D \cap H \in S(D)$. Since $\bar{\rho}$ is invariant under Δ and $\bar{\rho} \in \mathcal{F}_G^{n.o}(\mathbb{F})$,

$$(\text{Inv}) \qquad\qquad \bar{\rho}_{D',i}^{[\sigma]} = \bar{\rho}_{D',i} \text{ in } gr(V) \text{ for all } i \text{ and } \sigma \in D.$$

For $\rho \in \mathcal{F}_X^{n.o}(A)$, we have $g_D \in \widehat{GL}_n(A)$ such that $\rho(D) \subset g_D^{-1} P_D(A) g_D$. This implies that $V(\rho)$ has a filtration $F_D(\rho) : \{0\} = F_0(\rho) \subset F_1(\rho) \subset \cdots \subset F_r(\rho) = V(\rho)$ stable under D such that $F_i(\rho)$ is a direct A–summand of $V(\rho)$ for all i and $F_D(\rho) \otimes_A \mathbb{F} = F_D$. We write $\rho_{D,i}$ for the representation of D on $F_i(\rho)/F_{i-1}(\rho)$. Now suppose $\rho \in \mathcal{F}_H^{\Delta,n.o}(A)$ and $[\rho] = 0$ in $H^2(\Delta, \widehat{\mathbb{G}}_m(B))$ for a flat A–algebra B. Then we find an extension $\pi : G \to GL_n(B)$ of ρ. Let $\sigma \in D$ and $D' = H \cap D$. Thus $\pi(\sigma)\rho(d')\pi(\sigma)^{-1} = \rho(\sigma d'\sigma^{-1}) \in g_{D'}^{-1} P_D(A) g_{D'}$ for all $d' \in D'$ and hence $\pi(\sigma)\rho(d')\pi(\sigma)^{-1}\rho(d')^{-1} \in g_{D'}^{-1} P_D(A) g_{D'}$. From this and (Reg$_{D'}$), it follows that $\pi(\sigma) \in g_{D'}^{-1} P_D(B) g_{D'}$ for $\sigma \in D$ [see [T] Proof of Proposition 6.2]. Thus, taking $g_D = g_{D'}$, we confirm that $\pi \in \mathcal{F}_G^{n.o}(A)$. Since $\mathcal{F}_G^{n.o}$ is stable under the action of $\widehat{\Delta}$, all the arguments given for $\mathcal{F}_X$ in the previous paragraph are valid for $\mathcal{F}_X^{n.o}$. Writing $(R_X^{n.o}, \rho_X^{n.o})$ for the universal couple representing $\mathcal{F}_X^{n.o}$, we conclude

THEOREM A.2.2. — *Suppose* (Z_H), *(Reg$_D$) for all* $D \in S_H$ *and that* n *is prime to* p. *Then we have the equality* $R_G^{n.o} = \mathrm{Im}(\alpha^{n.o})\Lambda_G^{n.o}$, *where* $\alpha^{n.o} :$ $R_H^{n.o} \to R_G^{n.o}$ *is an* $\mathcal{O}$–*algebra homomorphism given by* $\alpha^{n.o}\rho_H^{n.o} \approx \rho_G^{n.o}|_H$ *and* $\Lambda_G^{n.o}$ *is the image of* $\mathcal{O}[[G_p^{ab}]]$ *in* $R_G^{n.o}$. *Moreover we have*

$$(R_G^{n.o}, \rho_G^{n.o}) \cong (\mathrm{Im}(\alpha^{n.o}) \otimes_{\Lambda_H} \Lambda_G^{n.o}, \alpha^{n.o}\rho_H^{n.o} \otimes \det(\rho_G^{n.o}))$$
$$\cong (\mathrm{Im}(\alpha^{n.o}) \otimes_{\mathcal{O}[[H_p^{ab}]]} \mathcal{O}[[G_p^{ab}]], \alpha^{n.o}\rho_H^{n.o} \otimes \kappa).$$

A.2.3. — Ordinary deformations

Fix a normal closed subgroup $I = I_D$ of each $D \in S$. For $D' = gDg^{-1} \cap H \in S(D)$, we put $I_{D'} = gI_Dg^{-1} \cap H$. We call $\rho \in \mathcal{F}_X^{n.o}(A)$ ordinary if ρ satisfies the following conditions :

$$(\text{Ord}_X) \qquad \rho_{D,1} \text{ is of rank } 1 \text{ over } A \text{ and } I \subset \mathrm{Ker}(\rho_{D,1}) \text{ for every } D \in S_X.$$

We then consider the following subfunctor $\mathcal{F}_X^{ord}$ of $\mathcal{F}_X^{n.o}$:

$$\mathcal{F}_X^{ord}(A) = \{\rho \in \mathcal{F}_X^{n.o}(A) \mid \rho \text{ is ordinary}\}.$$

It is easy to see that the functor $\mathcal{F}_X^{ord}$ is representable by $(R_X^{ord}, \rho_X^{ord})$ under (Reg_D) for every $D \in S_X$.

Let $\rho \in \mathcal{F}_H^{ord}(A)$. Suppose $[\rho] = 0$ in $H^2(\Delta, \widehat{\mathbb{G}}_m(B))$ for a flat A-algebra B. Then we have at least one extension π of ρ in $\mathcal{F}_X^{n.o}(B)$. We consider $\pi_{D,1} : D \to A^{\times}$ for $D \in S$. We suppose one of the following two conditions for each $D \in S$:

(TR_D) $|I_D/I_D \cap H|$ is prime to p;
(Ex_D) Every p-power order character of $I_D/I_D \cap H$ can be extended to a character of Δ having values in a flat extension B' of B so that it is trivial on $I_{D'}$ for all $D' \in S$ different from D.

Under (TR_D), as a homomorphism of groups, $\pi_{D,1}$ restricted to I_D factors through $\bar{\rho}_{D,1}$ which is trivial on I. Thus $\pi_{D,1}$ is trivial on I_D. We note that $\pi_{D,1}$ is of p-power order on $I_D/H \cap I_D$ because $\bar{\rho}_{D,1}$ is trivial on I_D and $\rho_{D,1}$ is trivial on $I_D \cap H$. Thus we may extend $\pi_{D,1}$ to a character η of Δ congruent 1 modulo $m_{B'}$. Then we twists π by η^{-1}, getting an extension $\pi' = \pi \otimes \eta^{-1}$ such that $\pi'_{D,1}$ is trivial on I_D. Repeating this process for the D's satisfying (Ex_D), we find an extension $\pi \in \mathcal{F}_G^{ord}(B)$ for a flat extension B of A. We now consider

$$\mathcal{F}_{G,H}^{ord}(A) = \{\rho|_H \in \mathcal{F}_H^{ord}(A) \mid \rho \in \mathcal{F}_G^{ord}(B) \text{ for a flat extension } B \text{ of } A\}.$$

In the same manner as in Section 2, if either n is prime to p or $[\alpha^{ord}\rho_H^{ord}] = 0$ in $H^2(\Delta, \widehat{\mathbb{G}}_m(B))$ for a flat extension B of $\text{Im}(\alpha^{ord})$ in $CNL_{\mathcal{O}}$, we know that $\mathcal{F}_{G,H}^{ord}$ is represented by $(\text{Im}(\alpha^{ord}), \alpha^{ord}\rho_H^{ord})$, where $\alpha^{ord} : R_H^{ord} \to R_G^{ord}$ is an $\mathcal{O}$-algebra homomorphism given by $\alpha^{ord}\rho_H^{ord} \approx \rho_G^{ord}|_H$.

Let $\rho \in \mathcal{F}_{G,H}^{ord}(A)$ and π be its extension in $\mathcal{F}_G^{ord}(B)$ for a flat A-algebra B in $CNL_{\mathcal{O}}$. The character $\det(\pi)$ is uniquely determined by ρ on the subgroup of G_p^{ab} generated by all $I_{D,p}$, because another choice is $\pi \otimes \chi$ for a character χ of Δ and $(\pi \otimes \chi)_{D,1} = \chi$ on $I_{D,p}$. If G_p^{ab} is generated by the $I_{D,p}$'s and H_p, $\det(\pi)$ is uniquely determined by ρ. Thus assuming that n is prime to p, π itself is uniquely determined by ρ. Therefore the natural transformation : $\mathcal{F}_G^{ord} \to \mathcal{F}_{G,H}^{ord}$ given by $\rho \longmapsto \rho|_H$ identifies $\mathcal{F}_G^{ord}$ with a subfunctor of $\mathcal{F}_{G,H}^{ord}$, inducing a surjective $\mathcal{O}$-algebra homomorphism $\beta : \text{Im}(\alpha^{ord}) \to R_G^{ord}$ such that $\rho_G^{ord}|_H = \beta\alpha\rho_H^{ord}$. Since $\rho_G^{ord}|_H = \alpha\rho_H^{ord}$, β is the identity on $\text{Im}(\alpha^{ord})$, and we conclude that $\text{Im}(\alpha^{ord}) = R_G^{ord}$. This implies

THEOREM A.2.3. — *Suppose* (Z_H), (Reg_D) *for* $D \in S_H$, *either* (TR_D) *or* (Ex_D) *for each* $D \in S$ *and that* n *is prime to* p. *Suppose further that the* $I_{D,p}$'s *for all* $D \in S$ *and* H_p *generate* G_p^{ab}. *Then we have* $\text{Im}(\alpha^{ord}) = R_G^{ord}$. *In particular, for any deformation* $\rho \in \mathcal{F}_{G,H}^{ord}(A)$, *there is a unique extension* $\pi \in \mathcal{F}_G^{ord}(A)$ *such that* $\pi|_H = \rho$. *If further* $[\rho_H^{\Delta,ord}] = 0$ *in* $H^2(\Delta, \widehat{\mathbb{G}}_m(B))$ *for a flat extension* B *of* $R_{H,\Delta}^{ord}$, *then* $R_{H,\Delta}^{ord} \cong \text{Im}(\alpha^{ord}) = R_G^{ord}$, *where* $R_{H,\Delta}^{ord} = R_H^{ord}/\Sigma_{\sigma \in \Delta} R_H^{ord}([\sigma] - 1)R_H^{ord}$.

A.2.4. — Deformations with fixed determinant

We take a character $\chi : G \to \mathcal{O}^\times$ such that $\chi \equiv \det(\overline{\rho}) \bmod m_{\mathcal{O}}$. We then define $\mathcal{F}_X^{\chi,?}(A) = \{\rho \in \mathcal{F}_X^?(A) \mid \det(\rho) = \chi|_X\}$. Supposing the representability of $\mathcal{F}_X^?$, it is easy to check that $\mathcal{F}_X^{\chi,?}$ is representable. Since the determinant is already fixed and can be extended to G, by the argument in the previous sections shows that if n is prime to p,

$$\mathcal{F}_H^{\chi,?,\Delta} = \mathcal{F}_{G,H}^{\chi,?} = \mathcal{F}_G^\chi .$$

Write $(R_X^{\chi,?}, \rho_X^{\chi,?})$ for the universal couple representing $\mathcal{F}_X^{\chi,?}$ and define $\alpha^{\chi,?} : R_H^{\chi,?} \to R_G^{\chi,?}$ so that $\alpha^{\chi,?}\rho_H^{\chi,?} \approx \rho_G^{\chi,?}$. Then we have

PROPOSITION A.2.3. — *Suppose* (Z_H), (Reg_D) *for* $D \in S_H$ *and that* n *is prime to* p. *Then we have*

$$R_H^{\chi,?}/\Sigma_{\sigma \in \Delta} R_H^{\chi,?}([\sigma] - 1)R_H^{\chi,?} = R_{G,H}^{\chi,?} \cong \text{Im}(\alpha^{\chi,?}) = R_G^{\chi,?} ,$$

where $R_G^{\chi,?}$ *is either* R_G^χ, $R_G^{\chi,n.o}$ *or* $R_G^{\chi,ord}$.

For each $\rho \in \mathcal{F}_H^{n.o}(A)$, we decompose $\det(\rho) = \chi\xi$ so that ξ is a p–power order. If n is prime to p, there is a unique character $\xi^{1/n} : H \to \widehat{\mathbb{G}}_m(A)$ exists. Then we define $\rho^\chi = \rho \otimes \xi^{-1/n}$, which is an element of $\mathcal{F}_H^{\chi,n.o}(A)$. Writing f_H for the deformation functor $\mathcal{F}_H$ for $\chi|H$ in place of $\overline{\rho}|_H$, we have a natural transformation : $\mathcal{F}_H^{n.o} \to \mathcal{F}_H^{\chi,n.o} \times f_H$ given by $\rho \longmapsto (\rho^\chi, \det(\rho))$. If $(\rho^\chi, \det(\rho)) = (\rho'^\chi, \det(\rho'))$, then

$$\rho = \rho^\chi \otimes (\det(\rho)/\chi)^{1/n} = \rho'^\chi \otimes (\det(\rho')/\chi)^{1/n} = \rho' .$$

Thus the transformation is a monomorphism. For a given $(\rho^\chi, \det(\rho))$, we can recover ρ as above. Thus we get $\mathcal{F}_H^{n.o} \cong \mathcal{F}_H^{\chi,n.o} \times f_H$. Since $(\mathcal{O}[[H_p^{ab}]], \kappa)$ represents f_H, we see, if n is prime to p,

$$(R_H^{n.o}, \rho_H^{n.o}) \cong (R_H^{\chi,n.o}[[H_p^{ab}]], \varepsilon_H^{n.o} \otimes \kappa^{1/n}) .$$

Similarly we get

$$(R_H, \rho_H) \cong (R_H^\chi[[H_p^{ab}]], \varepsilon_H \otimes \kappa^{1/n}).$$

Note that, if n is prime to p,

$$\mathcal{F}_H^{?,\Delta} \cong \mathcal{F}_H^{\chi,?,\Delta} \times f_H^\Delta = \mathcal{F}_{G,H}^{\chi,?} \times f_H^\Delta \text{ and } \mathcal{F}_{G,H}^? \cong \mathcal{F}_{G,H}^{\chi,?} \times \mathcal{F}_{G,H}.$$

Thus $\alpha^? = \alpha^{\chi,?} \times \alpha'$ for α' as in the end of the paragraph A.2.1. This shows that

THEOREM A.2.4. — *Suppose* (Z_H), (Reg_D) *for* $D \in S_H$ *and that* n *is prime to* p. *Then if* $\mathcal{O}$ *contains a primitive* $|\Delta_p^{ab}|$*-root of unity, we have*

$$\text{Im}(\alpha^?) = R_G^{\chi,?} \widehat{\otimes}_\mathcal{O} (\mathcal{O}[[G_p^{ab}]])^{\widehat{\Delta}(\mathcal{O})},$$

where $R_G^{\chi,?}$ *is either* R_G^χ *or* $R_G^{\chi,n.o}$.

Manuscrit reçu le 23 octobre 1995

Corrections to "On Λ-adic forms of half integral weight for $SL(2)_{/\mathbf{Q}}$"

by Haruzo Hida

in "Number Theory, Paris 1992-93" Lecture Note Series **215**, 139-166

p. 145 line 15 : "$U(p^\alpha) = \left\{ s \in \mathbb{S} \mid s_p \equiv \begin{pmatrix} * & * \\ 0 & 1 \end{pmatrix} \bmod p^\alpha \right\}$" should read

$$U(p^\alpha) = \left\{ s \in U \mid s_p \equiv \begin{pmatrix} * & * \\ 0 & 1 \end{pmatrix} \bmod p^\alpha \right\}$$

p. 145 line 1 from the bottom : "for $\omega_{1/2}^{\otimes k} \otimes \omega^\circ \mid_{U_\alpha}$" should read "for $\omega_{1/2}^{\otimes k} \otimes \omega^\circ \mid_{U_{\alpha'}}$ where ω° is the dualizing sheaf on $X_{\alpha/\mathbb{Z}_p}$".

p. 146 line 8 : "the first horizontal map" should read "the first vertical map".

p. 146 line 10 : "the first row" should read "the second column".

p. 146 line 14 : "second row" should read "first column".

p. 146 line 14 : "the vertical maps" should read "the horizontal maps".

p. 146 line 15 : "rows" should read "columns"

The second diagram in p. 146 should be replaced by the following :

$$
\begin{array}{ccc}
0 & & 0 \\
\downarrow & & \downarrow \\
H^0(U_\gamma, \omega(k + \tfrac{3}{2})) \otimes \mathbb{Z}/p^\beta\mathbb{Z} & \xleftarrow{\ E_\alpha\ } & H^0(U_\gamma, \omega(k + \tfrac{1}{2}) \otimes \mathbb{Z}/p^\beta\mathbb{Z}) \\
\downarrow & & \downarrow \\
H^0(U_\gamma, \omega(k + 2)) \otimes \mathbb{Z}/p^\beta\mathbb{Z} & \xleftarrow{\ E_\alpha\ } & H^0(U_\gamma, \omega(k + 1) \otimes \mathbb{Z}/p^\beta\mathbb{Z}) \\
\downarrow & & \downarrow \\
H^0(U_\gamma, \mathcal{O}(D)) \otimes \mathbb{Z}/p^\beta\mathbb{Z} & = & H^0(U_\gamma, \mathcal{O}(D) \otimes \mathbb{Z}/p^\beta\mathbb{Z})
\end{array}
$$

The diagram in p. 147 should be replaced by the following :

$$
\begin{array}{ccc}
0 & & 0 \\
\downarrow & & \downarrow \\
H^0(U_\infty, \omega(k+\tfrac{3}{2})) \otimes \mathbb{Z}/p^\beta\mathbb{Z} & \xleftarrow{\ E_\alpha\ } & H^0(U_\infty, \omega(k+\tfrac{1}{2}) \otimes \mathbb{Z}/p^\beta\mathbb{Z} \\
\downarrow & & \downarrow \\
H^0(U_\infty, \omega(k+2)) \otimes \mathbb{Z}/p^\beta\mathbb{Z} & \xrightarrow{\ \sim\ } & H^0(U_\infty, \omega(k+1) \otimes \mathbb{Z}/p^\beta\mathbb{Z}) \\
\downarrow & & \downarrow \\
H^0(U_\infty, \mathcal{O}(D)) \otimes \mathbb{Z}/p^\beta\mathbb{Z} & = & H^0(U_\infty, \mathcal{O}(D) \otimes \mathbb{Z}/p^\beta\mathbb{Z})
\end{array}
$$

The second formula in (4.1) : "$a(n, f \mid T(q^2)) = a(p^2 n, f)$ if $q \mid Np^\alpha$", should read

$$\text{"}a(n, f \mid T(q^2)) = a(q^2 n, f) \text{ if } q \mid Np^\alpha\text{"}$$

p. 149 line 9 from the bottom "$P^{r(P)-1}$" should read $p^{r(P)-1}$".

In the formula of Theorem 3 in p. 153 : "$\psi_p(n + m)$" should read "$\psi_p(n/m)$".

At several places in pp. 155–157, "$\mathbb{Q}_e ll$" should read "$\mathbb{Q}_\ell$".

In the proof of Lemma 3, $(k/2)$ should read $k + (1/2)$ (thus $(k/2) - 1$ is replaced by $(2k - 1)/2$).

p. 157 line 5 from the bottom : "$\mu^2 \neq \alpha$" should read "$\mu^2 = \alpha$".

References

[C] H. CARAYOL. — *Formes modulaires et représentations galoisiennes à valeurs dans un anneau local compact*, Contemporary Math. **165** (1994), 213–237.

[CR] C.W. CURTIS and I. REINER. — *Methods of representation theory*, John Wiley and Sons, New York, 1981.

[DHI] K. DOI, H. HIDA and H. ISHII. — *Discriminant of Hecke fields and the twisted adjoint L-values for GL(2)*, preprint, 1995.

[F] M. FLACH. — *A finiteness theorem for the symmetric square of an elliptic curve*, Inventiones Math. **109** (1992), 307–327.

[G] R. GREENBERG. — *Iwasawa theory and p-adic deformation of motives*, Proc. Symp. Pure Math. **55** Part 2 (1994), 193–223.

[H] H. HIDA. — *Elementary Theory of L-functions and Eisenstein series*, 1993, Cambridge University Press.

[H1] H. HIDA. — *Galois representations into $GL_2(\mathbb{Z}_p[[X]])$ attached to ordinary cusp forms*, Inventiones Math. **85** (1986), 545–613.

[H2] H. HIDA. — *Modules of congruence of Hecke algebras and L-functions associated with cusp forms*, Amer. J. Math. **110** (1988), 323–382.

[H3] H. HIDA. — *Hecke algebras for GL_1 and GL_2*, Sém. Théorie des Nombres de Paris, 1984–85, **63** (1986), 131–163.

[H4] H. HIDA. — *On the search of genuine p-adic modular L-functions for $GL(n)$*, preprint 1995.

[HT] H. HIDA and J. TILOUINE. — *On the anticyclotomic main conjecture for CM fields*, Inventiones Math. **117** (1994), 89–147.

[M] B. MAZUR. — *Deforming Galois representations*, in "Galois group over $\mathbb{Q}$", MSRI publications, Springer, New York.

[MT] B. MAZUR and J. TILOUINE. — *Représentations Galoisiennes, différentielles de Kähler et "conjectures principales"*, Publ. IHES **71** (1990), 65–103.

[Sch] M. SCHLESSINGER. — *Functors of Artin rings,*, Trans. Amer. Math. Soc. **130** (1968), 208–222.

[T] J. TILOUINE. — *Deformation of Galois representations and Hecke algebras,* Lecture notes at MRI (Allahabad, India).

[W] A. WILES. — *Modular elliptic curves and Fermat's last theorem,* Ann. of Math. **142** (1995), 443–551.

[EGA] A. GROTHENDIECK. — *Eléments de géométrie algébrique IV,* Publ. IHES, vol. 20, 1964.

Haruzo HIDA
Department of Mathematics,
UCLA, Los Angeles,
Ca 90095–1555,
U.S.A.

Algèbres de Hecke et corps locaux proches

(une preuve de la conjecture de Howe pour $GL(N)$ en caractéristique > 0)

Bertrand Lemaire

1. — Introduction

Enoncé de la conjecture. Soient

- F un corps local non archimédien de corps résiduel fini [rappelons les deux possibilités : ou bien F est une extension finie d'un corps p–adique $\mathbb{Q}_p$, ou bien F est isomorphe à un corps de séries formelles $k((\varpi))$ en une indéterminée ϖ sur un corps fini k], $\mathcal{O}_F$ l'anneau des entiers de F et $\mathcal{P}_F$ l'idéal maximal de $\mathcal{O}_F$.
- G le groupe des F–points d'un groupe réductif connexe **G** défini sur F, muni de la topologie totalement discontinue héritée de F.
- $\mathcal{H}(G)$ l'algèbre de Hecke des fonctions $G \to \mathbb{C}$ localement constantes à support compact, munie du produit de convolution $*$ induit par une mesure de Haar dg sur G.
- K un sous–groupe ouvert compact de G, $e_K = \mathrm{vol}(K, dg)^{-1}\mathbf{1}_K$ l'idempotent de $\mathcal{H}(G)$ associé à K et $\mathcal{H}(G, K) = e_K * \mathcal{H}(G) * e_K$ la sous–algèbre des fonctions K–biinvariantes.
- $J_G \subset \mathcal{H}(G)^*$ l'espace des distributions $\mathrm{Ad}G$–invariantes sur G.
- Ω une partie *compacte modulo conjugaison* dans G (i.e. fermée, $\mathrm{Ad}G$–invariante et contenue dans $\mathrm{Ad}G(W)$ pour une partie compacte W de G).
- $J_G(\Omega) \subset J_G$ le sous–espace des distributions à support dans Ω.

CONJECTURE DE HOWE. — *Pour tous G, K et Ω comme ci–dessus,*

$$\dim_{\mathbb{C}}\left(J_G(\Omega)\Big|_{\mathcal{H}(G,K)} \right) < \infty\,.$$

Commentaires.

Pour F de caractéristique nulle, cette conjecture a été prouvée par Clozel pour n'importe quel groupe réductif connexe **G** défini sur F ([Cl 2]).

La conjecture de Howe est le type même de résultat local de finitude induisant une algébrisation toujours plus profonde de la théorie des représentations automorphes (cf. les derniers papiers d'Arthur où l'ingrédient "conjecture de Howe" est presque toujours présent). Dans le même ordre d'idée, il est significatif de constater le glissement progressif d'une analyse harmonique basée sur l'espace de Schwartz (Harish–Chandra) vers une analyse harmonique basée sur l'algèbre de Hecke $\mathcal{H}(G)$ (Bernstein/Kazhdan), même si l'espace de Schwartz conserve toute son utilité dans certaines situations (cf. [Cl 3]).

Quid de la caractéristique > 0? Pour F de caractéristique > 0 et $\mathbf{G} = GL(N)$, cette conjecture est prouvée dans [Le] Chap. 4 ; c'est cette preuve que nous présentons ici.

Essentiellement deux obstacles s'opposent à la généralisation de la démonstration de Clozel à la caractéristique > 0 :

1. Le nombre de classes de conjugaison de sous-groupes de Cartan, contrairement à la caractéristique nulle, peut être infini [exemple : si $k = \mathbb{Z}/2\mathbb{Z}$, il y a une infinité de classes d'isomorphisme d'extensions quadratiques séparables de $F = k((\varpi))$, chacune fournissant une classe de conjugaison de sous-groupes de Cartan (en l'occurence elliptiques) de $GL(2, F)$]. On ne peut donc, après réduction de la conjecture de Howe à une conjecture portant sur les intégrales orbitales de G, raisonner comme le fait Clozel en ne s'intéressant qu'aux éléments appartenant à la trace $\Omega \cap \Gamma$ de Ω sur un sous-groupe de Cartan fixé Γ de G.

2. L'intégrabilité locale des caractères.

Rappels : on appelle représentation *lisse* de G un triplet (π, G, V) où V est un $\mathbb{C}$–espace–vectoriel (de dimension le plus souvent infinie) et $\pi : G \to \mathrm{Aut}_{\mathbb{C}}(V)$ un homomorphisme de groupes tel que pour tout $\nu \in V$, le stabilisateur $G_\nu = \{g \in G, \pi(g) \cdot \nu = \nu\}$ de ν est ouvert dans G. Notant (π^*, G, V^*) la représentation duale de (π, G, V) donnée par

$$< \pi(g) \cdot \nu^*, \nu > = < \nu^*, \pi(g^{-1}) \cdot \nu > \quad (\nu^* \in V^*, \ g \in G, \ \nu \in V),$$

on définit la *représentation contragrédiente* $(\widetilde{\pi}, G, \widetilde{V})$ de (π, G, V) comme la partie lisse de (π^*, G, V^*) ; le choix d'un couple $(\nu, \widetilde{\nu}) \in V \times \widetilde{V}$ définit alors un *coefficient* de π

$$\pi_{\widetilde{\nu}, \nu}(g) = < \widetilde{\nu}, \pi(g) \cdot \nu > \quad (g \in G).$$

Une représentation lisse (π, G, V) est dite *admissible* si pour tout sous–groupe ouvert compact H de G, $V^H = \{\nu \in V, \pi(h) \cdot \nu = \nu \text{ pour tout } h \in H\}$ est un sous–espace de dimension finie de V.

Une représentation lisse (π, G, V) induit un homomorphisme d'algèbres $\pi : \mathcal{H}(G) \to \mathrm{End}_{\mathbb{C}}(V)$ donné par $\pi(\varphi) = \int_G \varphi(g)\pi(g)dg$ $(\varphi \in \mathcal{H}(G))$. Si (π, G, V) est admissible, alors l'opérateur $\pi(\varphi)$ $(\varphi \in \mathcal{H}(G))$ est à valeur dans un sous–espace de dimension finie de V, et l'on peut définir le *caractère-distribution* $\Theta_\pi \in J_G$ de π par

$$< \Theta_\pi, \varphi > = \mathrm{trace}\big(\pi(\varphi)\big) \quad \big(\varphi \in \mathcal{H}(G)\big).$$

Pour F de caractéristique nulle, on sait grâce à Harish–Chandra ([HC 1]) que si (π, G, V) est admissible et *irréductible*[1] (*i.e.* s'il n'existe aucun sous-espace propre non trivial de V stable sous l'action de G), alors Θ_π est une distribution localement intégrable au sens où il existe une fonction $\gamma_\pi \in L^1_{\mathrm{loc}}(G)$ telle que $< \Theta_\pi, \varphi > = \int_G \varphi(g)\gamma_\pi(g)dg$ $(\varphi \in \mathcal{H}(G))$.

Pour F de caractéristique > 0, cette intégrabilité locale des caractères est dans le cas général (c'est–à–dire pour n'importe quel groupe réductif connexe $\mathbf{G}$ défini sur F) encore conjecturale. D'où l'impossibilité d'utiliser la formule d'intégration de Weyl, pas plus que le(s) procédé(s) de troncature de la trace de Clozel.

Pourquoi se limiter à $GL(N)$?

L'application exponentielle, outil fréquemment utilisé en caractéristique nulle pour remonter du groupe à son algèbre de Lie, n'est pas définie en caractéristique > 0. Mais pour $GL(N)$, on dispose de la miraculeuse application $x \longmapsto 1 + x$ et des résultats de Bushnell–Kutzko sur l'entrelacement des strates simples (cf. la section **3** ci–après).

La présence d'éléments inséparables (*i.e.* dont une au moins des composantes irréductibles du polynôme minimal à travers un plongement $\rho : G \to GL(d, F)$ est inséparable sur F), et les phénomènes d'instabilité des orbites à leur voisinage, impliquent de revoir l'analyse harmonique – et tout particulièrement la théorie des intégrales orbitales – développée en caractéristique nulle [exemples : dégénérescence des orbites inséparables par extension radicielle du corps de base (les orbites inséparables fermées de G ne sont plus fermées lorsque considérées dans $\mathbf{G}(\bar{F})$ pour une clôture algébrique $\bar{F}$ de F); absence, pour les éléments inséparables, d'une décomposition de Jordan en parties semi–simple[2] et unipotente commutant entre elles] ; on a donc tout naturellement commencé par nettoyer $GL(N)$, cf. [Le] Chap. 3.

---

[1] Et même seulement admissible de longueur finie.

[2] On entend par *semi–simple* un élément $s \in G$ tel que le polynôme minimal de $\rho(s)$ est produit de composantes irréductibles (*non nécessairement séparables*) sur F apparaissant chacune avec une multiplicité 1.

Désormais et jusqu'à la fin de l'exposé $\mathbf{G} = GL(N)$ (N entier ≥ 2), et soient

- $K_F = \mathbf{G}(\mathcal{O}_F)$ et K_F^m (m entier ≥ 1) les sous–groupes de congruence modulo $\mathcal{P}_F^m$ de K_F.
- B_F le sous–groupe d'Iwahori standard de $\mathbf{G}(F)$ i.e. l'image réciproque par la projection $K_F \to \mathbf{G}(\mathcal{O}_F/\mathcal{P}_F)$ du sous–groupe de Borel de $\mathbf{G}(\mathcal{O}_F/\mathcal{P}_F)$ formé des matrice triangulaires supérieures, et B_F^m (m entier ≥ 1) les sous–groupes de congruence modulo $\mathcal{P}_F^m$ de B_F.

2. — L'idée : comparer l'analyse harmonique des groupes $\mathbf{G}(F)$ et $\mathbf{G}(E)$ pour F de caractéristique > 0 et E de caractéristique nulle.

Résultat de Kazhdan sur les corps locaux proches.

DÉFINITION. — *On dit que deux corps locaux non archimédiens E et E' sont m–proches pour un entier $m \geq 1$ si les anneaux locaux tronqués $\mathcal{O}_E/\mathcal{P}_E^m$ et $\mathcal{O}_{E'}/\mathcal{P}_{E'}^m$ sont isomorphes.*

THÉORÈME[3] ([Ka]). — *Soit n un entier ≥ 1. Alors il existe un entier $m = m(n) \geq n$ tel que pour tout corps local E m–proche de F, les algèbres de Hecke $\mathcal{H}(\mathbf{G}(F), K_F^n)$ et $\mathcal{H}(\mathbf{G}(E), K_E^n)$ sont isomorphes.*

CONJECTURE ([Ka]). — *L'entier $m(n) = n$ convient.*

Côté galoisien, Deligne, inspiré par le résultat de Kazhdan et par la philosophie de Langlands, obtient un résultat du même genre, quoique beaucoup plus précis. Notons $\mathcal{E}(F)$ la catégorie des extensions finies séparables de F et $\mathcal{E}_m(F)$ (m entier ≥ 1) la sous–catégorie pleine de $\mathcal{E}(F)$ formée des extensions F'/F telles que le m–ième groupe de ramification en notation supérieure $\mathrm{Gal}(F''/F)^m$ de la clôture normale F''/F de F'/F est trivial.

"THÉORÈME" ([De]). — *Soit n un entier ≥ 1. Alors la catégorie $\mathcal{E}_n(F)$ "ne dépend"[4] que du triplet $Tr_n(F) = (\mathcal{O}_F/\mathcal{P}_F^n, \mathcal{P}_F/\mathcal{P}_F^{n+1}, \varepsilon_{F,n})$, où $\varepsilon_{F,n}$ désigne le morphisme de $\mathcal{O}_F/\mathcal{P}_F^n$–modules $\mathcal{P}_F/\mathcal{P}_F^{n+1} \to \mathcal{O}_F/\mathcal{P}_F^n$ induit par l'inclusion $\mathcal{P}_F \to \mathcal{O}_F$ par passage au quotient.*

Retour côté automorphe où l'on traduit sur les isomorphismes d'algèbres de Hecke de Kazhdan la précision obtenue par Deligne côté galoisien.

---

[3] Enoncé et montré par Kazhdan pour n'importe quel $\mathbb{Z}$–groupe réductif déployé.

[4] Plus précisément, Deligne donne une notion générale de *triplet*, organise ces triplets en une catégorie, puis construit une catégorie $\mathcal{E}(Tr_n(F))$ des "triplets au-dessus de $Tr_n(F)$" et une équivalence de catégorie canonique $\mathcal{E}_n(F) \to \mathcal{E}(Tr_n(F))$.

Fixons un entier $n \geq 1$, un corps local E n–proche de F et un isomorphisme d'anneaux $\lambda : \mathcal{O}_F/\mathcal{P}_F^n \to \mathcal{O}_E/\mathcal{P}_E^n$. Fixons aussi un jeu d'uniformisantes (ϖ_F, ϖ_E) respectivement de F et E, compatible avec λ au sens où $\lambda(\varpi_F + \mathcal{P}_F^n) = \varpi_E + \mathcal{P}_E^n$. Via la décomposition de Bruhat-Tits $\mathbf{G}(F) = \coprod_w B_F w B_F$ où w parcourt les éléments du groupe de Weyl affine $W^a(\varpi_F) \cong S_N \times \mathbb{Z}^N$, on construit aisément un isomorphisme d'*espaces vectoriels*

$$\zeta = \zeta(\lambda, \varpi_F, \varpi_E) : \mathcal{H}(\mathbf{G}(F), B_F^n) \xrightarrow{\cong} \mathcal{H}(\mathbf{G}(E), B_E^n) .$$

PROPOSITION 1. — *ζ est un isomorphisme d'algèbres (et même d'anneaux).*

Preuve : *cf.* la présentation de $\mathcal{H}(\mathbf{G}(F), B_F^n)$ par générateurs et re-lations donnée dans [Ho]. L'application ζ induisant une bijection entre l'ensemble des générateurs de $\mathcal{H}(\mathbf{G}(F), B_F^n)$ défini par ϖ_F et l'ensemble des générateurs de $\mathcal{H}(\mathbf{G}(E), B_E^n)$ défini par ϖ_E, il suffit de vérifier qu'elle préserve les relations.

COROLLAIRE. — *Pour tout sous–groupe ouvert compact H de $\mathbf{G}(F)$ con-tenant B_F^n, ζ induit par restriction un isomorphisme d'algèbres $\mathcal{H}(\mathbf{G}(F), H) \to \mathcal{H}(\mathbf{G}(E), \zeta(H))$ où $\zeta(H)$ est le sous–groupe ouvert compact de $\mathbf{G}(E)$ contenant B_E^n défini par $\mathbf{1}_{\zeta(H)} = \zeta(\mathbf{1}_H)$.*

Preuve : après avoir vérifié que la partie $\zeta(H)$ ainsi définie est bien un groupe, on conclut grâce à l'égalité $\mathcal{H}(\mathbf{G}(F), H) = e_H * \mathcal{H}(\mathbf{G}(F), B_F^n) * e_H$.

On a en fait légèrement mieux. La donnée du jeu d'uniformisantes (ϖ_F, ϖ_E) permet de relever l'isomorphisme λ en un isomorphisme de groupes $\widetilde{\lambda} : \mathcal{P}_F/\mathcal{P}_F^{n+1} \to \mathcal{P}_E/\mathcal{P}_E^{n+1}$ défini par $\widetilde{\lambda}(\varpi_F x) = \varpi_E \lambda(x)$ $(x \in \mathcal{O}_F/\mathcal{P}_F^n)$, compatible avec les structures de modules sur $\mathcal{O}_F/\mathcal{P}_F^n$ et $\mathcal{O}_E/\mathcal{P}_E^n$ au sens où $\widetilde{\lambda}(ax) = \lambda(a)\widetilde{\lambda}(x)$ $(a \in \mathcal{O}_F/\mathcal{P}_F^n, x \in \mathcal{P}_F/\mathcal{P}_F^{n+1})$, et induisant un diagramme commutatif

$$
\begin{array}{ccc}
\mathcal{O}_F/\mathcal{P}_F^n & \xrightarrow{\;\;\lambda\;\;} & \mathcal{O}_E/\mathcal{P}_E^n \\
\Big\downarrow{\scriptstyle \varepsilon_{F,n}} & \widetilde{} & \Big\downarrow{\scriptstyle \varepsilon_{E,n}} \\
\mathcal{P}_F/\mathcal{P}_F^{n+1} & \xrightarrow[\;\widetilde{\lambda}\;]{} & \mathcal{P}_E/\mathcal{P}_E^{n+1}
\end{array}
$$

Alors $\zeta(\lambda, \varpi_F, \varpi_E) = \zeta(\lambda, \widetilde{\lambda})$ [traduction : la classe d'isomorphisme de l'algèbre de Hecke $\mathcal{H}(\mathbf{G}(F), B_F^n)$ "ne dépend" que de celle du triplet $Tr_n(F)$].

Conséquence immédiate (et séduisante) quant à la théorie des représen-tations : le foncteur $(\pi, \mathbf{G}(F), V) \longmapsto (\pi_n, \mathcal{H}(\mathbf{G}(F), B_F^n), V_n)$ (où V_n désigne le sous–espace des vecteurs $\nu \in V$ tels que $\pi(b) \cdot \nu = \nu$ pour tout $b \in B_F^n$) de la catégorie des représentations lisses de $\mathbf{G}(F)$ dans celle des $\mathcal{H}(\mathbf{G}(F), B_F^n)$–modules, induit une bijection $'\pi' \longmapsto \zeta('\pi')$ entre l'ensemble $\varepsilon_n(\mathbf{G}(F))$ des classes de représentations admissibles irréductibles[5] de $\mathbf{G}(F)$ ayant un vecteur non nul fixé par B_F^n et l'ensemble $\varepsilon_n(\mathbf{G}(E))$ [en fait, on obtient même une équivalence entre la catégorie des représentations lisses de $\mathbf{G}(F)$ qui sont engendrées par leurs vecteurs fixes sous l'action de B_F^n et la catégorie correspondante pour E ([Le] Chap. 2)]. Comme toute représentation lisse de $\mathbf{G}(F)$ possède un vecteur non nul fixé par B_F^n pour n suffisamment grand, et comme on peut approcher un corps local de caractéristique > 0 d'aussi près que l'on veut par un corps local non archimédien de caractéristique nulle [si $F \cong k((\varpi))$ avec $\#k = p^r$, alors toute extension de $\mathbb{Q}_p$ de degré résiduel r et d'indice de ramification $\geq n$, est n–proche de F], on peut résumer la situation en disant que pour F de caractéristique > 0, la théorie des représentations de $\mathbf{G}(F)$ est limite des théories des représentations de $\mathbf{G}(E)$, E extension finie de $\mathbb{Q}_p$ telle que $\mathcal{O}_E/\mathcal{P}_E \cong \mathcal{O}_F/\mathcal{P}_F$, quand l'indice de ramification absolu $e(E/\mathbb{Q}_p)$ tend vers l'infini.

3. — Une preuve de la conjecture de Howe pour $GL(N)$ en caractéristique > 0

Le résultat (comme chez Clozel) se montre par induction sur la dimension des sous–groupes de Levi de $\mathbf{G}$. On procède en trois étapes, les deux premières (indépendantes l'une de l'autre) permettant dans la troisième de traiter les éléments elliptiques, c'est–à–dire ceux intervenant dans le dernier cran de l'induction.

Première étape : un relèvement uniforme des intégrales orbitales elliptiques.

Soient n un entier ≥ 1 et y un élément *elliptique* (i.e. de polynôme caractéristique irréductible sur F) de $\mathbf{G}(F)$.

Notations : soient $x \in \mathbf{G}(F)$, dg_x une mesure de Haar sur le centralisateur $\mathbf{G}(F)_x$ de x dans $\mathbf{G}(F)$ et $\varphi \in \mathcal{H}(\mathbf{G}(F))$. On définit *l'intégrale orbitale de φ au point x dans $\mathbf{G}(F)$* par

$$I^{\mathbf{G}(F)}(\varphi, x, dg_x) = \int_{\mathbf{G}(F)_x \backslash \mathbf{G}(F)} \varphi(g^{-1}xg) \frac{dg}{dg_x} ,$$

où dg est la mesure de Haar sur $\mathbf{G}(F)$ arbitrairement normalisée par $\mathrm{vol}(K_F, dg) = 1$. Pour tout x, cette intégrale est absolument convergente. Soit $\mathbf{G}(F)_e$ l'ensemble (ouvert) des éléments elliptiques de $\mathbf{G}(F)$.

---

[5] On sait (grâce à Jacquet) qu'une représentation lisse irréductible est admissible.

Normalisation : si de plus $x \in \mathbf{G}(F)_e$, soit $I^{\mathbf{G}(F)}(\cdot, x)$ l'intégrale orbitale au point x définie par la mesure dg_x sur $\mathbf{G}(F)_x$ telle que $\mathrm{vol}(\mathcal{O}_{F[x]}^{\times}, dg_x) = 1$.

THÉORÈME 1. — *Il existe un voisinage $\mathcal{V} = \mathcal{V}(y, n)$ de y dans $\mathbf{G}(F)_e$ tel que $I^{\mathbf{G}(F)}(\varphi, x) = I^{\mathbf{G}(F)}(\varphi, y)$ pour toute fonction $\varphi \in \mathcal{H}(\mathbf{G}(F), K_F^n)$ et tout élément $x \in \mathcal{V}$.*

Quitte à remplacer y par un de ses conjugués, on suppose que y est en *position standard*, *i.e.* que le sous-groupe parahorique H de $\mathbf{G}(F)$ normalisé par $F[y]^{\times}$ satisfait la double inclusion $B_F \subset H \subset K_F$.

THÉORÈME 2. — *Il existe un entier $r = r(y, n) \geq n$ tel que pour tout corps local E r–proche de F et tout isomorphisme d'algèbres $\zeta : \mathcal{H}(\mathbf{G}(F), B_F^r) \to \mathcal{H}(\mathbf{G}(E), B_E^r)$ comme dans la Proposition 1, le voisinage $\mathcal{V}$ du Théorème 1 est B_F^r-biinvariant, la partie $\zeta(\mathcal{V})$ définie par $\mathbf{1}_{\zeta(\mathcal{V})} = \zeta(\mathbf{1}_{\mathcal{V}})$ est contenue dans $\mathbf{G}(E)_e$ et $I^{\mathbf{G}(E)}(\zeta(\varphi), x') = I^{\mathbf{G}(F)}(\varphi, y)$ pour toute fonction $\varphi \in \mathcal{H}(\mathbf{G}(F), K_F^n)$ et tout élément $x' \in \varphi(\mathcal{V})$.*

Principe de la preuve : l'idée est, grâce au *principe de submersion* d'Harish–Chandra [Ha 2] appliqué à l'application partout submersive $\delta_{P,y}$: $\mathbf{G}(F) \times P \to \mathbf{G}(F)$, $(g, p) \longmapsto g^{-1}ygp$ où P désigne un sous–groupe de Borel de $\mathbf{G}(F)$, d'écrire l'intégrale orbitale $I^{\mathbf{G}(F)}(\varphi, y)$ d'une fonction $\varphi \in \mathcal{H}(\mathbf{G}(F), K_F^n)$ comme combinaison linéaire infinie (indéxée par les éléments w du groupe de Weyl affine) d'intégrales de la forme $\int_{\mathbf{G}(F)} \alpha_{P,y}(g)\varphi_w(g)dg$ ($\varphi_w \in \mathcal{H}(\mathbf{G}(F))$), puis de "calculer" cette application $\alpha_{P,y} \in \mathcal{H}(\mathbf{G}(F))$ issue du principe de submersion en la cassant en une combinaison linéaire (finie) de fonctions caractéristiques d'ouverts compacts de $\mathbf{G}(F)$ (ce qui revient à contrôler l'image des ouverts de $\mathbf{G}(F) \times P$ par la submersion $\delta_{P,y}$); le voisinage $\mathcal{V}(y, n)$ du Théorème 1 comme l'entier $r(y, n)$ du Théorème 2, se lisent alors directement sur la formule obtenue. Ces deux résultats – "calcul" des intégrales orbitales elliptiques et leur transport – constituent la partie la plus délicate de la démonstration, les arguments utilisés étant pour l'essentiel puisés dans le travail de Bushnell–Kutzko [BK], cf. ci–dessous.

Soit x un élément *irréductible* (i.e. de polynôme minimal irréductible sur F) de $\mathbf{G}(F)$. Alors $F[x] = L$ est une extension de F contenue dans $\mathfrak{g} = M(N, F)$ et l'on note $\mathfrak{b} \subset \mathfrak{g}$ le commutant de L dans $\mathfrak{g}$. Identifiant de manière standard $\mathfrak{g}$ et $\mathfrak{b}$ avec leur dual de Pontrjagin, on appelle *corestriction modérée dans $\mathfrak{g}$ relative à L/F* un homomorphisme de $(\mathfrak{b}, \mathfrak{b})$–bimodules $s : \mathfrak{g} \to \mathfrak{b}$ réalisant la restriction des caractères[6], c'est-à-dire tel que $\psi_F \circ \mathrm{tr}_{\mathfrak{g}/F}(gb) = \psi_L \circ \mathrm{tr}_{\mathfrak{b}/L}(s(g)b)$ $((g, b) \in \mathfrak{g} \times \mathfrak{b})$ pour des caractères additifs

---

[6] Pour nous, un *caractère* est simplement un homomorphisme continu dans $\mathbb{C}^{\times}$.

$\psi_F : F \to \mathbb{C}^\times$ et $\psi_L : L \to \mathbb{C}^\times$ de conducteurs respectifs $\mathcal{P}_F$ et $\mathcal{P}_L$. Si $\mathcal{G}$ est un $\mathcal{O}_F$–ordre héréditaire dans $\mathfrak{g}$ normalisé par $L^\times$, alors $s(\mathcal{G}) = \mathfrak{b} \cap \mathcal{G}$ est un $\mathcal{O}_L$–ordre héréditaire dans $\mathfrak{b}$ et la classe $x + \mathcal{G}^k$ $(k \in \mathbb{Z})$, appelée *strate* de $\mathfrak{g}$ et notée $[\mathcal{G}, -\nu_{\mathcal{G}}(x), -k, x]$ où $\nu_{\mathcal{G}}$ désigne la "valuation" sur $\mathfrak{g}$ induite par les puissances du radical de Jacobson de $\mathcal{G}$, est une *strate simple* si x minimise le degré des extensions $F[g]/F$ $(g \in x + \mathcal{G}^k$, irréductible). Dans cette situation, on a une relation explicite entre la corestriction modérée s et l'application adjointe $[x, \cdot] : \mathfrak{g} \to \mathfrak{g}$, relation permettant à Bushnell–Kutzko de calculer l'*entrelacement* $\{g \in \mathbf{G}(F), \ g^{-1}(x + \mathcal{G}^k)g \cap (x + \mathcal{G}^k) \neq \varnothing\}$ de la strate $[\mathcal{G}, -\nu_{\mathcal{G}}(x), -k, x]$; en particulier si $x \in \mathbf{G}(F)_e$, cet entrelacement est compact modulo le centre et contenu dans le normalisateur de $\mathcal{G}$ dans $\mathbf{G}(F)$. C'est ce calcul de l'entrelacement des strates simples elliptiques que nous utilisons pour contrôler l'image des ouverts de $\mathbf{G}(F) \times P$ par l'application $\delta_{P,y}$ puis pour transporter la formule obtenue de $\mathbf{G}(F)$ à $\mathbf{G}(E)$.

Deuxième étape : une majoration du résultat de Clozel indépendante de la ramification.

Soient n un entier ≥ 1, $q = p^r$, $E_1/\mathbb{Q}_p$ une extension non ramifiée de degré r, et $\mathcal{F}$ une famille finie de suites ordonnées $(\alpha) = (\alpha_1 \leq \cdots \leq \alpha_N)$ $(\alpha_i \in \mathbb{Z})$. Pour chaque extension finie E/E_1, on définit la partie $X_{E,\mathcal{F}} = \mathrm{Ad}\mathbf{G}(E)(\coprod_{(\alpha) \in \mathcal{F}} K_E \varpi_E^{(\alpha)} K_E)$ de $\mathbf{G}(E)$ où (ϖ_E désignant une quelconque uniformisante de E)

$$\varpi_E^{(\alpha)} = \mathrm{diag}(\varpi_E^{\alpha_1}, \ldots, \varpi_E^{\alpha_N}) \quad ((\alpha) \in \mathbb{Z}^N).$$

THÉORÈME 3. — *Il existe une constante* $c = c(q, n, \mathcal{F}) > 0$ *telle que pour toute extension finie totalement ramifiée* E/E_1 *et toute partie* Ω_E *de* $\mathbf{G}(E)$ *fermée* $\mathrm{Ad}\mathbf{G}(E)$*-invariante et contenue dans* $X_{E,\mathcal{F}}$,

$$\dim_{\mathbb{C}} \left(J_{\mathbf{G}(E)}(\Omega_E) \Big|_{\mathcal{H}(\mathbf{G}(E), K_E^n)} \right) \leq c.$$

Principe de la preuve : on reprend le(s) article(s) de Clozel en contrôlant à chaque étape de la démonstration que la majoration qu'il obtient ne dépend pas de la ramification. L'idée centrale de la preuve de Clozel consiste à casser l'expression trace$(\pi(\varphi))$ (où π est une représentation admissible irréductible de $\mathbf{G}(E)$ et $\varphi \in \mathcal{H}(\mathbf{G}(E))$). Les deux versions qu'il donne de sa démonstration correspondent à deux troncations différentes de cette expression. Dans la première version ([Cl 1]), il coupe cette trace en "trace elliptique" et "trace non elliptique" ce qui l'amène à considérer *toutes* les (classes de) représentations elliptiques[7] de $\mathbf{G}(E)$, ceci bien que la ramification des fonctions considérées (*i.e.* le niveau n du sous–groupe de congruence K_E^n) soit fixée, et donc à travailler modulo une hypothèse conjecturale

---

[7] Une représentation admissible irréductible π de G est dite *tempérée* s'il

de finitude des exposants spéciaux de la série discrète[8]; cette hypothèse a été vérifiée par Clozel pour le groupe linéaire ([Cl 1]) [notons qu'en théorie des formes automorphes, cette hypothèse de finitude correspondrait au fait que les pôles des séries d'Eisenstein construites à partir des formes cuspidales relativement à un sous–groupe parabolique donné, appartiennent à un ensemble fini, indépendant de la forme cuspidale induisante; c'est une conjecture très forte, pas même vérifiée pour le groupe linéaire]. Dans la deuxième version ([Cl 2]), la trace est coupée suivant des termes indéxés par les classes de conjugaison de sous–groupes paraboliques de $\mathbf{G}(E)$ et correspondant à la stratification de Deligne–Casselman; Clozel établit ainsi une formule remarquable exprimant la trace complète $\mathrm{trace}(\pi(\varphi))$ en termes des "traces compactes" des modules de Jacquet de π relatifs à un système de représentants des classes de conjugaison de sous–groupes paraboliques de $\mathbf{G}(E)$ (nous utilisons en fait la version duale de cette formule, montrée par Clozel dans [Cl 3], exprimant la trace compacte de $\pi(\varphi)$ en terme des traces complètes des modules de Jacquet de π relatifs à ces mêmes sous–groupes paraboliques).

Nous raisonnons essentiellement sur la seconde troncature de Clozel, utilisant néanmoins la propriété de finitude des exposants spéciaux de la série discrète puisqu'on est amenés à compter (grâce à la paramétrisation de Zelevinski) les orbites du groupe des caractères *non ramifiés* (*i.e.* les caractères triviaux sur le groupe $\mathbf{G}(E)_0 = \{g \in \mathbf{G}(E), \det(g) \in \mathcal{O}_E^\times\}$) de $\mathbf{G}(E)$ agissant par torsion sur l'ensemble des classes de représentations elliptiques de $\mathbf{G}(E)$ ayant un vecteur non nul fixé par K_E^n.

Troisième étape.

On suppose F de caractéristique > 0. Soient K un sous–groupe ouvert compact de $\mathbf{G}(F)$ et Ω une partie compacte modulo conjugaison dans $\mathbf{G}(F)$. Soient n un entier ≥ 1 tel que $K_F^n \subset K$ et $\mathcal{F}$ une famille finie de suites ordonnées $(\alpha) = (\alpha_1 \leq \cdots \leq \alpha_N)$ $(\alpha_i \in \mathbb{Z})$ telle que $\Omega \subset X_{F,\mathcal{F}}$.

---

existe un caractère ω de G tel que les coefficients de la représentation tordue $\omega \otimes \pi$ sont dans l'espace de Schwartz. Une représentation tempérée π de G est dite *elliptique* si son caractère–distribution Θ_π n'est pas identiquement nul sur l'ouvert des éléments elliptiques de G.

[8] Une représentation admissible irréductible π de G est dite *essentiellement de carré intégrable* s'il existe un caractère ω de G tel que les coefficients de la représentation tordue $\omega \otimes \pi$ sont de carré intégrable modulo le centre de G. La *série discrète* de G est l'ensemble des classes de représentations essentiellement de carré intégrable de G. Noter que pour $GL(N)$, une représentation est essentiellement de carré intégrable si et seulement si elle est elliptique (Jacquet). Pour l'hypothèse sur les exposants spéciaux, on renvoie à la Définition 1 de [Cl 1].

On raisonne par induction sur la dimension des sous–groupes de Levi de $\mathbf{G}$. On suppose la conjecture de Howe vraie pour tous les sous–groupes de Levi standards $\mathbf{M}$ de $\mathbf{G}$ (elle est trivialement vraie pour le tore diagonal $\mathbf{A}_0$, une partie compacte modulo conjugaison dans $\mathbf{A}_0(F)$ étant tout simplement compacte). La propriété bien connue de descente des intégrales orbitales non–elliptiques, jointe à un argument de descente des parties compactes modulo conjugaison dans $\mathbf{G}(F)$, entraînent que les fonctionnelles linéaires $\varphi \longmapsto I^{\mathbf{G}(F)}(\varphi, x, dg_x)$ ($x \in \Omega$ non elliptique, $\varphi \in \mathcal{H}(\mathbf{G}(F), K_F^n)$) engendrent un sous–espace vectoriel de dimension finie du dual de $\mathcal{H}(\mathbf{G}(F), K_F^n)$. Grâce aux étapes 1 et 2, on montre (petit raisonnement par l'absurde) que les fonctionnelles linéaires $\varphi \longmapsto I^{\mathbf{G}(F)}(\varphi, x, dg_x)$ ($x \in \Omega$ elliptique, $\varphi \in \mathcal{H}(\mathbf{G}(F), K_F^n)$) engendrent elles aussi un sous–espace vectoriel de dimension finie du dual de $\mathcal{H}(\mathbf{G}(F), K_F^n)$. On conclut grâce à la propriété de densité des intégrales orbitales $I^{\mathbf{G}(F)}(\varphi, x, dg_x)$ ($x \in \Omega$) dans l'espace $J_{\mathbf{G}(F)}(\Omega)$, lequel s'obtient grâce aux résultats de Gelfand–Kazhdan en rangeant les nappes de Dixmier de $\mathbf{G}(F)$ par dimension d'orbites croissante, *cf.* [Le] Chap. 3.

4. — Conclusion

L'intégrabilité locale des caractères en caractéristique > 0, si elle reste conjecturale dans le cas général, est désormais montrée pour le groupe linéaire (*cf.* [le] Chap. 5, à paraitre dans Compositio Math.). On peut dès lors suivre d'encore plus près la démonstration de Clozel et montrer cette conjecture de Howe pour $GL(N)$ en caractéristique > 0 directement, c'est–à–dire sans passer par la comparaison des analyses harmoniques [la démonstration, non encore rédigée, repose alors sur le théorème de densité des caractères-distributions des représentations tempérées dans l'espace $J_{\mathbf{G}(F)}$, dont la preuve ne dépend pas de la caractéristique du corps de base]. Cette seconde approche se prêtant nettement plus facilement que la première à une généralisation à n'importe quel groupe réductif connexe, il est naturel de commencer par essayer de montrer l'intégrabilité locale des caractères dans le cas général. A suivre donc.

Manuscrit reçu le 22 mars 1994

Bibliographie

[BK] C.J. BUSHNELL et P.C. KUTZKO. — *The admissible dual of GL(N) via compact open subgroups*, Ann. of Math. Studies **129**, Princeton U. Press, Princeton, New Jersey, 1993.

[Cl 1] L. CLOZEL. — *Sur une conjecture de Howe I*, Compositio Math. **56** (1985), 87–110.

[Cl 2] L. CLOZEL. — *Orbital integrals on p–adic groups : a proof of the Howe conjecture*, Ann. of Math. **129** (1989), 237–251.

[Cl 3] L. CLOZEL. — *The fundamental lemma for stable base change*, Duke Math. J. **61** (1990), 225–302.

[De] P. DELIGNE. — *Les corps locaux de caractéristique p comme limites de corps locaux de caractéristique 0* dans Représentations des groupes réductifs sur un corps local, Travaux en cours, Hermann, Paris (1984), 119–157.

[Ha 1] HARISH–CHANDRA. — *Admissible invariant distributions on reductive p–adic groups*, Queen's Papers in Pure and applied Math. **48** (1978), 377–380.

[Ha 2] HARISH–CHANDRA. — *A submersion principle and its application* in Papers dedicated to the memory of V.K. Patodi, Indian Academy of Sciences, Bangalore, and the Tata Institute for Fundamental Research, Bombay (1980), 95–102.

[Ho] R. HOWE. — *Harish–Chandra homomorphism for p–adic groups*, CBMS Regional Conf. Series in Math. **59**, Amer. Math. Soc., Providence, Rhode Island, 1985.

[Ka] D. KAZHDAN. — *Representations of groups over close local fields*, J. Analyse Math. **47** (1986), 175–179.

[Le] L. LEMAIRE. — *These*, Univ. de Paris-Sud, 8 février 1994.

Bertrand LEMAIRE
Université Paris–Sud
Département de Mathématiques
Bâtiment 425
91405 ORSAY CEDEX

Aspects expérimentaux de la conjecture abc

Abderrahmane Nitaj

1. — Introduction

Si n est un entier non nul, le produit de ses différents facteurs premiers, noté $r(n)$, sera appelé radical de n (certains auteurs utilisent le terme support, conducteur ou noyau), avec par convention $r(1) = 1$. La conjecture abc de J. Oesterlé et D.W. Masser est liée à la notion de radical. Cette conjecture date de 1985 et est devenue maintenant classique.

CONJECTURE (abc). — *Soit $\varepsilon > 0$. Il existe une constante $c(\varepsilon) > 0$ telle que, pour tout triplet (a, b, c) d'entiers positifs vérifiant $a + b = c$ et $(a, b) = 1$ on ait :*

$$c \leq c(\varepsilon)\, (r(abc))^{1+\varepsilon} ,$$

où $r(abc)$ est le radical de abc.

La conjecture abc a des conséquences très étonnantes. Elle implique en particulier le théorème de Fermat asymptotique (*i.e.* vrai pour les grands exposants), le théorème de Faltings (née Conjecture de Mordell) et s'applique bien aux équations diophantiennes. On peut trouver certaines de ses applications dans $[2, 3, 5, 6, 11, 12, 13, 14, 15, 16, 17, 19]$.

L'inégalité de la conjecture abc implique que, pour tout triplet d'entiers positifs (a, b, c), vérifiant $a + b = c$, $(a, b) = 1$ et $c \geq 3$, on a :

$$(1.1) \qquad \frac{\log c}{\log r(abc)} \leq 1 + \varepsilon + \frac{\log c(\varepsilon)}{\log r(abc)} \leq 1 + \varepsilon + \frac{\log c(\varepsilon)}{\log 6}.$$

Ainsi, le rapport

$$(1.2) \qquad \alpha = \alpha(a, b, c) = \frac{\log c}{\log r(abc)}$$

est borné. Elle implique de même que le rapport

(1.3) $$\rho = \rho(a,b,c) = \frac{\log abc}{\log r(abc)}$$

est borné car $\alpha(a,b,c) \leq \rho(a,b,c) \leq 3\alpha(a,b,c)$. Expérimentalement, les plus grandes valeurs connues pour les rapports (1.2) et (1.3) proviennent respectivement d'un exemple de E. Reyssat (1987), et d'un exemple de l'auteur (1992) :

(1.4) $$\begin{cases} 2 + 3^{10}.109 = 23^5 & \alpha = 1,62991, \\ 13.19^6 + 2^{30}.5 = 3^{13}.11^2.31, & \rho = 4,41901. \end{cases}$$

Le meilleur résultat prouvé provient d'un travail de C.L. Stewart et K. Yu, qui ont établit le théorème suivant :

THÉORÈME 1.5 (Stewart-Yu). — *Soit* $\varepsilon > 0$. *Il existe une constante* $c_1(\varepsilon) > 0$ *telle que, pour tout triplet* (a,b,c) *d'entiers positifs vérifiant* $a + b = c$ *et* $(a,b) = 1$, *on ait :*

$$\log c \leq c_1(\varepsilon)\,(r(abc))^{2/3+\varepsilon}.$$

L'inégalité de ce théorème est cependant assez loin de celle de la conjecture *abc*.

Dans la partie 2 de cet article, on étudie les deux rapports (1.2) et (1.3), et dans la partie 3, on décrit des algorithmes qui permettent de "tester" numériquement la conjecture *abc*. Ces résultats numériques nous permettront de proposer des conjectures plus faibles que la conjecture *abc*, mais dans lesquelles les constantes sont explicites.

2. — Etude des rapports (1.2) et (1.3)

Les inégalités (1.1) impliquent que les rapports (1.2) et (1.3) vérifient :

(2.1) $$\limsup_{c\to\infty} \alpha(a,b,c) = 1, \qquad \limsup_{c\to\infty} \rho(a,b,c) = 3,$$

où les limites sont prises pour des triplets d'entiers positifs vérifiant $a+b = c$ et $(a,b) = 1$. En effet, le théorème de Mahler assure que

$$\lim_{c\to\infty} r(abc) = +\infty.$$

En particulier, les limites (2.1) montrent qu'il ne peut y avoir qu'un nombre fini de triplets d'entiers positifs (a,b,c), vérifiant $a + b = c$, $(a,b) = 1$ et admettant des rapports (1.2) ou (1.3) respectivement supérieurs à $1,62991$ ou à $4,41901$, et donc meilleurs que les exemples (1.4). D'autre part, indépendamment de la conjecture *abc*, nous démontrons la proposition suivante.

PROPOSITION 2.2. — *Pour tout réel $\alpha_0 > 0$, il n'y a qu'un nombre fini de triplets (a, b, c) d'entiers positifs vérifiant $a + b = c$, $(a, b) = 1$ et tels que $\alpha(a, b, c) = \alpha_0$.*

Preuve : soient (x_i, y_i, z_i), $i = 1, 2$, deux triplets d'entiers positifs différents vérifiant pour $i = 1, 2$, $x_i + y_i = z_i$, $(x_i, y_i) = 1$ et tels que $\alpha(x_i, y_i, z_i) = \alpha_0$. On pose $r_i = r(x_i y_i z_i)$, $i = 1, 2$.

Si $r_1 = r_2$, alors $z_1 = z_2$ et l'équation $x + y = z_1$ n'admet qu'un nombre fini de solutions telles que $r(xyz_1) = r_1$, où r_1 est fixé.

Supposons maintenant que $r_1 \neq r_2$. Soit (x_3, y_3, z_3) un triplet d'entiers positifs vérifiant $x_3 + y_3 = z_3$, $(x_3, y_3) = 1$ et tel que $\alpha(x_3, y_3, z_3) = \alpha_0$. Soit $r_3 = r(x_3 y_3 z_3)$. Si $r_3 = r_i$, $i = 1, 2$, alors $z_3 = z_i$. Supposons donc que $r_3 \neq r_i$, $i = 1, 2$. Ainsi

$$\frac{\log z_1}{\log r_1} = \frac{\log z_2}{\log r_2} = \frac{\log z_3}{\log r_3} = \alpha_0,$$

et α_0 n'est pas rationnel puisque $r(z_i) \neq r_i$, $i = 1, 2, 3$, en excluant le triplet trivial $(1, 1, 2)$. Alors, le théorème de Lang (voir [20], p. 51) implique que $\log r_1$, $\log r_2$ et $\log r_3$ sont $\mathbb{Q}$-linéairement dépendants. Soient a_i, $i = 1, 2, 3$ trois entiers positifs non nuls tels que $(a_1, a_2, a_3) = 1$ et $a_3 \log r_3 = a_1 \log r_1 + a_2 \log r_2$. Ainsi $r_3^{a_3} = r_1^{a_1} r_2^{a_2}$ et r_i, $i = 1, 2, 3$, est sans facteurs carrés. Si $(r_1, r_2) = 1$ alors $a_3 = a_1 = a_2$ et donc $r_3 = r_1 r_2$. Si $(r_1, r_2) \neq 1$, alors $a_3 = a_1 + a_2$, $a_3 = a_i$, avec $i = 1$ ou $i = 2$, et donc $r_3 = r_2$ ou $r_3 = r_1$, ce qui est impossible. Ainsi, pour α_0 donné, il n'existe au plus que trois familles, finies, de triplets (a, b, c) d'entiers tels que $\alpha(a, b, c) = \alpha_0$. Chaque famille est représentée par le même radical $r(abc)$. ∎

Remarques :

1) Si la conjecture de Lang, appelée conjecture des quatre exponentielles (voir [20], p. 59) est vraie, on peut montrer que pour chaque réel α_0 il ne peut y avoir au plus qu'une seule famille, finie, de triplets (a, b, c) d'entiers, ayant le même radical et tels que $\alpha(a, b, c) = \alpha_0$.

2) Une famille de triplets intéressante est la suivante. Pour un entier $n \geq 2$, on pose :

$$a_n = 2(2^{n-1} - 1), \quad b_n = (2^n - 1)^2, \quad x_n = 2^n - 1, \quad y_n = 2^{n+1}(2^{n-1} - 1).$$

Alors $a_n + b_n = x_n + y_n$, $r(a_n b_n) = r(x_n y_n)$ et donc $\alpha(a_n, b_n, a_n + b_n) = \alpha(x_n, y_n, x_n + y_n)$.

3) Voici d'autres exemples de triplets qui ne sont pas de la forme ci-dessus, ayant le même rapport (1.2), avec :

– Deux représentants : $1 + 2.3 = 2^2 + 3 = 7$;
– Trois représentants : $3 + 2^2.5 = 5 + 2.3^2 = 2^3 + 3.5 = 23$;
– Quatre représentants : $2 + 3.5^2 = 2.5^2 + 3^3 = 2^5 + 3^2.5 = 7.11$;
– Cinq représentants : $2^5.5.7 + 3^2 = 2^3.3^3.5 + 7^2 = 2^2.5^2 + 3.7^3 = 2^{10} + 3.5.7 = 1129$.

Nous allons maintenant montrer que les triplets (a, b, c), d'entiers positifs vérifiant $(a, b) = 1$, $a+b = c$ et pour lesquels $\alpha(a, b, c) > 1$ et $\rho(a, b, c) > 3$ sont nombreux. Soit (X, Y, Z) un triplet d'entiers vérifiant $(X, Y) = 1$, $X \neq Y$, $X + Y = Z$ et Z pair. Ecrivons $X = Ax^3$, $Y = By^3$ et $Z = Cz^3$ où A, B et C sont les parties sans facteur cube de X, Y et Z. Définissons les suites (x_n), (y_n) et (z_n) par $x_0 = x, y_0 = y, z_0 = z$ et par les relations de récurrence :

$$(2.3) \qquad \begin{cases} dx_{n+1} = x_n(Ax_n^3 + 2By_n^3), \\ dy_{n+1} = -y_n(2Ax_n^3 + By_n^3), \\ dz_{n+1} = z_n(Ax_n^3 - By_n^3), \end{cases}$$

où d désigne le pgcd des trois termes de droite.

PROPOSITION 2.4. — *Soient (x_n), (y_n) et (z_n) les suites définies par* (2.3). *Alors pour tout* $n \geq 0$, $Ax_n^3 + By_n^3 = Cz_n^3$, $(Ax_n^3, By_n^3) = 1$ *et* $2^n|z_n$. *En particulier les trois suites sont infinies.*

Preuve : par définition des suites (x_n), (y_n) et (z_n), on a $Ax_0^3 + By_0^3 = Cz_0^3$ avec $(Ax_0^3, By_0^3) = 1$. Supposons ceci vrai pour les termes x_n, y_n, z_n et que $2^n|z_n$. Alors, par les relations (2.3), on a $Ax_{n+1}^3 + By_{n+1}^3 = Cz_{n+1}^3$. Puisque $(x_{n+1}, y_{n+1}, z_{n+1}) = 1$ et C est sans facteur cube, alors $(x_{n+1}, y_{n+1}) = 1$. On a de même $(x_{n+1}, z_{n+1}) = (y_{n+1}, z_{n+1}) = 1$. Ainsi $(Ax_{n+1}^3, By_{n+1}^3) = (A, y_{n+1}^3)(x_{n+1}^3, B)$. Si p est premier et si $p|(Ax_{n+1}^3, By_{n+1}^3)$, alors $p|Cz_{n+1}^3$ et $p|(A, y_{n+1}^3)$ ou $p|(x_{n+1}^3, B)$, ce qui n'est pas possible.

Soit d comme dans (2.3). On a

$$\left(x_n(Ax_n^3 + 2By_n^3), y_n(2Ax_n^3 + By_n^3)\right) = (Ax_n^3 + 2By_n^3, 2Ax_n^3 + By_n^3)$$
$$= (3Ax_n^3, 3By_n^3).$$

Alors $d = 1$ ou $d = 3$. Comme Ax_n^3 et By_n^3 sont impairs, alors $2|(Ax_n^3 - By_n^3)$ et donc $2^{n+1}|z_{n+1}$. La conclusion de la proposition découle du fait que les termes de la suite (z_n) ne s'annulent jamais. ∎

On pose maintenant pour tout $n \geq 0$:

$$(2.5) \qquad \begin{cases} a_n = \min\left(|Ax_n^3|, |By_n^3|, |Cz_n^3|\right), \\ c_n = \max\left(|Ax_n^3|, |By_n^3|, |Cz_n^3|\right), \\ b_n = c_n - a_n, \end{cases}$$

PROPOSITION 2.6. — *Soient* (x_n), (y_n) *et* (z_n) *les suites définies par* (2.3). *Alors il existe un entier* n_0 *tel que pour tout* $n \geq n_0$, *on ait :*

$$\alpha(a_n, b_n, c_n) > 1, \qquad \rho(a_n, b_n, c_n) > 3.$$

Preuve : pour tout $n \geq 0$, les triplets (a_n, b_n, c_n) vérifient $a_n + b_n = c_n$ et $(a_n, b_n) = 1$. Alors $r(a_n b_n c_n) = r\left(|ABC| |x_n y_n z_n|^3\right)$ et

$$r\left(|ABC| |x_n y_n z_n|^3\right) = r\left(2|ABC x_n y_n| \frac{|z_n|}{2^n}\right) \leq |ABC x_n y_n z_n| / 2^{n-1}.$$

Soit $n_0 \geq 1$ un entier tel que $|ABC| / 2^{n_0 - 1} < |ABC|^{1/3}$. Alors pour tout $n \geq n_0$, on a :

$$r(a_n b_n c_n) < |ABC|^{1/3} |x_n y_n z_n| = (a_n b_n c_n)^{1/3} < c_n,$$

et donc $\alpha(a_n, b_n, c_n) > 1$ et $\rho(a_n, b_n, c_n) > 3$. ∎

La constante $c(\varepsilon)$ de la conjecture *abc* agit comme un terme correcteur. Nous pouvons utiliser les suites définies par les relations (2.3) pour montrer que la condition $\varepsilon > 0$ est obligatoire dans la conjecture *abc*.

PROPOSITION 2.7. — *Soit* $\varepsilon \longmapsto c(\varepsilon)$ *une application vérifiant la conjecture* abc. *Alors*
$$\lim_{\varepsilon \to 0} c(\varepsilon) = +\infty.$$

Preuve : On reprend les suites (x_n), (y_n) et (z_n) définies par les relations (2.3). Soit n_0 un entier tel que $|ABC| / 2^{n_0 - 1} < |ABC|^{1/3}$. Appliquons la conjecture *abc* aux triplets (a_n, b_n, c_n) définis par (2.5) avec $n \geq n_0$:

$$c_n \leq c(\varepsilon) \left(r(a_n b_n c_n)\right)^{1+\varepsilon} \leq c(\varepsilon) \left(\frac{c_n}{2^{n-n_0}}\right)^{1+\varepsilon}.$$

Alors $c(\varepsilon) \geq c_n^{-\varepsilon} 2^{(n-n_0)(1+\varepsilon)}$ et donc

$$\liminf_{\varepsilon \to 0} c(\varepsilon) \geq 2^{n-n_0},$$

ce qui implique que $\lim_{\varepsilon \to 0} c(\varepsilon) = +\infty$. ∎

3. — Recherche de bons triplets pour la conjecture *abc*.

Nous dirons qu'un triplet (a, b, c) d'entiers positifs vérifiant $a + b = c$, $(a, b) = 1$ est *bon* pour la conjecture *abc* si l'un de ses rapports $\alpha(a, b, c)$ ou

$\rho(a, b, c)$ est assez grand par rapport aux valeurs conjecturales 1 et 3. Ceci ne peut se produire que si c ou le produit abc est assez grand par rapport au radical $r(abc)$. Nous résumons dans cette partie une étude amplement détaillée dans [9, 10].

Soit $n \geq 2$ un entier et soient $A > 0$, $B \neq 0$ et $C > 0$ des entiers premiers entre eux deux à deux. Notre recherche de bons triplets pour la conjecture abc va être basée sur la résolution de l'équation diophantienne :

$$(3.1) \qquad\qquad Ax^n - By^n = Cz,$$

avec $(y, C) = 1$. Cette équation a des solutions si et seulement si la congruence

$$(3.2) \qquad\qquad At^n \equiv B \pmod{C},$$

a une solution t avec $0 < |t| < C/2$. Les entiers a, b et c dont on calculera les rapports $\alpha(a, b, c)$ et $\rho(a, b, c)$ seront pris parmi $|Ax^n|$, $|By^n|$ et $|Cz|$, en les réduisant par leur pgcd et pour les quels $|z|$ est assez petit. La recherche de solutions pour l'équation (3.1) avec $|z| = 1$ est basée sur les théorèmes suivants.

THÉORÈME 3.3. — *Soient $B < 0$ et n pair. Si $(x, y, 1)$ est une solution de l'équation (3.1) avec $(y, C) = 1$, alors il existe une solution t de (3.2) avec $0 < |t| < C/2$ et une réduite u/y de la fraction continue de t/C telles que $x = ty - Cu$.*

Supposons maintenant $B > 0$ ou n impair. On peut alors définir les quantités :

$$\delta = (B/A)^{1/n},$$

$$y_0 = \left(\frac{2^n}{An\delta^{n-1}}\right)^{1/(n-2)} \quad \text{si } n \geq 3.$$

Pour un réel x, $[x]$ désigne sa partie entière.

THÉORÈME 3.4. — *Soient B et n deux entiers tels que $B > 0$ où n est impair. Soit $(x, y, \pm 1)$ une solution de l'équation (3.1) et soit*

$$\varepsilon = \begin{cases} 1 & \text{si } x\delta > 0, \\ \cos \cdot \left(2\pi \frac{[(n-1)/2]}{n}\right) & \text{si } x\delta < 0. \end{cases}$$

Si $n = 2$ et $AB \geq 4$ ou si $n \geq 3$ et $y \geq y_0$, alors il existe une solution t de (3.2) avec $0 < |t| < C/2$ et une réduite u/y de la fraction continue de $(t - \varepsilon\delta)/C$ telles que $x = ty - Cu$.

Les théorèmes (3.3) et (3.4) nous permettent d'écrire deux algorithmes pour chercher de bons triplets pour la conjecture abc. Pour cela, il faut choisir des entiers A, B et C tels que $r(ABC)$ soit petit par rapport à $|ABC|$. On peut choisir par exemple $|AB|$ petit et C de la forme p^e où p est premier et où e est un entier assez grand. Les deux algorithmes ont le même principe :

• Déterminer les solutions t de la congruence (3.2) en utilisant une des différentes méthodes connues (voir [4, 22]).

• Déterminer les réduites u/y de $(t - \Delta)/C$ où $\Delta = 0$ si $B < 0$ et n pair et $\Delta = \varepsilon\delta$ sinon.

• Poser $a_0 = A\,(ty - Cu)^n$, $b_0 = -By^n$ et $c_0 = a_0 + b_0$ et les diviser par leur $pgcd$.

• Poser $a = \min\,(|a_0|, |b_0|, |c_0|)$, $c = \max\,(|a_0|, |b_0|, |c_0|)$ et $b = c - a$.

• Calculer $\alpha(a, b, c)$ et $\rho(a, b, c)$ par (1.2) et (1.3). Si l'un de ces rapports est assez grand, enregistrer le triplet (a, b, c).

Nous avons appliqué ces deux méthodes dans les cas suivants :
1) $n = 2$, $1 \leq A \leq |B| \leq 300$, $C = p^e$, où $p \leq 31$ est premier et e est entier vérifiant $p^e \leq 2^{60}$ pour $B < 0$ et $p^e \leq 2^{40}$ pour $B > 0$.
2) $n = 3, 5$, $1 \leq A \leq B \leq 200$, $C = p^e$, où $p \leq 31$ est premier et e est entier vérifiant $p^e \leq 2^{40}$.

Cette recherche a donné 86 triplets (a, b, c) vérifiant $\alpha(a, b, c) \geq 1,4$ sur 115 connus et 103 autres triplets vérifiant $\rho(a, b, c) \geq 3,8$ sur 140 connus. La table (3.6) liste les triplets connus avec $\alpha \geq 1.49$ et la table (3.7) ceux avec $\rho \geq 4,00$, suivant les ordres décroissants de α et ρ.

On peut utiliser une autre méthode pour chercher de bons triplets pour la conjecture abc. Cette méthode repose sur la résolution de l'équation diophantienne linéaire :

$$(3.5) \qquad\qquad Ax - By = Cz,$$

où $A > 0$, $B \neq 0$, $C > 0$ sont premiers entre eux et où $|xyz|$ est petit par rapport à $|ABC|$. Cela peut se faire en déterminant les réduites u/y de $(at - b)/(ac)$ avec $t \equiv ba^{-1} \pmod{C}$, $0 < |t| < C/2$ et en prenant $x = ty - Cu$. On peut augmenter les chances de déterminer de bons triplets en prenant par exemple $A = p_1^{e_1}$, $B = p_2^{e_2}$ et $C = p_3^{e_3}$ où pour $i = 1, 2, 3$, p_i est un nombre premier et où $e_i \geq 1$ est assez grand.

La recherche de bons triplets pour la conjecture abc, basée sur la détermination des petites solutions de l'équation (3.5) n'est pas complètement achevée. Elle a permis cependant la découverte de certains bons triplets (notés "N. (1994)" dans les tables (3.6) et (3.7)).

Dans les deux tables ci-dessous, les auteurs des différents exemples sont désignés par leurs initiales :

B.-B. : J. Browkin et J. Brzezinski,

G. : X. Gang,

M.-R. : P. Montgomery et H. te Riele,

N. : A. Nitaj,

R. : E. Reyssat,

W. : B.M.M. de Weger.

Table 3.6				
a	b	c	α	Auteur
2	$3^{10}.109$	23^5	$1,62991$	R. (1987)
11^2	$3^2.5^6.7^3$	$2^{21}.23$	$1,62599$	W. (1985)
19.1307	$7.29^2.31^8$	$2^8.3^{22}.5^4$	$1,62349$	B.-B. (1992)
283	$5^{11}.13^2$	$2^8.3^8.17^3$	$1,58076$	B-B, N. (1992)
1	2.3^7	$5^4.7$	$1,56789$	W. (1985)
7^3	3^{10}	$2^{11}.29$	$1,54708$	W. (1985)
$7^2.41^2.311^3$	$11^{16}.13^2.79$	$2.3^3.5^{23}.953$	$1,54443$	N. (1994)
5^3	$2^9.3^{17}.13^2$	$11^5.17.31^3.137$	$1,53671$	R.-M. (1994)
13.19^6	$2^{30}.5$	$3^{13}.11^2.31$	$1,52700$	N. (1992)
$3^{18}.23.2269$	$17^3.29.31^8$	$2^{10}.5^2.7^{15}$	$1,52216$	N. (1994)
239	$5^8.17^3$	$2^{10}.37^4$	$1,50284$	B.-B., N. (1992)
$5^2.7937$	7^{13}	$2^{18}.3^7.13^2$	$1,49762$	W. (1985)
$2^2.11$	$3^2.13^{10}.17.151.4423$	$5^9.139^6$	$1,49243$	N. (1992)
73	$2^{13}.7^7.941^2$	$3^{16}.103^3.127$	$1,49159$	N. (1992)

Table 3.7				
a	b	c	ρ	Auteur
13.19^6	$2^{30}.5$	$3^{13}.11^2.31$	$4,41901$	N. (1992)
$2^5.11^2.19^9$	$5^{15}.37^2.47$	$3^7.7^{11}.743$	$4,26801$	N. (1994)
$2^{19}.13.103$	7^{11}	$3^{11}.5^3.11^2$	$4,24789$	W. (1985)
$2^{35}.7^2.17^2.19$	$3^{27}.107^2$	$5^{15}.37^2.2311$	$4,23069$	N. (1994)
$3^{18}.23.2269$	$17^3.29.31^8$	$2^{10}.5^2.7^{15}$	$4,22979$	N. (1994)
$17^4.79^3.211$	$2^{29}.23.29^2$	5^{19}	$4,22960$	N. (1994)
$5^{14}.19$	$2^5.3.7^{13}$	$11^7.37^2.353$	$4,22532$	N. (1994)
3^{21}	$7^2.11^6.199$	$2.13^8.17$	$4,20094$	N. (1992)
$5^{18}.6359$	$3^2.47^6.73^3$	$2^7.19^{10}.79$	$4,14883$	N. (1994)
$11^3.31^5.101.479$	107^8	$2^{31}.3^4.5^6.7$	$4,13000$	N. (1994)
$7.11^6.43$	$3^{11}.5^4$	$2^{17}.17^3$	$4,10757$	G. (1986)
$3^6.5^{11}.41$	$2^9.7^9.283$	$13^{10}.53$	$4,09700$	N. (1994)
$2^{16}.41.71$	$3^{15}.7^2$	19^7	$4,09655$	N. (1992)
$3^{12}.5^6$	$7^9.31^2$	$2^9.11^5.571$	$4,09647$	N. (1992)
$7^8.19$	$2^{15}.5^2.37^2$	3.17^7	$4,09080$	N. (1992)
$79^5.677$	2^{42}	$3^{12}.7.13^4$	$4,08331$	N. (1994)
$2^{24}.3^5$	$5.19^5.59^2$	$7^{10}.167$	$4,07114$	N. (1992)
$2.11^2.107^4.359.20947$	$5^3.7^3.29^{11}$	$3^{35}.23^3$	$4,07038$	N. (1994)
$19.47.71^6$	$3^{21}.193^2$	$2^7.5^{12}.127^2$	$4,06347$	N. (1994)
$3^6.157^3.283$	23^{10}	$2^{30}.5^2.11^2.13$	$4,05990$	B.-B., N. (1992)
$7^8.17^3$	$2^{32}.3^2$	$11.13^5.23^2.31$	$4,05301$	N. (1994)
$2^{13}.3^{13}.11^3$	$13.29.43^6.673$	$5^{20}.17$	$4,04710$	N. (1992)
$5^{13}.13$	$2^{17}.19^3.23$	$3^{17}.283$	$4,04498$	N. (1992)
$3^7.5^{14}.7^2$	$2^{51}.11^2$	$29^5.73.419^2.1039$	$4,03039$	N. (1994)
$3^2.5^7.79$	$2^{29}.13$	$11^7.19^2$	$4,02943$	N. (1992)
$3^6.19^{12}$	$2^5.37^3.59^4.19603$	$7^{17}.11^2.71$	$4,02904$	N. (1994)
$2^{41}.97^2$	$73^8.2347$	$3^2.5^5.11^{10}.43.61$	$4,01847$	N. (1994)
2.5^9	3^{14}	$7^5.11.47$	$4,01342$	N. (1992)
2^{37}	$3^{20}.853$	$5^2.7^6.29.191^2$	$4,01312$	N. (1994)
$7^2.41^2.311^3$	$11^{16}.13^2.79$	$2.3^3.5^{23}.953$	$4,00968$	N. (1994)
$2^2.73.89^5.347.997$	$7^4.23^{11}$	$3^{36}.19$	$4,00764$	N. (1994)
$2^{22}.11.43$	$3^2.17^7$	$7^2.41^5$	$4,00751$	N. (1994)
$2^{10}.19^{10}$	$5^6.13^4.29^5$	$3^{20}.4425749$	$4,00292$	N. (1992)
$2^{11}.11^4.13^2.23$	3^{25}	$73.29^5.137$	$4,00238$	N. (1994)
$5^3.11^4.31^2$	$3^{17}.7^2$	$2^{25}.241$	$4,00087$	N. (1992)

Plusieurs autres méthodes ont été mises au point pour la recherche de bons triplets pour la conjecture abc, avec plus ou moins de succès.

- En 1985, de Weger [21], a donné dans sa thèse une première table, en résolvant l'équation $x + y = z$ avec $r(xyz)$ fixé.

- En 1992, Brzezinski et Browkin [1] ont utilisé une méthode basée sur la détermination des réduites des nombres $d^{1/n}$ pour d et n donnés, en prenant les entiers a, b, c parmi $|x^n|$, $|y^n d|$ et $|x^n - dy^n|$.

- En 1992, Elkies et Kanapka ont testé tous les triplets (a, b, c) avec $c < 2^{32}$ et tentent de mener leur recherche en poussant la borne à 2^{40} (communication privée).

- En 1994, Montgomery et te Riele ont cherché de bons triplets en utilisant une variante de l'algorithme LLL (Communication privée).

Malgré toutes ces recherches, les exemples (1.4) restent les meilleurs pour les rapports (1.2) et (1.3). Il est à noter que les exemples (1.4) correspondent à des triplets (a, b, c) dans lesquels c est relativement petit à l'intérieur du domaine exploré. D'autre part, le théorème de Mason (voir [8]) empêche l'existence de bonnes familles infinies pour la conjecture abc, qui soient paramétrables polynômialement. Ceci nous amène tout naturellement à proposer :

CONJECTURE 3.8. — *Soit (a, b, c) un triplet d'entiers positifs vérifiant $a + b = c$ et $(a, b) = 1$. Alors :*

$$c < r(abc)^{1,63}, \qquad abc < r(abc)^{4,42}.$$

Les tables (3.6) et (3.7) nous renseignent aussi sur l'expression de la constante $c(\varepsilon)$ de la conjecture abc. Pour en donner une estimation, il faut tenir compte de la proposition (2.7), mais aussi du théorème suivant (voir [7]).

THÉORÈME 3.9 (Masser). — *Pour tout $\delta > 0$, il existe un triplet (a, b, c) d'entiers positifs vérifiant $a + b = c$, $(a, b) = 1$ et tel que*

$$abc \geq r^3 exp\{(12 - \delta)(\log r)^{1/2}(\log \log r)^{-1}\},$$

où $r = r(abc)$.

Ce théorème implique en particulier que l'inégalité de la conjecture abc ne peut pas être de la forme $c \leq k_1 r(abc) \log r(abc)^{k_2}$ où k_1 et k_2 sont des constantes. Par conséquent, la constante $c(\varepsilon)$ ne peut pas être de la forme $c(\varepsilon) = (1/\varepsilon)^{k_3}$ où k_3 est une constante.

Manuscrit reçu le 20 décembre 1994

BIBLIOGRAPHIE

[1] J. BROWKIN and J. BRZEZINSKI. — *Some remarks on the abc-conjecture*, Math. Comp. **62** (1994), 931–939.

[2] N.D. ELKIES. — *ABC implies Mordell*, Intern. Math. Res. Notices **7** (1991), 99–109.

[3] G. FREY. — *Links between elliptic curves and solutions of $A - B = C$*, Number Theory, Ulm 1987, Lect. Notes in Math. **1380**.

[4] K. HARDY, J.B. MUSKAT and K.S. WILLIAMS. — *A deterministic algorithm for solving $n = fu^2 + gv^2$ in coprime integers u and v*, Math. Comp. **55** (1990), 327–343.

[5] S. LANG. — *Old and new conjectured diophantine inequalities*, Bull. Amer. Math. Soc. **23** (1990), 37–75.

[6] M. LANGEVIN. — *Cas d'égalité pour le théorème de Mason et applications de la conjecture (abc)*, C. R. Acad. Sci. Paris, t. 317 (1993), 441–444.

[7] D.W. MASSER. — *Note on a conjecture of Szpiro*, Astérisques **183** (1990), 19–23.

[8] R.C. MASON. — *Diophantine equations over Function Fields*, LMS Lecture Notes **96**, Cambridge University Press 1984.

[9] A. NITAJ. — *An algorithm for finding good abc-examples*, C. R. Acad. Sci. Paris, t. 317 (1993), 811–815.

[10] A. NITAJ. — *Algorithms for finding good examples for the abc and the Szpiro conjectures*, Experimental Math. **2** (1993), 223–230.

[11] A. NITAJ. — *La conjecture abc*, L'Ens. Math. à paraître.

[12] J. OESTERLE. — *Nouvelles approches du théorème de Fermat*, Séminaire Bourbaki, 1987–88, n° 694. (Astérisque, vols. 161-162, 165–186) Paris, Soc. Math. Fr. 1988.

[13] M. OVERHOLT. — *The diophantine equation $n! + 1 = m^2$*, Bull. London Math. Soc. **25** (1993), 104.

[14] D. RICHARD. — *Equivalence of some questions in mathematical logic with some conjectures in Number Theory*, Number Theory and Applications, R. A. Mollin ed. , NATO-ASI Ser. (C–265), Kluwer, (1989), 529–545.

[15] W.M. SCHMIDT. — *Diophantine Approximations and Diophantine Equations*, Lect. Notes in Math. **1467**, Springer-Verlag 1991.

[16] J.H. SILVERMAN *Wieferich's criterion and the abc–conjecture*, J. Number Theory **30** (1988), 226–237.

[17] C.L. STEWART and R. TIJDEMAN, *On the Oesterlé-Masser conjecture*, Monatsh. Math. **102** (1986), 251–257.

[18] C.L. STEWART and K. YU. — *On the abc–conjecture*, Math Ann. **291** (1991), 225-230.

[19] P. VOJTA. — *Diophantine Approximation and Value Distribution Theory*, Lect. Notes in Math. **1239**, Springer-Verlag 1987.

[20] M. WALDSCHMIDT *Nombres Transcendants*, Lect. Notes in Math. **402**, Springer–Verlag 1974.

[21] B.M.M. de WEGER. — *Algorithms for diophantine equations*, CWI Tract, Centr. Math. Comput. Sci. , Amsterdam 1989.

[22] K.S. WILLIAMS and K. HARDY. — *A refinement of H.C. Williams'qth root algorithm*, Math. Comp. **61** (1993), 475–483.

Abderrahmane Nitaj
Université de Caen
Département de Mathématiques
14032 Caen Cedex
(email : nitaj@math.unicaen.fr)

Heights of points on subvarieties of $\mathbb{G}_m^n$

Wolfgang M. Schmidt[*]

1. — Introduction

Beginning with the deep 1992 work of Shouwu Zhang [14], much work has recently been done on the height of algebraic points on Zariski-closed subsets V of $\mathbb{G}_m^n$, where $\mathbb{G}_m = \mathbb{G}_m(\mathbb{C})$ is the multiplicative group of the complex numbers. Let $h(x)$ be the absolute logarithmic height (whose definition will be recalled in Section 5) of an algebraic number x. Let $\mathcal{A}_m = \mathbb{G}_m(\overline{\mathbb{Q}})$ be the multiplicative group of algebraic numbers. A very special case of Zhang's results is that when $(x, y) \in \mathcal{A}_m^2$ is on the line

$$(1.1) \qquad x + y = 1,$$

but is not of the type (ω, ω^2) where ω is a sixth root of 1, then

$$(1.2) \qquad h(x) + h(y) \geqq c_1 > 0$$

with an absolute constant c_1. Then Zagier [13] gave an elegant proof of this inequality and determined the best possible constant : $c_1 = \frac{1}{2} \log(\frac{1}{2}(1 + \sqrt{5}))$. Next, Schlickewei and Wirsing [11] used a method inspired by Zagier's to show that the algebraic solutions to (1.1) are in some sense well spaced with respect to the height function, and this in turn [9], [10] led to bounds $c_2(r)$ depending only on r for the number of solutions of (1.1) with (x, y) in a subgroup Γ of $\mathbb{G}_m^2$ of rank r.

From now on, a *point* x will be understood to lie in $\mathbb{G}_m^n$, but whenever heights are involved, it will be understood to lie in $\mathcal{A}_m^n$. Given a point $x = (x_1, \ldots, x_n)$, set

$$h_s(x) = \sum_{i=1}^{n} h(x_i).$$

[*] Supported in part by DMS–9401426.

Given two points x, y set

$$\delta(x, y) = h_s(xy^{-1}),$$

with the product and inverse taken in $\mathcal{A}_m^n$. Then
(i) $\delta(x, y) \geq 0$, with equality precisely when $xy^{-1} \in U^n$, where U is the group of roots of 1,
(ii) $\delta(x, y) = \delta(y, x)$,
(iii) $\delta(x, z) \leq \delta(x, y) + \delta(y, z)$.
Therefore δ is a semidistance on $\mathcal{A}_m^n$, and it induces a distance on $\mathcal{A}_m^n/U^n$.
Given $\varepsilon > 0$, call x, y *neighbors of distance* $< \varepsilon$ if $\delta(x, y) < \varepsilon$.

Carrying the ideas of Zagier, and of Schlickewei and Wirsing further, the author [12] showed that for certain curves $V \subset \mathbb{G}_m^2$, there are constants $q > 0$, $\varepsilon > 0$ depending on V in a rather simple way, such that a point $x \in V$ has at most q neighbors of distance $< \varepsilon$ in V. Very soon afterwards, Bombieri and Zannier [1] went even further. They gave an elementary proof of theorems of Zhang [14], [15] on heights of points on varieties $V \subset \mathbb{G}_m^n$, with bounds depending only on n and the degree of V. They showed in particular that for a certain subvariety V^a of V, there are numbers $q > 0$, $\varepsilon > 0$, such that a point $x \in \mathbb{G}_m^n$ has at most q neighbors of distance $< \varepsilon$ in $V \backslash V^a$. Here $q = q(n, d)$ and $\varepsilon = \varepsilon(n, d)$ when V is defined by polynomial equations of degree $\leq d$. Zagier [13] and the author [12] had used the involution $x \mapsto x^{-1}$, but Bombieri and Zannier [1], inspired by work of Dobrowolski [4], used the isogeny $x \mapsto x^p$ in their arguments, where p is a sufficiently large prime number.

Whereas the constants $q(n, d)$ and $\varepsilon(n, d)$ are explicitly computable by the method of [1], Bombieri and Zannier did not carry out such a computation, and indeed some estimates obtained by their method would involve n exponentiations. Our goal here will be to obtain explicit and representable (although still quite large or quite small) values for all the constants, and to make some other improvements. In order to achieve this, we will avoid as much as possible automorphisms of $\mathbb{G}_m^n$, resultants and induction on n.

In contrast to [1], we will work both with the partial degrees, as well as the total degree of a polynomial. A polynomial in n variables of total degree $\leq d$ has up to

$$(1.3) \qquad\qquad N(d) = \binom{n+d}{d} \leq (2n)^d$$

nonzero coefficients.

An *algebraic subgroup* of $\mathbb{G}_m^n$ is a subgroup which is an algebraic variety, i.e., it is Zariski-closed. A *torus* is an algebraic subgroup which is (absolutely) irreducible as a Zariski-closed set. By *coset* we will understand a coset gH where H is an algebraic subgroup. A *torus coset* will be a coset gH where H is a torus. A *torsion coset* will be a coset uH where u is a torsion point of $\mathbb{G}_m^n$, i.e., $u \in U^n$. Now let $V \subset \mathbb{G}_m^n$ be an algebraic variety. A coset gH contained in V will be called a *maximal* coset in V if there is no coset gH' with $gH \subsetneq gH' \subset V$. Let $V^a(H)$ be the union of all cosets gH contained in V, and $V_1^a(H)$ the union of all maximal cosets gH contained in V. Similarly define $V^u(H)$ and $V_1^u(H)$ as the union of all torsion cosets uH contained in V and the union of the maximal torsion cosets uH contained in V. Finally set

$$V^a = \bigcup_{\substack{H \\ \dim H > 0}} V^a(H), \qquad V^u = \bigcup_H V^u(H),$$

the unions being over algebraic subgroups H of $\mathbb{G}_m^n$.

We will suppose throughout that

$$\boxed{n \geq 2.}$$

THEOREM 1. — *For given d, there are algebraic subgroups $H_1, \ldots, H_m$ depending only on n and d, each a union of at most $(2d)^n$ tori, and with*

$$(1.4) \qquad\qquad m < (11d)^{n^2},$$

having the following property. When V is defined by polynomial equations of total degree $\leq d$, then

$$(1.5) \qquad V^a = \bigcup_{\substack{i=1 \\ \dim H_i > 0}}^m V^a(H_i) = \bigcup_{\substack{i=1 \\ \dim, H_i > 0}}^m V_1^a(H_i),$$

$$(1.6) \qquad V^u = \bigcup_{i=1}^m V^u(H_i) = \bigcup_{i=1}^m V_1^u(H_i).$$

THEOREM 2. — *Let $V \subset \mathbb{G}_m^n$ be defined by polynomial equations of total degree $\leq d$. Then*
  (i) *Each $V^a(H)$ is Zariski-closed and defined by polynomial equations of degree $\leq d$.*
  (ii) *When V is defined by equations with rational coefficients, then each $V^a(H)$ is defined by equations of degree $\leq d$ with rational coefficients.*

(iii) *Each* $V_1^u(H)$ *is the union of fewer than*

(1.7) $\exp(4N(d)!)$

   torsion cosets uH.

(iv) *When* V *is defined by equations with rational coefficients, then each*
   $V_1^u(H)$ *is the union of fewer than*

(1.8) $\exp(3N(d)^{3/2} \log N(d))$

   torsion cosets.

   In Laurent [6] and again in Bombieri–Zannier [1] it had been shown
that $V_1^u(H)$ is the union of at most $c(d, n, [K : \mathbb{Q}], M)$ cosets uH when the
defining polynomials of V had coefficients in a number field K and had
heights $\leqq M$.

   In Section 5 we will define a certain height $H'(f)$ for nonzero polynomials f with algebraic coefficients.

   THEOREM 3. — *Suppose* $V \subset \mathbb{G}_m^n$ *is defined by polynomial equations*
$f_\ell = 0$ $(\ell = 1, \ldots, t)$ *with total degree* $\leqq d$, *with coefficients in a normal*
number field K *of degree* k, *and with heights* $H'(f_\ell) \leqq M$ $(\ell = 1, \ldots, t)$.
Then every $x \in V \backslash V^u$ *has*

(1.9) $h_s(x) > 1/(dkN(dk) \cdot e^{3M_1})$

where

$$M_1 = k! M^{k^2}.$$

   In particular, $M_1 = M$ when $K = \mathbb{Q}$.

   We will also point out how Sieve Methods can be used to replace (1.9)
by the alternative estimates

(1.9') $h_s(x) > c_0(N(dk)) M_1^{-3N(dk)-10}$

where $c_0(N(dk))$ depends only on $N(dk)$ (hence only on n, d, k), and

(1.9'') $h_s(x) > \exp(-3000 N(dk) \log^2 N(dk)) M_1^{-3000 N(dk)}$.

   THEOREM 4. — *Suppose* $V \subset \mathbb{G}_m^n$ *is defined by polynomial equations*
$f_\ell = 0$ $(\ell = 1, \ldots, t)$ *of total degree* $\leqq d$. *Then every* $x \in V \backslash V^a$ *with at most*

(1.10) $q := \exp((4n)^{2dN(d)})$

exceptions, has

(1.11) $h_s(x) > \varepsilon := q^{-1}$.

COROLLARY. — *A point $y \in \mathbf{G}_m^n$ has at most q neighbors of distance $< \varepsilon$ in $V \backslash V^a$.*

For when $x \in V \backslash V^a$, then $z = xy^{-1} \in y^{-1}V \backslash y^{-1}V^a = y^{-1}V \backslash (y^{-1}V)^a$, and by Theorem 4 applied to $y^{-1}V$, all but at most q points z in $y^{-1}V \backslash (y^{-1}V)^a$ have $h_s(z) > \varepsilon$.

THEOREM 5. — *Let V be as above. Let $\Gamma \subset A_m^n$ be a group containing at most r multiplicatively independent elements. Then given $C \geq 1$, there are at most*

$$(1.12) \qquad\qquad q(qC)^r$$

points $x \in \Gamma \cap V \backslash V^a$ with $h_s(x) \leqq C$.

Of particular interest is the case $d = 1$. The variety V then is of the type

$$(1.13) \qquad\qquad \sum_{i=0}^{n} a_{\ell i} x_i = 0 \qquad (\ell = 1, \dots, t)$$

where we have set $x_0 = 1$, and where $x = (x_1, \dots, x_n) \in \mathbf{G}_m^n$. When $\mathcal{P}$ is a partition of $\{0, 1, \dots, n\}$, let $H_{\mathcal{P}}$ consist of the points x having $x_i = x_j$ for any i, j lying in the same subset λ of $\mathcal{P}$. Then $H_{\mathcal{P}}$ is a torus, and it turns out (see Section 3) that in Theorem 1 we may replace $\{H_1, \dots, H_m\}$ by the set of groups $H_{\mathcal{P}}$ where $\mathcal{P}$ is any partition. There are $\leqq (n+1)^{n+1}$ such groups. A coset $gH_{\mathcal{P}}$ consists of points x having $x_i = g_i x_\lambda$ (for $1 \leqq i \leqq n$, $i \in \lambda$ and λ a set of $\mathcal{P}$) with arbitrary x_λ (λ in $\mathcal{P}$), except that $x_\lambda = 1$ for the set λ containing 0. Further $V^a(H_{\mathcal{P}})$ consists of points x having

$$\sum_{i \in \lambda} a_{\ell i} x_i = 0 \qquad (\ell = 1, \dots, t)$$

for every set λ of the partition. Since $\dim H_{\mathcal{P}} > 0$ except when $\mathcal{P}$ is the trivial partition with the only set $\lambda = \{0, \dots, n\}$, we see that $V \backslash V^a$ consists of $x \in V$ such that there is no subset $\lambda \neq \emptyset$ of $\{1, \dots, n\}$ such that

$$\sum_{i \in \lambda} a_{\ell i} x_i = 0 \qquad (\ell = 1, \dots, t).$$

Solutions $x \in V \backslash V^a$ are usually called *non-degenerate*. When (1.13) is a single non-homogeneous equation, Schlickewei [8] proved that there are at most $2^{4(n+1)!}$ non-degenerate solutions $x \in U^n$.

In the present situation $N(d) = n + 1$, and since $n \geq 2$, Theorems 4, 5 hold with

$$q = q_0(n) := \exp((4n)^{3n}).$$

2. — Lattices and Algebraic Groups

In our usual pedantic manner we begin by recalling basic facts about algebraic groups $H \subset \mathbb{G}_m^n$. When $X = (X_1, \ldots, X_n)$ is a variable vector and $i = (i_1, \ldots, i_n) \in \mathbb{Z}^n$, set

$$X^i = X_1^{i_1} \cdots X_n^{i_n}.$$

Similarly define x^i when x is a point. Given i, the map $x \mapsto x^i$ is a character χ on $\mathbb{G}_m^n$. Let $e_1 = (1, 0, \ldots, 0), \ldots, e_n = (0, \ldots, 0, 1)$ be the standard basis vectors. When $\tau \in GL(n, \mathbb{Z})$, say $\tau e_i = (a_{i1}, \ldots, a_{in})$, let φ_τ be the map $\mathbb{G}_m^n \to \mathbb{G}_m^n$ defined by

$$\varphi_\tau(x) = (x^{\tau(e_1)}, \ldots, x^{\tau(e_n)}) = (x_1^{a_{11}} \cdots x_n^{a_{1n}}, \ldots, x_1^{a_{n1}} \cdots x_n^{a_{nn}}).$$

When also $\sigma \in GL(n, \mathbb{Z})$, then $\varphi_{\sigma\tau}(x) = \varphi_\tau \varphi_\sigma(x)$. It is now clear that φ_τ is an automorphism of $\mathbb{G}_m^n$. When $V \subset \mathbb{G}_m^n$ is Zariski-closed, then so is $\varphi_\tau V$, and when V is irreducible (as a Zariski-closed set), then so is $\varphi_\tau V$. For algebraic subgroups H_1, H_2 of $\mathbb{G}_m^n$, write

$$H_1 \sim H_2$$

if $H_2 = \varphi_\tau H_1$ for some $\tau \in GL(n, \mathbb{Z})$.

When $a = (a_1, \ldots, a_n) \in \mathbb{Z}^n$, then $\tau(\mathbf{a}) = a_1\tau(e_1) + \cdots + a_n\tau(e_n)$, so that

$$x^{\tau(a)} = (x^{\tau(e_1)})^{a_1} \cdots (x^{\tau(e_n)})^{a_n} = (\varphi_\tau((x))^a.$$

Applying this with τ^{-1}, $\tau(a)$ in place of τ, a, we obtain

$$(2.1) \qquad\qquad x^a = x^{\tau^{-1}\tau(a)} = (\varphi_{\tau^{-1}}(x))^{\tau(a)}.$$

By *lattice* we will understand a subgroup of $\mathbb{Z}^n$. A *full* lattice is a subgroup of rank n. When A is a lattice, let $S(A) \subset \mathbb{R}^n$ be the space spanned by A, and set $\tilde{A} = S(A) \cap \mathbb{Z}^n$. Then $\tilde{A}$ is a lattice containing A, and the index $\rho(A)$ of A in $\tilde{A}$ is finite. The lattice is called *primitive* if $A = \tilde{A}$, i.e., if $\rho(A) = 1$.

When A is a lattice, let H_A consist of $x \in \mathbb{G}_m^n$ with $x^a = 1$ for every $a \in A$. Then H_A is an algebraic variety defined over $\mathbb{Q}$. In fact it is an algebraic subgroup of $\mathbb{G}_m^n$ since $(xy)^a = x^a \cdot y^a$. In view of (2.1) we have

$$(2.2) \qquad\qquad \varphi_{\tau^{-1}} H_A = H_{\tau(A)}.$$

LEMMA 1. — *Suppose A is a lattice of rank r. Then*

(2.3)
$$H_A = F \times M^{n-r},$$

where the product $\times$ of groups is direct, F is finite of order $\rho(A)$, and $M^{n-r} \sim H^{n-r}$, where H^{n-r} consists of x with $x_1 = \cdots = x_r = 1$. Therefore H_A has $\rho(A)$ irreducible components, and in particular H_A is a torus precisely when A is primitive. When B is a lattice of rank r containing A, we have

(2.4)
$$H_B = F' \times M^{n-r},$$

where F' is of order $\rho(B)$, and is contained in F.

Remark. F is not uniquely determined by A, but M^{n-r} is, consisting of elements m of H_A which are "divisible" by any integer $q \neq 0$, i.e., $m = m_1^q$ where $(m_1 \in H_A$.

Proof : pick $\tau \in GL(n, \mathbb{Z})$ such that $\tau(\tilde{A})$ consists of the vectors $(b', 0)$ where $b' \in \mathbb{Z}^r$ and 0 is the origin in $\mathbb{Z}^{n-r}$. Then $\tau(A) = (A', 0)$, where A' is a full lattice in $\mathbb{Z}^r$ with $\rho(A') = \rho(A)$. Thus $H_{\tau(A)}$ consists of $(x', x'') \in \mathbb{G}_m^r \times \mathbb{G}_m^{n-r} = \mathbb{G}_m^n$ with $x' \in H_{A'}$ and x'' arbitrary in H^{n-r}. Since A' is full in $\mathbb{Z}^r$, the group $H_{A'}$ is easily seen to be finite of order $\rho(A') = \rho(A)$. We have $H_{\tau(A)} = H_{A'} \times H^{n-r}$. Therefore by (2.2), H_A is of the type (2.3).

The other assertions of the lemma follow easily.

LEMMA 2. — *The map $A \mapsto H_A$ sets up a bijection between lattices and algebraic subgroups of $\mathbb{G}_m^n$.*

Proof : we begin by showing that the map is injective. Suppose $H_A = H_B$. Then when $x \in H_A$, we have $x^a = 1$ for every $a \in A$, further $x^b = 1$ for $b \in B$, therefore $x^c = 1$ for every $c \in A + B$. Therefore $H_A \subseteq H_{A+B}$. Since the reverse inclusion is obvious, $H_A = H_{A+B}$. It now follows from Lemma 1 that rank $A = \text{rank}(A + B)$ and $\rho(A) = \rho(A + B)$. Therefore $A = A + B$, so that $B \subseteq A$. By symmetry $A = B$.

It remains to be shown that every algebraic subgroup H equals H_A for some lattice A. Since H is Zariski-closed, this will follow from Lemma 4 in the next section.

LEMMA 3. — *Let A, B be lattices. Then*

(2.5)
$$H_A H_B = H_{A \cap B}.$$

Proof : we first will show that a product HH' of algebraic groups H, H' in $\mathbb{G}_m^n$ is an algebraic group. Since HH' is obviously a group, we need only show that it is Zariski-closed. Since each group H or H' is a finite union of torus cosets, it will suffice to show that a product of two such cosets is closed, hence suffice to show that a product of two tori is closed. In fact we will show that the product of two tori H, H' is a torus. By Lemma 1, $H = \tau(H^b)$ with $\tau \in GL(n, \mathbb{Z})$ and some b, so that H consists of points $(y^{b_1}, \ldots, y^{b_n})$ with fixed $b_1, \ldots, b_n \in \mathbb{Z}^b$, and where y runs through $\mathbb{G}_m^b$. Similarly, H' consists of points $(z^{c_1}, \ldots, z^{c_n})$ where z runs through $\mathbb{G}_m^c$, say. Therefore HH' consists of points $(y^{b_1}z^{c_1}, \ldots, y^{b_n}z^{c_n})$. It will suffice to show that when $g > 0$ and $d_1, \ldots, d_n \in \mathbb{Z}^g$ are arbitrary, the set T of points

$$(w^{d_1}, \ldots, w^{d_n}) \in \mathbb{G}_m^n$$

as w runs through $\mathbb{G}_m^g$ is a torus.

Let A consist of vectors $a = (a_1, \ldots, a_n) \in \mathbb{Z}^n$ having $a_1d_1 + \cdots + a_nd_n = 0$. Then A is a primitive lattice, and $T \subset H_A$. We claim that $T = H_A$. Without loss of generality, $d_1, \ldots, d_r$ are independent, and we have relations

$$ad_i = a_{i1}d_1 + \cdots + a_{ir}d_r \quad (r < i \leqq n)$$

with integer coefficients and $a > 0$. As w runs through $\mathbb{G}_m^g$, then $z = (w^{d_1}, \ldots, w^{d_r})$ runs through $\mathbb{G}_m^r$. Now T is "divisible" by every $a > 0$, so that it consists of elements x^a with $x \in T$, i.e., of elements

(2.6) $$\left(z_1^a, \ldots, z_r^a, z_1^{a_{r+1,1}} \cdots z_r^{a_{r+1,r}}, \ldots, z_1^{a_{n1}} \cdots z_r^{a_{nr}} \right)$$

with $z \in \mathbb{G}_m^r$. The vectors

$$a_i = (a_{i1}, \ldots, a_{ir}, 0, \ldots, -a, \ldots, 0) \quad (r < i \leqq n)$$

generate a lattice $B \subset A$ with $\tilde{B} = A$. Since T contains the points (2.6), we have $H_B \subset T$. Therefore $H_A^\rho \subset H_B \subset T$, where $\rho = \rho(B)$ is the index of B in A. Since H_A, being a torus, is divisible by ρ, indeed $H_A \subset T$.

The proof of the lemma is now completed as follows. We have already shown that $H_A H_B$ is an algebraic group. It is the smallest group containing both H_A and H_B. The 1–1 correspondence between lattices and algebraic groups reverses inclusion relations. Since $A \cap B$ is the largest lattice contained in both A and B, the assertion (2.5) is established.

3. — Proof of Theorem 1

Let $V \subset \mathbb{G}_m^n$ be Zariski-closed and defined by polynomials f_ℓ ($\ell = 1, \ldots, t$). We may write

$$f_\ell = \sum_{i \in I} a_{\ell i} X^i \qquad (\ell = 1, \ldots, t),$$

where I is a finite set of n-tuples i. Let $D(I)$ be the difference set, consisting of points $i - j$ with $i, j \in I$.

An algebraic subgroup $H \subset V$ is called a *maximal* subgroup contained in V if there is no subgroup H' with $H \subsetneq H' \subset V$.

LEMMA 4. — *Let H be a maximal algebraic group contained in V. Then $H = H_A$ where A is generated by vectors in $D(I)$.*

Proof : we follow Laurent [6] and Bombieri–Zannier [1]. Given i, the restriction of the map $x \mapsto x^i$ to H is a character χ_i on H. When χ is any character on H, let I_χ be the set

$$I_\chi = \{i \in I \quad \text{with} \quad \chi_i = \chi\}.$$

On H we have the relations

$$\sum_\chi \left(\sum_{i \in I_\chi} a_{\ell i} \right) \chi = 0 \quad (\ell = 1, \ldots, t).$$

By Artin's Theorem on linear relations on characters, these must be trivial relations, so that

(3.1) $$\sum_{i \in I_\chi} a_{\ell i} = 0 \qquad (\ell = 1, \ldots, t; \text{ any } \chi).$$

For $x \in H$,

(3.2) $$x^i = x^j \quad \text{if} \quad i, j \in I_\chi \quad \text{for some} \quad \chi.$$

Therefore $H \subset H_A$, where A is the lattice generated by the vectors $i - j$ with $i, j \in I_\chi$ for some χ. When $x \in H_A$, then (3.2) implies that x^i has a common value for every $i \in I_\chi$, say the value x_χ, and then

$$f_\ell(x) = \sum_\chi \left(\sum_{i \in I_\chi} a_{\ell i} \right) x_\chi = 0 \quad (\ell = 1, \ldots, t)$$

by (3.1). Therefore $H_A \subset V$. Since H was maximal, $H = H_A$.

When V is defined by equations of total degree $\leqq d$, we may take

$$I = I(d),$$

where $I(d)$ consists of n-tuples of nonnegative integers whose sum is $\leqq d$. Therefore $I \subseteq d\mathcal{I}$ where $\mathcal{I}$ consists of the points $\xi \in \mathbb{R}^n$ with L_1-norm $\leqq 1$. Then $D(I) \subset dD(\mathcal{I}) = 2d\mathcal{I}$. A lattice A with generators in $2d\mathcal{I}$ does not necessarily have a basis in $2d\mathcal{I}$. However, when rank $A = r$, say, then there is a basis $a_1, \ldots, a_r$ with $a_j \in \max(1, j/2) \cdot 2d\mathcal{I}$ $(j = 1, \ldots, r)$; see, e.g., Lemma 8 in [2, Ch. V]. So when $n > 1$ and A is of any rank, it is generated by n vectors in $nd\mathcal{I}$. The number of integer points in $nd\mathcal{I}$ with nonnegative components is $N(nd)$. Considering the 2^n quadrants, the number of integer points in $nd\mathcal{I}$ is

$$\leqq 2^n N(nd) = 2^n \binom{nd + n}{n} \leqq 2^n (2dn)^n / n! < (4ed)^n < (11d)^n,$$

since $n! > (n/e)^n$. The number of choices for A then is $< (11d)^{n^2}$. In view of Lemma 4, every subgroup H_A obtained in this way is in a set $\{H_1, \ldots, H_m\}$ where m satisfies (1.4). Further when A of rank r has r linearly independent vectors in $2d\mathcal{I}$, these vectors will have Euclidean norm $\leqq 2d$, hence will span a parallelepiped of r-dimensional volume $\leqq (2d)^r \leqq (2d)^n$. Therefore

(3.3) $$\rho(A) \leqq (2d)^n,$$

and H_A is the union of at most $(2d)^n$ tori.

LEMMA 5. — *Every algebraic subgroup $H \subset V$ is contained in a maximal algebraic subgroup contained in V.*

Proof : the argument for Lemma 4 shows that $H \subset H_A \subset V$ where A is generated by vectors in $D(I)$. Since $D(I)$ is finite, there will be a group H_A which is maximal in V.

The proof of Theorem 1 follows immediately : every coset $xH \subset V$ has $H \subset x^{-1}V$, hence by Lemma 5 is contained in a coset $xH' \subset V$ where H' is maximal in $x^{-1}V$. Thus H' is among the groups $H_1, \ldots, H_m$ constructed above.

When V is defined by linear equations, I consists of the $n + 1$ vectors $e_0 := 0, e_1, \ldots, e_n$, hence $D(I)$ of the differences $e_i - e_j$. Given a lattice A generated by some vectors from $D(I)$, write $i \sim j$ if $e_i - e_j \in A$. Then H_A consists of $x = (x_1, \ldots, x_n)$ having $x_i = x_j$ when $i \sim j$, where we have set $x_0 = 1$. The equivalence relation $\sim$ defines a partition $\mathcal{P}$ of $\{0, 1, \ldots, n\}$, and $H_A = H_{\mathcal{P}}$ consists of $x \in \mathbb{G}_m^n$ having $x_i = x_j$ when i, j lie in the same set of the partition. This verifies a claim made in the Introduction.

4. — Proof of Theorem 2

(i) By Lemma 1, $H = F \times \tau(H^{n-r})$ with $\tau \in GL(n, \mathbb{Z})$, so that H consists of elements $v\tau(y)$ where $v \in F$, $y \in H^{n-r}$. When V is defined by polynomial equations $f_\ell = 0$ $(\ell = 1, \ldots, t)$, then $xH \subset V$ precisely if

$$(4.1) \qquad f_\ell(xv\tau(y)) = 0 \qquad (\ell = 1, \ldots, t)$$

for every $v \in F$ and every $y = (1, \ldots, 1, y_{r+1}, \ldots, y_n) \in H^{n-r}$. Each expression on the left hand side is a polynomial in $y_{r+1}, \ldots, y_n$. The coefficients of these polynomials are certain polynomials g_s^* in xv $(s = 1, \ldots, t')$, and we need that

$$g_s^*(xv) = 0 \qquad (s = 1, \ldots, t')$$

for every $v \in F$. But this means that x has to satisfy certain polynomial equations

$$(4.2) \qquad g_m(x) = 0 \qquad (m = 1, \ldots, t'').$$

$V^a(H)$ consists precisely of the points x satisfying the equations (4.2), which are clearly of total degree $\leq d$.

(ii) Now the polynomials f_ℓ defining V and the polynomials g_s^* introduced in (i) have rational coefficients. Consider a particular such $g_s^* = g^*$, say, and the system of equations

$$(4.3) \qquad g^*(xv) = 0 \qquad (v \in F).$$

When $v \in F$ is fixed, (4.3) implies the system

$$g^*(v^k x) = 0 \qquad (k \in \mathbb{Z}).$$

The components of v are powers of some root of 1, say powers of ζ where ζ is a primitive m-th root of 1. The automorphisms of $\mathbb{Q}(\zeta)$ are the $\mathbb{Q}$-linear maps σ_k with $\sigma_k(\zeta) = \zeta^k$ where $(k, m) = 1$. Applying these automorphisms to the coefficients of $f(X) := g^*(vX)$ we obtain the polynomials $f^{(k)}(X) = g^*(v^k X)$. Therefore (4.3) implies that x not only satisfies $f(x) = 0$, but also the $\phi(m)$ conjugate equations $f^{(k)}(x) = 0$. These equations are equivalent to a system of $\phi(m)$ equations with rational coefficients. Clearly all these equations have the same degree as g^*.

Before commencing with the proof of parts (iii), (iv), we need to insert the following.

LEMMA 6. — *Consider a system of linear equations*

(4.4)
$$\sum_{i=1}^{n} a_{\ell i} u_i = 0 \qquad (\ell = 1, \ldots, t).$$

A solution $u = (u_1, \ldots, u_n)$ *will be called non-degenerate if there is no subset* I *of* $\{1, \ldots, n\}$ *with* $0 < |I| < n$ *such that*

$$\sum_{i \in I} a_{\ell i} u_i = 0 \qquad (\ell = 1, \ldots, t).$$

Then up to proportionality, (4.4) has at most

$$\psi(n) := 2^{4n!}$$

non-degenerate solutions whose components are roots of 1.

In the special case when the coefficients $a_{\ell i}$ *in (4.4) are rational,* $\psi(n)$ *may be replaced by*
$$\kappa(n) := \exp(2n^{3/2} \log n).$$

Proof : let us first deal with the case of general coefficients. The case $t = 1$ is due to Schlickewei [8] (note that Schlickewei considers an inhomogeneous equation in n variables, which corresponds to a homogeneous equation in $n + 1$ variables). The general case is easily deduced from the case $t = 1$. For let $u_1, \ldots, u_N$ be distinct non-degenerate solutions. For each I with $0 < |I| < n$ and each m, $1 \leqq m \leqq N$, there is an ℓ in $1 \leqq \ell \leqq t$ such that $u_m = (u_{m1}, \ldots, u_{mn})$ has

$$\sum_{i \in I} a_{\ell i} u_{mi} \neq 0.$$

We therefore may pick $c_1, \ldots, c_t$ such that

$$\sum_{\ell=1}^{t} c_\ell \sum_{i \in I} a_{\ell i} u_{mi} \neq 0,$$

for every I with $0 < |I| < n$ and every m, $1 \leqq m \leqq N$. Then $u_1, \ldots, u_N$ are non-degenerate solutions of the single equation

$$\sum_{i=1}^{n} b_i u_i = 0$$

with $b_i = \sum_{\ell=1}^{t} c_\ell a_{\ell i}$. By Schlickewei's result $N \leq \psi(n)$.

Now consider the case of rational coefficients $a_{\ell i}$. By the argument given above it will suffice to deal with the situation when $t = 1$. By Conway and Jones [3], up to a factor of proportionality, the u_i will be m-th roots of 1 where m is of the type

$$m = p_1 p_2 \cdots p_\ell$$

with distinct primes $p_1, \ldots, p_\ell$ having

(4.5) $$\sum_{i=1}^{\ell} (p_i - 2) \leq n - 2.$$

Now when $p_1 < \cdots < p_\ell$, then $p_i \geq 2i - 1$, so that

(4.6) $$\sum_{i=1}^{\ell} (p_i - 2) \geq \sum_{i=1}^{\ell} (2i - 3) = \ell^2 - 2\ell,$$

and (4.5) yields $\ell < \sqrt{n} + 1 \leq 2\sqrt{n}$. Since each $p_i \leq n$, we have $m < n^{2\sqrt{n}} = \exp(2n^{1/2} \log n)$ [by the prime number theorem, $m < \exp((1+\varepsilon)(n \log n)^{1/2})$ when $n > n_0(\varepsilon)$, but we are interested in a simple and explicit estimate]. Given m, our equation has up to proportionality at most m^{n-1} solutions $(u_1, \ldots, u_n)$. Therefore the number of non-degenerate solutions is

$$< \exp(2n^{3/2} \log n) = \kappa(n).$$

Let $\mathcal{P}$ be a partition of $\{1, \ldots, n\}$. Consider the system of equations

(4.4 $\mathcal{P}$) $$\sum_{i \in \lambda} a_{\ell i} u_i = 0 \qquad (\ell = 1, \ldots, t; \ \lambda \in \mathcal{P}),$$

where $\lambda \in \mathcal{P}$ signifies that λ is a set of the partition $\mathcal{P}$. Let $\mathcal{S}(\mathcal{P})$ be the set of solutions $u \in U^n$ of (4.4) which satisfy (4.4 $\mathcal{P}$), but not (4.4 $\mathcal{Q}$) for any refinement $\mathcal{Q}$ of $\mathcal{P}$. Every solution of (4.4) lies in some (not necessarily unique) set $\mathcal{S}(\mathcal{P})$.

Consider the equations (4.4 $\mathcal{P}$) for fixed $\lambda \in \mathcal{P}$ and $\ell = 1, \ldots, t$. By Lemma 6, there are up to proportionality at most $\psi(|\lambda|)$ solutions in roots of unity u_i ($i \in \lambda$) for which the sums over a set $\lambda' \subset \lambda$ with $0 < |\lambda'| < |\lambda|$ do not all vanish. Denote these solutions by $\{u_i^{(p_\lambda)}\}_{i \in \lambda}$, where $1 \leq p_\lambda \leq q_\lambda$, with $0 \leq q_\lambda \leq \psi(|\lambda|)$. Thus, since a factor u_λ of proportionality is allowed, the solutions are $\{u_\lambda u_i^{(p_\lambda)}\}_{i \in \lambda}$.

This can be done for every $\lambda \in \mathcal{P}$. The solutions in $\mathcal{S}(\mathcal{P})$ therefore are of the type

(4.7) $u_i = u_\lambda u_i^{(p_\lambda)}$ when $i \in \lambda$ and $\lambda \in \mathcal{P}$,

where u_λ $(\lambda \in \mathcal{P})$ are arbitrary and $1 \leq p_\lambda \leq q_\lambda$ $(\lambda \in \mathcal{P})$. Let $\mathcal{S}(\mathcal{P}, \{p_\lambda\})$ for a given system $\{p_\lambda\}_{\lambda \in \mathcal{P}}$ be the solutions of (4.4) of the type (4.7) with the given values of p_λ.

The number of systems $\{p_\lambda\}$ is

$$\prod_{\lambda \in \mathcal{P}} q_\lambda \leq \prod_{\lambda \in \mathcal{P}} \psi(|\lambda|) \leq \psi(n),$$

since $\psi(a)\psi(b) \leq \psi(a+b)$. The number of partitions $\mathcal{P}$ is $\leq n^n$. Every solution $u \in U^n$ of (4.4) lies in one of the sets $\mathcal{S}(\mathcal{P}, \{p_\lambda\})$, and there are

(4.8) $\leq n^n \psi(n) = n^n \cdot 2^{4n!} \leq 2^{5n!} < \exp(4n!)$

such sets.

We now are ready to continue our proof of Theorem 2.

(iii) We have to show that for given H, there are fewer than (1.7) torsion cosets uH which are maximal in V. Now $uH \subset V$ precisely when u satisfies the equations (4.2).

Write

$$g_m(\mathbf{X}) = \sum_{i \in I(d)} b_{mi} \mathbf{X}^i \qquad (m = 1, \ldots, t''),$$

so that $uH \subset V$ precisely when

(4.9) $\sum_{i \in I(d)} b_{mi} u^i = 0 \qquad (m = 1, \ldots, t'').$

Set $u_i = u^i$; then $\{u_i\}_{i \in I(d)}$ satisfies the system of linear equations

(4.10) $\sum_{i \in I(d)} b_{mi} u_i = 0 \qquad (m = 1, \ldots, t'').$

This is exactly like the system (4.4), except that the set $\{1, \ldots, n\}$ has been replaced by $I(d)$, and the subscripts i by i. This time, by (4.8), the solutions fall into strictly less than

(4.11) $\exp(4|I(d)|!)$

sets $S(\mathcal{P}, \{p_\lambda\})$. Let $\mathcal{P}$ and $\{p_\lambda\}_{\lambda \in \mathcal{P}}$ be given; solutions in this set are

$$u_i = u_\lambda u_i^{(p_\lambda)} \quad \text{when} \quad i \in \lambda \quad \text{and} \quad \lambda \in \mathcal{P},$$

where the u_λ ($\lambda \in \mathcal{P}$) are arbitrary.

We have to go back from (4.10) to (4.9). We have to find $u \in U^n$ such that

$$u^i = u_\lambda u_i^{(p_\lambda)} \quad \text{when} \quad i \in \lambda \quad \text{and} \quad \lambda \in \mathcal{P},$$

where the u_λ are arbitrary. Possibly there is no solution for a given $\mathcal{P}$ and $\{p_\lambda\}_{\lambda \in \mathcal{P}}$. But if there is a solution u, the general solution will be uh where h^i for $i \in \lambda$ depends only on λ. In other words, $h^i = h^j$ if $i \overset{\mathcal{P}}{\sim} j$, where $\overset{\mathcal{P}}{\sim}$ is the equivalence relation induced by $\mathcal{P}$. Thus $h \in H_B$, where B is the lattice generated by the vectors $i - j$ with $i \overset{\mathcal{P}}{\sim} j$. Therefore if for given $\mathcal{P}$, $\{p_\lambda\}$ there is a solution u at all, the general solution will be

$$uu'$$

with $u' \in H_B \cap U^n$. But in fact for any $h \in H_B$, the point

$$uh$$

will be a zero of the polynomials g_m, so that $uhH \subset V$. But then $uH_BH \subset V$, where H_BH is an algebraic group by Lemma 3. Since we demanded that $uH \subset V$ be maximal, $H_B \subset H$. Therefore all the cosets $uu'H$ with $u' \in H_B \cap U^n$ will be the same.

Therefore for given $\mathcal{P}$, $\{p_\lambda\}$, there is at most one maximal torsion coset uH. *The number of maximal torsion cosets* $uH \subset V^u$ *therefore is bounded by* (4.11). Since $|I(d)| = N(d)$, (1.7) follows.

(iv) By part (ii), $uH \subset V$ precisely if u satisfies certain equations $g_m(u) = 0$ where each g_m has rational coefficients and is of total degree $\leq d$. We proceed as for (iii), but with κ in place of ψ, so that (4.8) is replaced by

(4.12) $\qquad \leq n^n \kappa(n) = \exp(n \log n + 2n^{3/2} \log n) \leq \exp(3n^{3/2} \log n),$

and (4.11) is replaced by

(4.13) $\qquad\qquad \exp(3|I(d)|^{3/2} \log |I(d)|).$

5. — Heights

When K is a number field, let M_K be its set of places, and for $v \in M_K$ let $|\cdot|_v$ be the absolute value belonging to v, normalized such that it extends the ordinary or a p-adic absolute value of $\mathbb{Q}$. Further set $\|x\|_v = |x|_v^{k_v/k}$ where k is the degree of K and k_v the local degree associated with v (caution : our $\|\cdot\|$ is denoted by $|\cdot|$ in [1]). The product formula

$$\prod_{v \in M_K} \|\alpha\|_v = 1$$

holds for $\alpha \in K^\times$. Given $x \in K^\times$, set

$$H(x) = \prod_{v \in M_K} \max(1, \|x\|_v).$$

This height, the *absolute multiplicative height* of x, is independent of the field K in which x is embedded. The absolute logarithmic height of x is

$$h(x) = \log H(x).$$

By the product formula,

(5.1) $$h(x) = \sum_{v \in M_K} \max(0, \log \|x\|_v) = \frac{1}{2} \sum_{v \in M_K} |\log \|x\|_v|.$$

One has $h(x) \geqq 0$, with equality precisely when $x \in U$, one has $h(1/x) = h(x)$ and $h(xy) \leqq h(x) + h(y)$. For points x, y with nonzero algebraic coordinates set again

$$h_s(\boldsymbol{x}) = \sum_{i=1}^{n} h(x_i),$$

$$\delta(\boldsymbol{x}, \boldsymbol{y}) = h_s(\boldsymbol{x}\boldsymbol{y}^{-1}).$$

Then δ clearly has the properties (i), (ii), (iii) stated in the Introduction.

We will now introduce a slightly different height, and we apologize to the reader for the inconvenience! When $\boldsymbol{x} \in K^n$, set

$$|\boldsymbol{x}|_v = \begin{cases} \max(|x_1|_v, \ldots, |x_n|_v) & \text{when } v \text{ is finite,} \\ |x_1|_v + \cdots + |x_n|_v & \text{when } v \text{ is infinite,} \end{cases}$$

and $\|\boldsymbol{x}\|_v = |\boldsymbol{x}|_v^{k_v/k}$. Further when $\boldsymbol{x} \neq 0$, set

$$H'(\boldsymbol{x}) = \prod_{v \in M_K} \|\boldsymbol{x}\|_v.$$

Then $H'(x)$ is independent of the field in which the coordinates of x are embedded. By the product formula, $H'(\alpha x) = H'(x)$ when $\alpha \neq 0$ is algebraic, so that H' depends only on the point $(x_1 : \cdots : x_n)$ in projective space $\mathbb{P}^{n-1}$. When f is a nonzero polynomial with coefficients in K, define $|f|_v$, $\|f\|_v$ and $H'(f)$ in terms of the coefficient vector.

When we substitute x into f, where f has partial degrees $\leqq d$, we obtain

(5.2)
$$|f(x)|_v \leqq |f|_v \prod_{i=1}^{n} \max(1, |x_i|_v)^d.$$

6. — Auxiliary Lemmas

When $\mathcal{P}$ is a partition of a set, let $\overset{\mathcal{P}}{\sim}$ be the equivalence relation associated with $\mathcal{P}$.

LEMMA 7. — *Suppose* $x_1, \ldots, x_n$ *are nonzero. Suppose they satisfy a system of equations*

(6.1)
$$\sum_{i=1}^{n} a_{\ell i} x_i^j = 0 \quad \text{for } 0 \leqq j \leqq n-1, \ 1 \leqq \ell \leqq t.$$

Then there is a partition $\mathcal{P}$ *of* $\{1, \ldots, n\}$ *such that*

$$x_i = x_j \quad \text{if} \quad i \overset{\mathcal{P}}{\sim} j$$

and

(6.2)
$$\sum_{i \in \lambda} a_{\ell i} = 0 \quad \text{for} \quad \lambda \in \mathcal{P}, \ 1 \leqq \ell \leqq t.$$

Proof : let $\mathcal{P}$ be the partition with $i \sim j$ precisely when $x_i = x_j$. For $\lambda \in \mathcal{P}$, write x_λ for the common value of x_i with $i \in \lambda$. The equations (6.1) become

(6.3)
$$\sum_{\lambda \in \mathcal{P}} \left(\sum_{i \in \lambda} a_{\ell i} \right) x_\lambda^j = 0 \quad \text{for} \quad 0 \leqq j \leqq n-1, \ 1 \leqq \ell \leqq t.$$

Since the x_λ with $\lambda \in \mathcal{P}$ are distinct, the Vandermonde determinant $|x_\lambda^j|$ with $\lambda \in \mathcal{P}$ and $0 \leqq j \leqq |\mathcal{P}| - 1$, where $|\mathcal{P}|$ is the number of sets $\lambda \in \mathcal{P}$, is nonzero. Then (6.3), applied with $0 \leqq j \leqq |\mathcal{P}| - 1$, shows that each "coefficient" is zero, i.e., that (6.2) holds.

LEMMA 8. — *Let $f \neq 0$ be a polynomial with coefficients in $\mathbb{Q}$, with partial degrees $\leq d$, and of height $H'(f)$. Let $p > H'(f)$ be a prime number. Then every algebraic point x with*

$$f(x) = 0, \qquad f(x^p) \neq 0$$

has

(6.4)
$$h_s(x) \geq \frac{1}{pd} \log(p/H'(f)).$$

Proof : this is essentially Lemma 1 of [1]. For completeness we will give a proof. We may suppose that f has coefficients in $\mathbb{Z}$, and these coefficients have no common factor > 1. By Fermat's Theorem,

$$f^p(x) = f(x^p) + pg(x),$$

where $g(X) \in \mathbb{Z}[X]$ has partial degrees $\leq pd$. Since $f(x) = 0$ by hypothesis,

(6.5)
$$f(x^p) = -pg(x).$$

Set $\alpha = f(x^p)$. When $v \mid p$ we have

$$|\alpha|_v \leq |p|_v \left(\prod_{i=1}^{n} \max(1, |x_i|_v) \right)^{pd} |f|_v$$

by (6.5) and since $|f|_v = 1$ in view of our convention on the coefficients of f. For the other v's we have

$$|\alpha|_v \leq \prod_{i=1}^{n} \max(1, |x_i|_v)^{pd} |f|_v$$

by (5.2). If we raise these inequalities to the exponent k_v/k, take the product over $v \in M_K$, observe that $\alpha = f(x^p) \neq 0$ satisfies the product formula, and that $\prod_{v \mid p} \|p\|_v = p^{-1}$, we obtain

$$1 \leq p^{-1} \left(\prod_{i=1}^{n} H(x_i) \right)^{pd} H'(f),$$

therefore

$$h_s(x) = \sum_{i=1}^{n} h(x_i) \geq \frac{1}{pd} \log(p/H'(f)).$$

LLEMMA 9. — *Let f be as above. Let m be an integer all of whose prime factors are*

$$> eH'(f).$$

Let x be a point with

(6.6) $$f(x) = 0, \qquad f(x^m) \neq 0.$$

Then

$$h_s(x) > 1/(md).$$

Proof : clearly $m > 1$. We will use induction on the number of prime factors of m. When m is a prime, the result follows from (6.4). When m is not a prime, write $m = pm'$ where p is a prime. Now if $f(x^{m'}) \neq 0$, our induction gives $h_s(x) > 1/(m'd) > 1/(md)$. But if $f(x^{m'}) = 0$, we apply Lemma 8 to $x^{m'}$, obtaining $h_s(x^{m'}) > 1/(pd)$, so that $h_s(x) = (1/m')h_s(x^{m'}) > 1/(md)$.

We will need arithmetic progressions of natural numbers

(6.7) $$a, a+b, \ldots, a+(n-1)b$$

whose members have no prime factor below some quantity r, and are small in size. When $a = 1$,

$$b = \prod_{p \leq r} p,$$

there are indeed no prime factors $p \leq r$ in (6.7). Moreover, $b = \exp(\vartheta(r))$ where ϑ is Chebyshev's function, which satisfies $\vartheta(r) < 1.017r$ (see, e.g., [7]). Therefore $b \leq e^{1.017r}$. We therefore obtain the almost trivial

LEMMA 10. — *There is an arithmetic progression of length n of natural numbers less than*

(6.8) $$n \cdot e^{1.017r},$$

all of whose prime factors exceed r.

Deeper results may be obtained by Sieve Methods. Professor Halberstam kindly pointed out to me that (6.8) may be replaced by

(6.8') $$c_3(n)r^{3n+10}.$$

On the other hand, following a suggestion by Professor P.D.T.A. Elliott, I could use the explicit sieve estimates in Chapter II of his treatise [5] to obtain the explicit bound

(6.8") $$\exp(90n \log^2 2n)r^{3000n}.$$

No doubt this can be much improved. No proofs of (6.8') or (6.8'') will be given here.

7. — Proof of Theorem 3 when $K = \mathbb{Q}$

Let $V \subset \mathbb{G}_m^n$ be defined by polynomial equations $f_\ell = 0$ $(\ell = 1, \ldots, t)$ with rational coefficients, of partial degrees $\leq d$ and heights $H'(f_\ell) \leq M$ $(\ell = 1, \ldots, t)$.

Set $r = [eM]$. By Lemma 10, but with n replaced by $N(d)$, there is an arithmetic progression

$$a, a + b, \ldots, a + (N(d) - 1)b$$

whose members are positive integers $< e^{1.017r}N(d)$, all of whose prime factors exceed eM. We may suppose that $\gcd(a, b) = 1$. Let $\boldsymbol{x} \in V$, so that $f_\ell(\boldsymbol{x}) = 0$ $(\ell = 1, \ldots, t)$.

Case (i)

$$f_\ell(\boldsymbol{x}^{a+jb}) \neq 0$$

for some ℓ in $1 \leq \ell \leq t$ and some j in $0 \leq j \leq N(d) - 1$. Since the prime factors of $a + jb$ exceed eM, and since $a + jb < e^{1.017r}N(d) < e^{3M}N(d)$, Lemma 9 yields

$$h_s(\boldsymbol{x}) > 1/(d \cdot e^{3M}N(d)),$$

so that (1.9) holds with $k = 1$.

If instead of (6.8) we had used (6.8') or (6.8''), the members of our arithmetic progression would have been bounded respectively by

$$c_3(N(d))r^{3N(d)+10} < c_4(N(d))M^{3N(d)+10}$$

or (since $N(d) \geq n + 1 \geq 3$) by

$$\exp(90N(d) \log^2(2N(d)))r^{3000N(d)}$$
$$< \exp(2900N(d) \log^2 N(d))M^{3000N(d)},$$

giving rise to (1.9') or (1.9'') with $k = 1$.

Case (ii)

$$f_\ell(\boldsymbol{x}^{a+jb}) = 0 \quad \text{for} \quad 1 \leq \ell \leq t, \ 0 \leq j \leq N(d) - 1.$$

So if $f_\ell = \sum_i a_{\ell i} X^i$, say, we have

$$\sum_i a_{\ell i}(\boldsymbol{x}^i)^{a+jb} = 0 \quad (1 \leq \ell \leq t, \ 0 \leq j \leq N(d) - 1),$$

which is the same as

$$\sum_i b_{\ell i} x_i^j = 0 \qquad (1 \leqq \ell \leqq t,\ 0 \leqq j \leqq N(d) - 1)$$

with

$$b_{\ell i} = a_{\ell i} (x^i)^a, \qquad x_i = (x^i)^b.$$

The sums are over $i \in I(d)$. Lemma 7, applied with $N(d)$ in place of n, gives a partition $\mathcal{P}$ of $I(d)$ such that

(7.2) $$(x^i)^b = (x^j)^b \quad \text{when} \quad i \overset{\mathcal{P}}{\sim} j,$$

and

(7.3) $$\sum_{i \in \lambda} a_{\ell i} (x^i)^a = \sum_{i \in \lambda} b_{\ell i} = 0 \quad (\lambda \in \mathcal{P},\ 1 \leqq \ell \leqq t).$$

By (7.2) we have

$$x^i = x_\lambda u_i \quad \text{for} \quad i \in \lambda,\ \lambda \in \mathcal{P},$$

where x_λ depends only on λ, and $u_i^b = 1$, so that u_i is a b-th root of 1. It lies in the field U_b generated by the b-th roots of 1. Now (7.3) may be rewritten as

(7.4) $$\sum_{i \in \lambda} a_{\ell i} u_i^a = 0 \qquad (\lambda \in \mathcal{P},\ 1 \leqq \ell \leqq t).$$

Since $\gcd(a, b) = 1$, there is an automorphism of $U_b/\mathbb{Q}$ which sends b-th roots of 1 to their a-th power. Since the coefficients $a_{\ell i}$ in (7.4) lie in $\mathbb{Q}$, we obtain

$$\sum_{i \in \lambda} a_{\ell i} u_i = 0 \qquad (\lambda \in \mathcal{P},\ 1 \leqq \ell \leqq t)$$

(this step is not needed in the construction of Lemma 10, for there we had $a = 1$). We may infer that

(7.5) $$\sum_{i \in \lambda} a_{\ell i} x^i = 0 \qquad (\lambda \in \mathcal{P},\ 1 \leqq \ell \leqq t).$$

Let B be the lattice in $\mathbb{Z}^n$ generated by the vectors $i - j$ with $i \overset{\mathcal{P}}{\sim} j$, and A the lattice generated by the vectors $bi - bj$ with $i \overset{\mathcal{P}}{\sim} j$. Then $A = bB \subseteq B$ and $H_B \subseteq H_A$. By (7.2), $x \in H_A$. Expressing H_A, H_B as in (2.3), (2.4), we see that

(7.6) $$x \in u H_B$$

for some $u \in F \subset U^m$.

Suppose $y \in uH_B$, so that $y = xh$ with $h \in H_B$. There are numbers $h_\lambda(\lambda \in \mathcal{P})$ such that $h^i = h_\lambda$ when $i \in \lambda$. Then

$$\sum_{i \in \lambda} a_{\ell i} y^i = h_\lambda \sum_{i \in \lambda} a_{\ell i} x^i = 0 \quad (\lambda \in \mathcal{P}, \ 1 \leqq \ell \leqq t)$$

by (7.5). We may conclude that $f_\ell(y) = 0$ $(\ell = 1, \dots, t)$, so that $y \in V$. Therefore $uH_B \subset V$, and by (7.6),

$$x \in V^u.$$

This completes the proof of Theorem 3 when $K = \mathbb{Q}$.

8. — Proof of Theorem 3 in general

Let W be an algebraic manifold defined by equations $f_\ell = 0$ $(\ell = 1, \dots, t)$ with coefficients in a normal number field K of degree k. Let $\sigma_1, \dots, \sigma_k$ be the embeddings of K into $\mathbb{C}$, and let $f_\ell^{(i)}$ be obtained from f_ℓ by applying σ_i to the coefficients of f_ℓ. Further let $W^{(i)}$ $(i = 1, \dots, k)$ be the manifold given by $f_\ell^{(i)}(x) = 0$ $(\ell = 1, \dots, t)$, so that $W^{(1)}, \dots, W^{(k)}$ are the "conjugates" of W.

Lemma 11. — *Let W as above be defined by equations $f_\ell = 0$ of partial degree $\leqq d$ and of height $H'(f_\ell) \leqq M$ $(\ell = 1, \dots, t)$. Then*

(8.1) $$W^* := \bigcup_{i=1}^{k} W^{(i)}$$

is defined by equations $g_m = 0$ $(m = 1, \dots, t')$ of partial degree $\leqq dk$ with coefficients in $\mathbb{Q}$ and with

$$H'(g_m) \leqq k! M^{k^2}.$$

Proof : set

$$F(T, X) = \prod_{i=1}^{k} \left(\sum_{\ell=1}^{t} T_\ell f_\ell^{(i)}(X) \right)$$

where $T_1, \dots, T_\ell$ are indeterminates. Expand F as a polynomial in $T_1, \dots, T_\ell$; the coefficients will be certain polynomials $g_m(X)$, and W^* is the zero set of these polynomials. They are of partial degree $\leqq dk$ and have coefficients in $\mathbb{Q}$. Say, without loss of generality, that g_m is the coefficient of

$$T_1^{j_1} \cdots T_q^{j_q}$$

with positive $j_1, \ldots, j_q$ and with $j_1 + \cdots + j_q = k$. Then g_m is the sum of polynomials

(8.2) $$f_1^{(i(1,1))} \cdots f_1^{(i(1,j_1))} \cdots f_q^{(i(q,1))} \cdots f_q^{(i(q,j_q))}$$

where $i(1,1), \ldots, i(q, j_q)$ is a permutation of $1, \ldots, k$.

We may suppose that each of the given polynomials f_ℓ has some coefficient equal to 1, and then $|f_\ell^{(i)}|_v \geqq 1$ for every ℓ, i and every $v \in M_K$. Then the polynomial (8.2) has absolute value (see the definition in Section 5!)
$$|\cdots|_v \leqq |f_1^{(1)}|_v \cdots |f_1^{(k)}|_v \cdots |f_q^{(1)}|_v \cdots |f_q^{(k)}|_v.$$
Write $\varepsilon_v = 0$ when v is finite and $\varepsilon_v = 1$ when v is infinite. Since g_m is a sum of at most $k!$ polynomials (8.2),
$$|g_m|_v \leqq k!^{\varepsilon_v} |f_1^{(1)}|_v \cdots |f_1^{(k)}|_v \cdots |f_q^{(1)}|_v \cdots |f_q^{(k)}|_v.$$

Raising this to the exponent k_v/k where k_v is the local degree, and taking the product over $v \in M_K$, we obtain

$$H'(g_m) \leqq k! M^{qk} \leqq k! M^{k^2}.$$

Remark : the lemma holds whether K is normal or not.

So now let V be Zariski-closed as in Theorem 3, defined by polynomial equations of degree $\leqq d$, with coefficients in a *normal* number field K, and with heights $\leqq M$. Let $V = V^{(1)}, \ldots, V^{(k)}$ be the conjugates of V. Let x be a point in V. Without loss of generality we may suppose that

(8.3) $$x \in V^{(1)}, \ldots, V^{(q)}, \quad \text{but} \quad x \notin V^{(q+1)}, \ldots, V^{(k)}.$$

Set
$$W = V^{(1)} \cap \cdots \cap V^{(q)}.$$

Then W is again defined by polynomial equations with coefficients in K, of degree $\leqq d$, and of height $\leqq M$. The variety W^* defined by (8.1), in view of Lemma 11, is defined by equations with coefficients in $\mathbb{Q}$, of degree $\leqq dk$, and with heights $\leqq M_1 = k! M^{k^2}$. By the case $K = \mathbb{Q}$ of Theorem 3 we have (1.9) (and also (1.9'), (1.9'')), unless $x \in W^{*u}$, so that $x \in uH$ where $u \in U^n$, H is a torus, and $uH \subseteq W^*$. Since uH is irreducible, $uH \subset W^{(i)}$ for some i. Here $W^{(i)} = V^{(r_1)} \cap \cdots \cap V^{(r_q)}$ with distinct $r_1, \ldots, r_q$. But in view of (8.3), the set $\{r_1, \ldots, r_q\}$ must be $\{1, \ldots, q\}$, so that $W^{(i)} = W$ and $uH \subset W \subset V^{(1)} = V$, therefore $x \in V^u$.

9. — A Variation

Suppose V is defined by equations $f_\ell = 0$ ($\ell = 1, \ldots, t$) where $f_\ell = \sum_{i \in I} a_{\ell i} X^i$. Then it is obvious from the proofs that $N(d)$ may be replaced by $|I| = \mathrm{card} I$ in Theorem 2 (iii), (iv) and in the case $k = 1$ of Theorem 3.

Now suppose that $W \subset \mathbb{G}_{in}^{ns}$ is defined by polynomial equations $f_\ell = 0$ ($\ell = 1, \ldots, t$) where $f_\ell = f_\ell(X_1, \ldots, X_s)$ has rational coefficients and is of total degree $\leqq d$ in each block of variables $X_i = (X_{i1}, \ldots, X_{in})$. Then in Theorem 1 we have to replace n, d by ns, ds, and in particular W^u is the union of

$$(9.1) \qquad\qquad < (11ds)^{n^2 s^2}$$

sets $W_1^u(H_i)$. The monomials occurring in f_ℓ are $X_1^{i_1} \cdots X_s^{i_s}$ with $(i_1, \ldots, i_s) \in I(d) \times \cdots \times I(d) = I$, say, where

$$(9.2) \qquad\qquad |I| = N(d)^s.$$

Therefore, when the coefficients are rational, we may conclude from what we have said above that each $W_1^u(H)$ is the union of fewer than

$$\exp(3|I|^{3/2} \log |I|) \leqq \exp(3|I|^2) = \exp(3N(d)^{2s})$$

torsion cosets. Combining this with (9.1) we see that W^u is the union of

$$(9.3) \qquad\qquad < (11ds)^{n^2 s^2} \exp(3N(d)^{2s})$$

torsion cosets. If s satisfies

$$s \leqq N(d),$$

then $\log(11ds) \leqq \log(11N^2(d)) < 2N(d)$ (since $N(d) \geqq n+1 \geqq 3$), and (9.3) becomes

$$
\begin{aligned}
(9.4) \qquad & < \exp(n^2 N^2(d) \cdot 2N(d) + 3N(d)^{2N(d)}) \\
& < \exp(2N(d)^5 + 3N(d)^{2N(d)}) < \exp(5N(d)^{2N(d)}) \\
& \leqq \exp(5 \cdot (2n)^{2dN(d)}).
\end{aligned}
$$

Assume furthermore that each defining polynomial has $H'(f_\ell) \leqq s! \leqq N(d)!$. Then by Theorem 3 with $k = 1$, with n, d replaced by $ns \leqq nN(d)$, $ds \leqq dN(d)$ respectively, and $N(d)$ replaced by $|I| = N(d)^s \leqq N(d)^{N(d)}$, we may infer that every $(x_1, \ldots, x_s) \in W \backslash W^u$ has

$$
\begin{aligned}
(9.5) \qquad h_s(x_1, \ldots, x_s) & > 1/(d \cdot N(d)N(d)^{N(d)} \cdot e^{3N(d)!}) \\
& > 1/(N(d)^{2N(d)} \exp(3N(d)!)) \\
& > \exp(-4N(d)^{N(d)}) \\
& \geqq \exp(-4(2n)^{dN(d)}).
\end{aligned}
$$

10. — Proof of Theorem 4

Let $\mathcal{L}(d)$ be the set of lattices which have a set of generators which lie in $D(d) := D(I(d))$. We had seen in (3.3) that every $A \in \mathcal{L}(d)$ has

$$\rho(A) \leqq (2d)^n.$$

LEMMA 12. — Let $A \subsetneqq A_1 \subsetneqq \cdots \subsetneqq A_\nu$ be lattices in $\mathcal{L}(d)$. Then

$$\nu \leqq 2n^2 d.$$

Proof : initially consider only chains of lattices which all have the same rank, say r. Then each lattice of the chain is contained in $\tilde{A}$, and the index of A in $\tilde{A}$ is $\rho(A)$. Since each index in our chain is $\geqq 2$, we obtain

$$\nu \leqq (\log \rho(A)) / \log 2 \leqq n(\log 2d) / \log 2 \leqq nd.$$

In general, the rank in the chain is between 0 and n, and can only increase. Only n such increases are possible. Since $\{0\}$ is the only lattice of rank 0, we obtain

$$\nu \leqq n \cdot nd + n \leqq 2n^2 d.$$

For any $A \in \mathcal{L}(d)$, let $\nu = \nu(A)$ be largest such that there is a chain as in the lemma. Then $\nu(\mathbb{Z}^n) = 0$, and $\nu(A) \leqq 2n^2 d$ in general.

PROPOSITION. — Let V be a variety defined by polynomial equations of total degree $\leqq d$. Suppose $V \subset gH_A$ where $A \in \mathcal{L}(d)$. Then every point $x \in V \backslash V^a$, with at most

$$(10.1) \qquad \exp(\nu(A) \cdot 5(2n)^{2dN(d)})$$

exceptions, has

$$(10.2) \qquad h_s(x) > 3\varepsilon = 3\exp(-(4n)^{2dN(d)}).$$

When $A = \{0\}$, so that $gH_A = g\mathbb{G}_m^n = \mathbb{G}_m^n$, the condition $V \subseteq gH_A$ is no restriction. Since $\nu(\{0\}) \leqq 2n^2 d$, we see that for any V defined by equations of total degree at most d, all but at most

$$\exp(2n^2 d \cdot 5(2n)^{2dN(d)}) < \exp((4n)^{2dN(d)}) = q$$

points $x \in V \backslash V^a$ have (10.2). Therefore the Proposition implies Theorem 4.

When $\nu(A) = 0$, then $A = \mathbb{Z}^n$, therefore $H_A = \{(1,\ldots,1)\}$, and $gH_A = \{g\}$ consists of a single element. The Proposition is true in this case. We will prove it in general by induction on $\nu(A)$.

So suppose $V \subset gH_A$ where $\nu(A) > 0$. Now A is generated by certain vectors $i - j$ with $i, j \in I(d)$, so that H_A is defined by certain equations $x^i = x^j$ with $i, j \in I(d)$. We obtain a partition $\mathcal{P}$ of $I(d)$ such that $i \overset{\mathcal{P}}{\sim} j$ precisely if $x^i = x^j$ is a valid relation in H_A, i.e., if the characters $\chi_i : x \mapsto x^i$ and $\chi_j : x \mapsto x^j$ are the same on H_A.

The coset gH_A consists of points x having

$$x^i = x_\lambda g^i \quad \text{when} \quad i \in \lambda, \ \lambda \in \mathcal{P},$$

with arbitrary x_λ ($\lambda \in \mathcal{P}$). Let $I_1 \subset I(d)$ consist of a set of representatives of the sets $\lambda \in \mathcal{P}$, so that I_1 contains exactly one element $i(\lambda) \in \lambda$ for every $\lambda \in \mathcal{P}$. By Artin's Theorem on characters, the monomials x^i with $i \in I_1$ are linearly independent in H_A.

When $V = gH_A$ where $\dim H_A > 0$, then $V^a = V$ and $V \backslash V^a = \emptyset$. When $V = gH_A$ where $\dim H_A = 0$, then $|V| = |H_A| = \rho(A) \leq (2d)^n$. We therefore may suppose that $V \subsetneq gH_A$. Since V is defined by equations of total degree $\leq d$, the monomials x^i with $i \in I_1$ are linearly dependent in V.

Let E be the set of points $x \in V \backslash V^a$ which violate (10.2). With $x \in E$ we associate the vector $v(x)$ with components x^i ($i \in I_1$). Thus $v(x)$ lies in the space $\mathbb{C}^{|I_1|}$ of dimension $|I_1|$. Let $T \subset \mathbb{C}^{|I_1|}$ be the subspace of $\mathbb{C}^{|I_1|}$ spanned by the vectors $v(x)$ with $x \in E$. Since the monomials x^i ($i \in I_1$) are linearly dependent on V, $\dim T < |I_1|$. Say $\dim T = s - 1$ where $1 \leq s \leq |I_1|$. A matrix with rows $v(x_1),\ldots,v(x_s)$ where $x_1,\ldots,x_s$ lie in E is an $(s \times |I_1|)$-matrix of rank $\leq s-1$. This means precisely that certain $s \times s$-determinants $\Delta_1,\ldots,\Delta_Q$ depending on $x_1,\ldots,x_s$ vanish. The equations

$$\Delta_\ell(x_1,\ldots,x_s) = 0 \qquad (\ell = 1,\ldots,Q)$$

define a Zariski-closed set $W \subset (\mathbb{G}_m^n)^s$. The polynomials Δ_ℓ have $H'(\Delta_\ell) = s!$, and they are of total degree $\leq d$ in each block of n variables. Thus W is of the type considered in Section 9. Therefore every point $(x_1,\ldots,x_s) \in W \backslash W^u$ satisfies (9.5). But this is impossible when $x_1,\ldots,x_s$ lie in E, for then

$$(10.3) \qquad h_s(x_1,\ldots,x_s) = \sum_{j=1}^{s} h_s(x_j) \leq 3s \exp(-(4n)^{2dN(d)})$$

$$< \exp(-4(2n)^{dN(d)}).$$

Therefore when $x_1,\ldots,x_s$ lie in E, we have

$$(x_1,\ldots,x_s) \in W^u.$$

Fix $x_1^*, \ldots, x_{s-1}^*$ in E with $v(x_1^*), \ldots, v(x_{s-1}^*)$ linearly independent.
Let us restrict to $x \in E$ such that

(10.4) $$(x_1^*, \ldots, x_{s-1}^*, x) \in \tilde{u}\tilde{H},$$

where $\tilde{u}\tilde{H}$ is a fixed maximal torsion coset in W^u. Here $\tilde{H} = \tilde{H}_B$, where by
Lemma 4, B is a lattice in $\mathbb{Z}^{ns}$ generated by certain vectors $\tilde{b} = (b_1, \ldots, b_s)$
where each $b_j \in D(d) = D(I(d))$; this holds since each Δ_ℓ has degree $\leq d$
in each of $x_1, \ldots, x_s$. The coset $\tilde{u}\tilde{H}$ is defined by certain equations

(10.5) $$x_1^{b_1} \cdots x_s^{b_s} = \delta,$$

for certain vectors $(b_1, \ldots, b_s) \in D(d) \times \cdots \times D(d)$. More precisely, since we
restricted to exponents $i \in I_1$, we have $(b_1, \ldots, b_s) \in D(I_1) \times \cdots \times D(I_1)$.

We claim that $b_s \neq 0$ for some of the relations (10.5). For otherwise, the
relation (10.4) would be independent of x. Thus if it holds for any x at all,
it holds for every x, so that

$$(x_1^*, \ldots, x_{s-1}^*, x) \in \tilde{u}\tilde{H} \subset W$$

for every x, and therefore $v(x_1^*), \ldots, v(x_{s-1}^*), v(x)$ would be linearly dependent for every x, contradicting the fact that even if we restrict to $x \in H_A$,
the vectors $v(x)$ span a space of dimension $|I_1| \geq s$.

Consider a particular relation (10.5) with $b_s \neq 0$. Applied to (10.4) we
obtain

$$x_1^{*b_1} \cdots x_{s-1}^{*b_{s-1}} x_s^{b_s} = \delta,$$

so that, since $x_1^*, \ldots, x_{s-1}^*$ are fixed, $x^{b_s} = \delta'$. Now if there is some x_0 at
all with $x_0 \in gH_A$, $x_0^{b_s} = \delta'$, the general x will be $x = x_0 h$ where $h^a = 1$
for $a \in A$, and $h^{b_s} = 1$, so that $h \in H_C$ where C is the lattice generated
by A and by b_s. Here $b_s \in D(I_1)$, say $b_s = i - j$ with $i \neq j$ in I_1, so that
$i \overset{p}{\sim} j$. Therefore $x^i = x^j$ is not a valid relation in H_A, and $b_s = i - j \notin A$.
Therefore $A \subsetneq C$. Since C is generated by A and b_s, we have $C \in \mathcal{L}(d)$ and
$\nu(C) \leq \nu(A) - 1$.

We have $x \in x_0 H_C$, so that $x \in \tilde{V}$ where

$$\tilde{V} = V \cap x_0 H_C.$$

Here V and H_C, therefore $\tilde{V}$, is defined by polynomial equations of degree
$\leq d$, and $\tilde{V} \subseteq x_0 H_C$. Further $x \notin V^u$ yields $x \notin \tilde{V}^a$. By induction, such x,
with at most

$$\exp((\nu(A) - 1) \cdot 5(2n)^{2dN(d)})$$

exceptions, have (10.2).

In order to deal with general $x \in V\backslash V^u$, we have to multiply by the number of cosets $\tilde{u}\tilde{H}$ making up W^u. This number is bounded by (9.4), so that indeed the number of exceptions to (10.2) is bounded by (10.1).

11. Proof of Theorem 5. We proceed as in [12]. Suppose initially that Γ is finitely generated and of rank r. There are $x_1, \ldots, x_r$ such that the elements of Γ are

$$(11.1) \qquad\qquad x = u x_1^{i_1} \cdots x_r^{i_r}$$

where $u \in U^n$ and $i = (i_1, \ldots i_r)$ runs through $\mathbb{Z}^r$. Here Γ, hence the components of $x_1, \ldots, x_r$ lie in a number field K. For $v \in M_K$ put

$$\alpha_{ijv} = \log \|x_{ij}\|_v \quad (1 \le i \le r,\ 1 \le j \le n),$$

$$\alpha_{jv}(\xi) = \sum_{i=1}^{r} \alpha_{ijv}\xi_i,$$

where $\xi = (\xi_1, \ldots, \xi_r) \in \mathbb{R}^r$. Set

$$\psi(\xi) = \frac{1}{2} \sum_{v \in M_K} \sum_{j=1}^{n} |\alpha_{jv}(\xi)|.$$

When $i \in \mathbb{Z}^r$, formula (5.1) yields

$$(11.2) \qquad \psi(i) = \sum_{j=1}^{n} h(x_{1j}^{i_1} \cdots x_{rj}^{i_r}) = \sum_{j=1}^{n} h(x_j) = h_s(x)$$

where x is given by (11.1). We have
(a) $\psi(\xi) \ge 0$.
(b) $\psi(\alpha\xi) = |\alpha|\psi(\xi)$ for $\alpha \in \mathbb{R}$.
(c) $\psi(\xi + \eta) \le \psi(\xi) + \psi(\eta)$.
(d) $\psi(i) \ge c(k) > 0$
for every $i \in \mathbb{Z}^r \backslash \{0\}$, where $k = \deg K$. The last of these assertions holds since x given by (11.1) does not lie in U^n unless $i = 0$, and by a well known theorem of Dobrowolski [4].

LEMMA 13. — *Suppose ψ satisfies* (a), (b), (c), (d). *Let $\mathcal{I} \subseteq \mathbb{R}^r$ be a set of points such that*

$$\psi(i - j) \ge \varepsilon > 0$$

for $i \neq j$ in $\mathcal{I}$. Then the number of $i \in \mathcal{I}$ with

$$\psi(i) \leqq C$$

is

$$\leqq ((2C/\varepsilon) + 1)^r.$$

Proof : this easy result is Lemma 4 of [12].

Define q, ε by (1.10), (1.11). In view of the Corollary to Theorem 4, an element $x \in V \backslash V^a$ has at most q neighbors in $V \backslash V^a$ of distance $< \varepsilon$. Let $\mathcal{S}$ be the set of points to be estimated in Theorem 5. Consider subsets $\mathcal{S}'$ of $\mathcal{S}$ no two of whose elements are neighbors of distance $< \varepsilon$. Let $\mathcal{S}'$ be a maximal set with this property, i.e., a set such that no set $\mathcal{S}'' \supsetneqq \mathcal{S}'$ has this property. It is easily seen that

$$|\mathcal{S}| \leqq q|\mathcal{S}'|.$$

When $x \neq y$ lie in $\mathcal{S}'$, we have $h_s(xy^{-1}) = \delta(x, y) \geqq \varepsilon$. When x is given by (11.1), and $y = u' x_1^{j_1} \cdots x_r^{j_r}$, then

(11.3) $$\psi(i - j) = h_s(xy^{-1}) \geqq \varepsilon.$$

Let $\mathcal{I}$ be the set of exponents $i \in \mathbb{Z}^r$ such that (11.1) holds for some $x \in \mathcal{S}'$. Then

$$|\mathcal{S}'| = |\mathcal{I}|,$$

and (11.3) holds for $i \neq j$ in $\mathcal{I}$. By Lemma 13,

$$|\mathcal{I}| \leqq ((2C/\varepsilon) + 1)^r < (3C/\varepsilon)^r,$$

therefore

$$|\mathcal{S}| \leqq q|\mathcal{S}'| = q|\mathcal{I}| \leqq q(3C/\varepsilon)^r.$$

Note that by (10.2), Theorem 4 holds with 3ε in place of ε. Therefore with $q = \varepsilon^{-1}$,

$$|\mathcal{S}| \leqq q(qC)^r.$$

In general, Γ is a union of finitely generated groups $\Gamma_1 \subseteq \Gamma_2 \subseteq \cdots$ of rank r. Our estimate holds for each Γ_i $(i = 1, 2, \ldots)$, hence also for Γ.

Manuscrit reçu le 30 janvier 1996

References.

[1] E. BOMBIERI and U. ZANNIER. — *Algebraic Points on Subvarieties of* $\mathbb{G}_m^n$, International Mathematics Research Notices **7** (1995), 333–347.

[2] J. W. S. CASSELS. — *An Introduction to the Geometry of Numbers*, Springer Grundlehren **99** (1959).

[3] J. H. CONWAY and A. J. JONES. — *Trigonometric diophantine equations (On vanishing sums of roots of unity)*, Acta Arith. **30** (1976), 229–240.

[4] E. DOBROWOLSKI. — *On a question of Lehmer and the number of irreducible factors of a polynomial*, Acta Arith. **34** (1979), 391–401.

[5] P.D.T.A. ELLIOTT. — *Probabilistic Number Theory I*, Springer Grundlehren **240** (1980).

[6] M. LAURENT. — *Equations diophantiennes exponentielles*, Invent. Math. **78** (1984), 299–327.

[7] J. B. ROSSER and L. SCHOENFELD. — *Approximate formulas for some functions of prime numbers*, Illinois J. of Math. **6** (1962), 64–94.

[8] H. P. SCHLICKEWEI. — *Equations in roots of unity*, Acta Arith. (to appear).

[9] H. P. SCHLICKEWEI. — *Equations* $ax + by = 1$, Annals of Math. (to appear).

[10] H. P. SCHLICKEWEI and W. M. SCHMIDT. — *Linear equations in variables which lie in a multiplicative group*, preprint.

[11] H. P. SCHLICKEWEI and E. WIRSING. — *Lower bounds for the heights of solutions of linear equations*, Invent. Math. (to appear).

[12] W. M. SCHMIDT. — *Heights of Algebraic Points Lying on Curves or Hypersurfaces*, Proc. A.M.S. (to appear).

[13] D. Zagier. — *Algebraic numbers close to both 0 and 1*, Math. Computation **61** (1993), 485–491.

[14] S. Zhang. — *Positive line bundles on arithmetic surfaces*, Annals of Math. **136** (1992), 569–587.

[15] S. Zhang. — *Positive line bundles on arithmetic varieties*, Journal A.M.S. **8** (1995), 187–221.

Wolfgang M. Schmidt
Department of Mathematics
University of Colorado, Boulder
Boulder, Colorado 80309–0395
U.S.A.

Some applications of Diophantine

approximations to Diophantine equations

T.N. Shorey

1. — In 1975, Erdös and Selfridge [5] confirmed an old conjecture by proving that a product of two or more consecutive postive integers is never a power. In other words, the equation

(1) $(m+1)\cdots(m+k) = y^{\ell}$ in integers $k \geq 2, \ell \geq 2, m \geq 0, y > 0$

has no solution. We shall consider a more general equation than (1). For introducing this equation, I need some notation. For an integer $\nu > 1$, we write $P(\nu)$ for the greatest prime factor of ν and we put $P(1) = 1$. Let $b \geq 1, k \geq 2, \ell \geq 3, m \geq 0, t \geq 2$ and $y > 0$ be integers such that $P(b) \leq k$. Let $d_1, \ldots, d_t$ be distinct integers in the interval $[1, k]$. We observe that

$$\{d_1, \ldots, d_t\} = \{1, 2, \ldots, k\} \text{ if } t = k.$$

We shall consider the equation

(2) $(m+d_1)\cdots(m+d_t) = by^{\ell}.$

Equation (2) with $t = k$ and $b = 1$ is equation (1). If $P(y) \leq k$, equation (2) asks for all the solutions of the inequality

$$P((m+d_1)\cdots(m+d_t)) \leq k$$

and we do not intend to consider this question in this talk. Therefore we make the assumption that $P = P(y) > k$ throughout section 1. Then there is unique i with $1 \leq i \leq t$ such that $P \mid (m+d_i)$. In fact $P^{\ell} \mid (m+d_i)$. Thus

$$m + k \geq m + d_i \geq P^{\ell} \geq (k+1)^{\ell}$$

which implies the following relation

$$m > k^\ell.$$

Before I state the results of this section, I wish to point out that I have taken $\ell \geq 3$ in equation (2). In fact equation (2) with $\ell = 2$ was the topic of my talk in this seminar in 1992.

The first result on equation (2) is due to Erdös in 1955.

THEOREM 1. ([4]). — Let $\epsilon > 0$. Equation (2) with

$$t \geq k - (1 - \epsilon)k\frac{\log \log k}{\log k}$$

implies that k is bounded by an effectively computable number depending only on ϵ.

Theorem 1 has been improved considerably by Shorey in 1986-87.

THEOREM 2. ([8], [9]). — (a) Equation (2) with

$$\ell \geq 3, t \geq \frac{1}{2}(1 + \frac{4\ell^2 - 8\ell + 7}{2(\ell - 1)(2\ell^2 - 5\ell + 4)})k = \nu'_\ell k$$

implies that k is bounded by an effectively computable number depending only on ℓ.

(b) There exist effectively computable absolute constants C_1 and C_2 such that equation (2) with

$$\ell > C_1, t \geq k\ell^{-1/11} + \pi(k) + 2$$

implies that $k \leq C_2$.

We observe that $\nu'_\ell > \frac{1}{2}$ for $\ell \geq 3$ and

$$\nu'_3 = \frac{47}{56}, \nu'_4 = \frac{135}{192}, \nu'_\ell \leq \frac{2}{3} \; for \; \ell \geq 5.$$

Theorem 2 has recently been sharpened for $\ell \geq 7$ by Nesterenko and Shorey. For stating this result, we define

$$\nu_\ell = \begin{cases} \frac{112\ell^2 - 160\ell + 29}{28\ell^3 - 76\ell + 29} & \text{if } \ell \equiv 1 \pmod 2 \\[2mm] \frac{112\ell^2 - 160\ell + 17}{28\ell^3 - 188\ell + 129} & \text{if } \ell \equiv 0 \pmod 2 \end{cases}$$

For $\ell \geq 7$, we observe that

$$\nu_\ell \leq \begin{cases} \frac{4}{\ell}\left(1 - \frac{1}{(.875)\ell}\right) & \text{if } \ell \equiv 1 \pmod 2 \\[2mm] \frac{4}{\ell}\left(1 - \frac{1}{(1.412)\ell}\right) & \text{if } \ell \equiv 0 \pmod 2 \end{cases}$$

and

$$\nu_7 \leq .4832, \nu_8 \leq .4556, \nu_9 \leq .3878, \nu_{10} \leq .3664,$$
$$\nu_{11} \leq .3243, \nu_{12} \leq .3076, \nu_{13} \leq .2787, \nu_{14} \leq .2655.$$

THEOREM 3. ([13]). — (a) *Equation (2) with*

$$\ell \geq 7, \ t \geq \nu_\ell k$$

implies that k is bounded by an effectively computable number depending only ℓ.

(b) *Let $\epsilon > 0$. There exist effectively computable numbers C_3 and C_4 depending only on ϵ such that equation (2) with*

$$\ell > C_3, \ t \geq k\ell^{-1+\epsilon} + \pi(k) + 2$$

implies that $k \leq C_4$.

The proofs of Theorem 2(b) and 3(b) depend on Baker's sharpening [3] in the theory of linear forms in logarithms. Linear forms in logarithms with $\alpha_i's$ close to one appear in the proofs and the best possible estimates of Shorey [8, Lemma 2], namely replacing $\log A$ for $\log A_1 \cdots \log A_n$ with $A = \max_{1 \leq i \leq n} A_i$, for these linear forms in logarithms are required. For Theorem 3(b), we utilise a refinement due to Loxton, Mignotte, van der Poorten and Waldschmidt [6] of the preceding result of Shorey. The proofs of Theorems 2(a) and 3(a) depend on the method of Roth and Halberstam on difference between consecutive ν-free integers and approximations of algebraic numbers by rationals proved by Pade approximations. For Theorem 3(a), we require the following result which is a consequence, as proved in [13], of a result of Baker [2].

LEMMA 1. — *Let A, B, K and n be positive integers such that $A > B, K < n, n \geq 3$ and $\omega = (B/A)^{1/n} \notin \mathbb{Q}$. For $0 < \phi < 1$, put*

$$\delta = 1 + \frac{2 - \phi}{K}, \ s = \frac{\delta}{1 - \phi},$$
$$\mu_1 = 40^{n(K+1)(s+1)/(Ks-1)}, \mu_2^{-1} = K2^{K+s+1}40^{n(K+1)}.$$

Assume that

(3) $$A(A-B)^{-\delta}\mu_1^{-1} > 1.$$

Then
$$\left| \omega - \frac{p}{q} \right| > \frac{\mu_2}{Aq^{K(s+1)}}$$

for all integers p and q with $q > 0$.

Proof : we may assume that

$$\left| \omega - \frac{p}{q} \right| < \frac{1}{2}.$$

Observe that
$$\delta > 1 + \frac{1}{K},$$

and
$$\frac{(K+1)s}{Ks-1} = \frac{(K+1)\delta}{K\delta-1+\phi} = \frac{(K+1)\delta}{K+2-\phi-1+\phi} = \delta.$$

Put :
$$\lambda_1 = 40^{n(K+1)}A, \lambda_2 = 40^{n(K+1)}(A-B)^{K+1}A^{-K}.$$

Observe that
$$\lambda_2 < 1$$

if
$$40^{n(1+\frac{1}{K})}(A-B)^{1+\frac{1}{K}}A^{-1} < 1$$

if
$$A(A-B)^{-(1+\frac{1}{K})}40^{-n(1+\frac{1}{K})} > 1$$

which is the case by (3).

Put
$$\Lambda = \frac{\log \lambda_1}{\log \lambda_2}.$$

Observe that
$$-\Lambda < s$$

i.e.
$$\lambda_1 < (\lambda_2^{-1})^s$$

i.e.
$$40^{n(K+1)(s+1)} < (A-B)^{-(K+1)s}A^{Ks-1}$$

i.e.
$$A(A-B)^{-\delta}\mu_1^{-1} > 1,$$

which is (3). For integers $r > 0, p$ and $q > 0$, Baker [2] proved in 1967 that there exists $P_r(X) \in \mathbf{Z}[X]$ satisfying

$$\text{(i)} \ \deg P_r \leq K \quad \text{(ii)} \quad H(P_r) \leq \lambda_1^r ;$$
$$\text{(iii)} \ P_r(\tfrac{p}{q}) \neq 0 \quad \text{(iv)} \ \mid P_r(\omega) \mid \leq \lambda_2^r.$$

Define r as the smallest positive integer such that

$$\lambda_2^r \leq \frac{1}{2q^K}.$$

Then :

$$\lambda_2^{r-1} > \frac{1}{2q^K}$$

i.e.

$$\lambda_2^r > \frac{\lambda_2}{2q^K}.$$

Observe that :

$$\lambda_1^r = (\lambda_2^r)^\wedge < \left(\frac{\lambda_2}{2q^K}\right)^\wedge = \lambda_1 2^{-\wedge} q^{-\wedge K} < \lambda_1 2^s q^{sK}.$$

We have :

$$\frac{1}{q^K} \leq \mid P_r(\tfrac{p}{q}) \mid \leq \mid P_r(\tfrac{p}{q}) - P_r(\omega) \mid + \mid P_r(\omega) \mid \leq \mid P_r(\tfrac{p}{q}) - P_r(\omega) \mid + \frac{1}{2q^K}.$$

Thus

$$\mid P_r(\tfrac{p}{q}) - P_r(\omega) \mid \geq \frac{1}{2q^K}.$$

On the other hand :

$$\mid P_r(\tfrac{p}{q}) - P_r(\omega) \mid \leq K 2^K \lambda_1^r \mid \omega - \frac{p}{q} \mid \leq \lambda_1 K 2^{K+s} q^{sK} \mid \omega - \frac{p}{q} \mid.$$

Thus :

$$\mid \omega - \frac{p}{q} \mid \geq \frac{\mu_2}{A q^{K(s+1)}}.$$

This completes the proof of Lemma 1.

We shall be applying Lemma 1 with $n = \ell \geq 7$ and

$$K = \begin{cases} (\ell - 3)/2 & \text{if} \ \ell \equiv 1 (\text{mod } 2) \\ (\ell - 4)/2 & \text{if} \ \ell \equiv 0 (\text{mod } 2). \end{cases}$$

Let us look at the exponent of irrationality

$$K(s+1) = \frac{K\delta}{1-\phi} + K = \frac{K+2-\phi}{1-\phi} + K$$

$$= K(\frac{1}{1-\phi}+1) + \frac{2 \div \phi}{1-\phi}$$

$$= (K+1)\left(\frac{2-\phi}{1-\phi}\right) \geq 2(K+1) \geq \ell - 2.$$

Thus the exponent of irrationality is not far off from the trivial exponent ℓ. In fact, the precision in the exponent of irrationality is not important for the application. Baker [1] proved earlier in 1964 irrationality measures with the assumption $\delta = 2$. The important feature of Lemma 1 is that we can take $\delta < 2$. The assumption $\delta = 2$ corresponds to Theorem 2(a). Finally, we close this section by giving a sketch of the proof of Theorem 1.

Proof of Theorem 1 : we may assume that k exceeds a sufficiently large effectively computable number depending only on ϵ. Recall that $m > k$. A prime divisor greater than k of the left hand side of equation (2) divides precisely one factor and therefore its exponent has to be congruent to zero mod ℓ. Thus

$$m + d_i = a_i x_i^\ell, \ 1 \leq i \leq t,$$

where

$$P(a_i) \leq k, (x_i, \prod_{p \leq k} p) = 1.$$

Put

$$S = \{a_1, \cdots, a_t\}.$$

We show that

(4) $$|S| = t.$$

For $i \neq j$, let

$$a_i = a_j.$$

We may assume that $x_i > x_j$ and

$$k > d_i - d_j = a_i(x_i^\ell - x_j^\ell) > a_i x_i^{\ell-1} \geq (a_i x_i^\ell)^{(\ell-1)/\ell} \geq m^{(\ell-1)/\ell} > k^{\ell-1}.$$

For every prime $p \leq k$, we take an $f(p) \in S$ such that p does not appear to a higher power in the factorisation of any other element of S. Put

$$S_1 = S - \{f(p) \mid p \leq k\}.$$

Thus

$$| S_1 | \geq t - \pi(k).$$

It is easy to see that

$$\prod_{a_i \in S_1} a_i \leq \prod_{p \leq k} p^{[\frac{k}{p}] + [\frac{k}{p^2}] + \cdots} = k! \leq k^k.$$

Then there exists a subset S_2 of S_1 such that :

$$| S_2 | \geq \epsilon k / 2$$

and

$$a_i \leq k (\log k)^{1 - \frac{\epsilon}{2}} \ if \ a_i \in S_2.$$

Further, as in the proof of (4), we show that

$$a_i a_j \ with \ a_i \in S_2, a_j \in S_2$$

are distinct. The primes satisfy this property and what Erdös has shown is that the estimate for the number of numbers satisfying the above property is like that of primes. Erdös showed that

$$| S_2 | \leq 3k (\log k)^{-\epsilon/2}.$$

Hence k is bounded.

2. — In this section, we mention another application of the theory of linear forms in logarithms and Lemma 1. Let b, d, k, ℓ, m and y be positive integers satisfying $P(b) \leq k, \gcd(m, d) = 1, k > 2, \ell \geq 2$ and

(5) $$m(m + d) \cdots (m + (k - 1)d) = by^\ell.$$

The first result on equation (5) is due to Fermat that there are no four squares in arithmetical progression. In view of the results of 1, we assume in this section that $d \geq 2$. Shorey and Tijdeman [14] showed that either $(m, d, k) = (2, 7, 3)$ or the left hand side of equation (5) is divisible by a prime exceeding k. Erdös conjectured that equation (5) implies that k is bounded by an effectively computable absolute constant. Marszalek [7] confirmed this conjecture in the case $b = 1$ and d fixed, Shorey [10] in the case $\ell > 2$ and d composed of fixed primes and Shorey and Tijdeman [15] in the case ℓ fixed and d composed of fixed number of prime divisors. Further Shorey and

Tijdeman [16] sharpened the result of Marszalek by showing that equation (5) implies that :

(6) $$d \geq k^{C_5 \log \log k}$$

where $C_5 > 0$ is an effectively computable absolute constant. Further Shorey [11] proved that there exist effectively computable absolute constants C_6 and C_7 such that equation (5) with $k \geq C_6$ implies that :

(7) $$m \geq d^{1-C_7 \Delta_\ell}$$

where
$$\Delta_\ell = \ell^{-1}(\log \ell)^2 \log \log (\ell + 1).$$

The proof depends on Baker's sharpening [3] and the best possible estimates of Shorey mentioned in 1 on linear forms in logarithms with $\alpha_i's$ close to one. The estimate (7) is trivial whenever the exponent $1 - C_7 \Delta_\ell$ is negative which is the case only if ℓ is bounded. For bounded ℓ, Shorey [12] derived from Lemma 1 that given $\epsilon > 0$ there exists an effectively computable number C_8 depending only on ϵ and ℓ such that equation (5) with $k \geq C_8$ implies that

(8) $$m \geq d^{1-(6+\epsilon)\ell^{-1}}.$$

Now we combine (7), (8) and (6) for deriving that equation (5) with $\ell \geq 7$ implies that
$$m \geq k^{C_9 \log \log k}$$

where $C_9 > 0$ is an effectively computable absolute constant. In particular, equation (5) with $\ell \geq 7$ implies that k is bounded by an effectively computable number depending only on m. In other words, the conjecture of Erdös mentioned above is confirmed whenever $\ell \geq 7$ and m fixed.

Manuscrit reçu le 5 juillet 1995

REFERENCES

[1] A. BAKER. — *Rational approximations to $\sqrt[3]{2}$ and other algebraic numbers*, Quart. J. Math. Oxford (2) **15** (1964), 375–383.

[2] A. BAKER. — *Simultaneous rational approximations to certain algebraic numbers*, Proc. Cambridge Philos. Soc. **63** (1967), 693–702.

[3] A. BAKER. — *The theory of linear forms in logarithms*, Transcendence Theory : Advances and Applications, ed. by A. Baker and D.W. Masser, Academic Press (1977), 1–27.

[4] P. ERDÖS. — *On the product of consecutive integers* III, Indag. Math. **17** (1955), 85–90.

[5] P. ERDÖS and J.L. SELFRIDGE. — *The product of consecutive integers is never a power*, Illinois J. Math. **19** (1975), 202–301.

[6] J. H. LOXTON, M. MIGNOTTE, A.J. van der POORTEN and M. WALDSCHMIDT. — *A lower bound for linear forms in the logarithms of algebraic numbers*, C.R. Math. Report Acad. Sci. Canada **11** (1987), 119–124.

[7] R. MARSZALEK. — *On the product of consecutive elements of an arithmetic progression*, Monatsh. für. Math. **100** (1985), 215–222.

[8] T.N. SHOREY. — *Perfect powers in values of certain polynomials at integer points*, Math. Proc. Camb. Phil. Soc. **99** (1986), 195–207.

[9] T.N. SHOREY. — *Perfect powers in products of integers from a block of consecutive integers*, Acta Arith. **49** (1987), 71–79.

[10] T.N. SHOREY. — *Some exponential Diophantine equations*, New advances in Transcendence Theory, ed. by A. Baker, Cambridge University Press (1988), 352–365.

[11] T.N. SHOREY. — *Some exponential Diophantine equations* (II), Number Theory and Related Topics, ed. by S. Raghavan, Tata Institute of Fundamental Research (1988), 217–229.

[12] T.N. SHOREY. — *Perfect Powers in products of arithmetical progressions with fixed initial term*, to appear.

[13] T.N. Shorey and Yu.V. Nesterenko. — *Perfect powers in products of integers from a block of consecutive integers* (II), to appear.

[14] T.N. Shorey and R. Tijdeman. — *On the greatest prime factor of an arithmetical progression*, A Tribute to Paul Erdös, ed. by A. Baker, B. Bollobas and A. Hajnal, Cambridge University Press (1990), 385–389.

[15] T.N. Shorey and R. Tijdeman. — *Perfect powers in products of terms in an arithmetical progression*, Compositio Math. **75** (1990), 307–344.

[16] T.N. Shorey and R. Tijdeman. — *Perfect powers in products of terms in an arithmetical progression* (II), Compositio Math. **82** (1992), 119–136.

T.N. Shorey
School of Mathematics
Tata Institute of
Fundamental Research
Homi Bhabha Road
Bombay 400 005
India

Fields containing values of algebraic functions and related questions

Umberto Zannier

1. — Introduction

The problems discussed below originate in the celebrated Hilbert Irreducibility Theorem (from now on H.I.T.) the simplest case of which states

Let $F \in \mathbb{Q}[x,y]$ be an irreducible polynomial. Then, for infinitely many natural numbers n, $F(n,y)$ is irreducible in $\mathbb{Q}[y]$.

(Many proofs and generalizations of such result in various directions are known. We quote for instance [De2], [De3], [La], [Sch], [Se], [SZ] for different points of view.)

We can restate H.I.T. as the assertion that the roots of $F(n,y) = 0$ have degree d over $\mathbb{Q}$ for infinitely many $n \in \mathbb{N}$ or even, letting θ be an algebraic function solution of $F(x, \theta(x)) = 0$, we may say that the degree over $\mathbb{Q}$ of the values of θ at infinitely many rational integers n equals the degree of θ over $\mathbb{Q}(x)$.[1]

In such a context the following problem seems natural to us : *to estimate the degree obtained adjoining to $\mathbb{Q}$ many values of type $\theta(n)$* . In other words we ask how independent are the values of an algebraic function at rational integral arguments.

The problem is clearly capable of generalizations. However we shall discuss here only the basic case just mentioned.

We shall sketch the proof of some results concerning the order of magnitude, for $n \to \infty$, of the degree over $\mathbb{Q}$ of fields of type $K^*(n) = \mathbb{Q}(\theta_1, \ldots, \theta_n)$ for a prescribed sequence $\{\theta_j\}$ satisfying $F(j, \theta_j) = 0$. We define, for given F

(1) $$D^*(n) := \min[K^*(n) : \mathbb{Q}]$$

---

[1] Of course the value of an algebraic function at n is not well defined, but in our context it turns out that the choice of the "branch" makes no difference.

where the minimum is taken over all sequences $\{\theta_j\}$ as above. Let $D :=$ $\deg_y F$.

Assuming that F is an absolutely irreducible polynomial with $D > 1$ (as we shall do in the sequel) it may be easily proved, using H.I.T. over general number fields, that $D^*(n) \to \infty$. In fact, under such assumptions, there exists, given any n, a natural number $n' > n$ such that $F(n', y)$ is irreducible over $K^*(n)$. Then $K^*(n')$ contains $K^*(n)$ properly.

It is in fact possible to improve substantially on such qualitative information.

For notational convenience we shall work from now on with the logarithm of $D^*(n)$. The obvious upper bound $n \log D$ for $\log D^*(n)$ is not always attained, even as an order of magnitude, as the most simple examples $F(x,y) = y^D - x$ show : in such cases $K^*(n)$ is contained in the compositum of the fields generated over $\mathbb{Q}(e^{\frac{2\pi i}{D}})$ by the D-th roots of the prime numbers up to n, whence $\log D^*(n) \ll \frac{n}{\log n}$, by Chebyshev's estimate. Such upper bound is anyway meaningful since it generally holds as a lower bound too, as far as the order of magnitude is taken into account. In fact we have

THEOREM 1. — *Let $F \in \mathbb{Q}[x,y]$ be an absolutely irreducible polynomial*[2] *with $\deg_y F > 1$. Then $\log D^*(n) \gg \frac{n}{\log n}$.*

In some cases we can actually prove better bounds. Before stating them we introduce a bit of notation, which shall be useful throughout. Let F be absolutely irreducible and define Σ as the splitting field of F over $\mathbb{Q}(x)$. Let $\mathcal{G}$ be the Galois group of $k\Sigma$ over $k(x)$, k being some number field containing the algebraic closure of $\mathbb{Q}$ in Σ. The function field $k\Sigma$ corresponds to a nonsingular curve $\mathcal{C}$ over k. Let $S(x)$ be a squarefree polynomial whose roots are the finite ramification points of x (viewed as function on $\mathcal{C}$). In view of our assumptions S is non constant (otherwise $\mathcal{C}$ would be a non-ramified covering of the Riemann sphere, whence the number of sheets, and so D, would be 1. The Hurwitz genus formula gives an even more precise information). Also, since F has coefficients in $\mathbb{Q}$, the same may be assumed to hold for S. The roots of the polynomial S are precisely the roots of the discriminant $G(x)$[3], say, of the integral closure of $\mathbb{Q}[x]$ in Σ.

Also, define $\Sigma(m)$ to be the splitting field over $\mathbb{Q}$ of the polynomial $F(m, y)$, and denote by $D(n)$ the absolute degree of the composite $K(m)$ of the fields $\Sigma(m)$ for $1 \le m \le n$. It will be seen that $D^*(n)$ and $D(n)$ have

---

[2] This could be replaced with the necessary condition that the absolutely irreducible factors of F have degrees ≥ 2 in y. The proof does not present new difficulties, however, and we restrict to the special case.

[3] defined up to multiplication by the square of a rational number

logarithms of the same order of magnitude, so we shall consider $D(n)$. We have

THEOREM 2. — *Assumptions being as in Theorem 1, let $S(x)$ have an irreducible factor of degree 2 or 3. Then $\log D(n) \gg n$.*

In both cases the implied constants may depend on F only.

Both results were obtained by R. Dvornicich and myself already in 1983 (see [DZ1] or [De4]), but for a number of reasons their publication has been delayed, so only very recently they appeared in [DZ2].

The proofs show that the condition *degree 2 or 3* in Theorem 2 could be replaced by *degree ≥ 2* provided natural (and classical) conjectures on the distribution of squarefree values of polynomials were true; the connection with such questions (answered at the moment only in quite special cases, see e.g. [Ho]) is somewhat intrinsic to the problem, as the examples $F(x,y) = y^2 - f(x)$ show. Concerning the exact order of $\log D(n)$, A. Schinzel has proposed the following general conjecture.

SCHINZEL'S CONJECTURE. — *We have either $\log D(n) \gg n$ or (i) $S(x)$ splits into linear factors over $\mathbb{Q}$ and (ii) $\mathcal{G}$ is abelian.*

By applying Kummer Theory it is easy to see that, conversely, conditions (i) and (ii) imply that Σ is contained in a composite of a finite number of fields of type $L(\sqrt[k]{x+b})$, where L is a suitable number field and $b \in \mathbb{Q}$, so practically we are reduced to the basic examples mentioned above, and in such cases $\log D(n) \ll \frac{n}{\log n}$.

We have just remarked that condition (i) is connected with classical problems in analytic number theory. To approach, so to say, the more algebraic part of the problem, one can assume (i) to be verified but $\mathcal{G}$ to be nonabelian. Recently in [DZ3] we have obtained in this direction the following

THEOREM 3. — *Let condition (i) in Schinzel's conjecture hold. Assume moreover that there exists a prime $p \| \#\mathcal{G}$ such that $\mathcal{G}$ has no normal subgroups of index p. Then $\log D(n) \geq \frac{6 \log 2}{\pi^2} n + o(n)$.*

(Contrary to the previous theorems, the constant in the lower bound is now absolute for large n). The result includes all nonabelian groups of squarefree order as well as the nonabelian S_n and A_n (use *Bertrand's postulate*). We also remark that it suffices if the assumption is verified by some quotient of $\mathcal{G}$ rather than by $\mathcal{G}$ itself.

In a somewhat different direction P.Dèbes has pointed out that, if one considers the fields $K(t, n) := \prod_{t < j \leq t+n} \mathbb{Q}(\theta_j)$ then, if t is sufficiently large with respect to n, an exponential bound always holds. In fact he has proved in [De5] that, for large t, $[K(t, n) : \mathbb{Q}] \geq c2^n$ where $c > 0$ depends only on D (Dèbes gives an explicit value for c; also, he works with polynomials over general number fields, and then c depends on the corresponding degree as well).

We shall sketch below the proofs of theorems 1,2,3. Some analytic number theory is used in both Theorems 1 and 2, while a simple sieve method, combined with Kummer theory and a somewhat striking result in field theory (Lemma 4 in [DZ3]) is crucial for the proof of Theorem 3. But the common and most essential ingredient in all such proofs is Lemma 2 below which, roughly speaking, constructs prime numbers ramified in $\Sigma(m)$ as certain prime divisors of $G(m)$, thus comparing (together with Lemma 1) the discriminant of the specialized field with the specialization of the discriminant. We give in [DZ1,2] a simple direct proof of such result, though it may be derived from the celebrated Weil's Decomposition Theorem (see [We1], [We2] Ch.1, or [La], p.263).

We have found further applications of such a principle. A first one allows to obtain, via Baker's fundamental theorems in effective diophantine analysis (see e.g. [Ba], Ch.4), an effective form of the classical Siegel's theorem on integral points on curves ([La], Ch.8 or [Se], Ch.8) in the special case the projection on the x-line induces a Galois cover : this corresponds to an effective answer to the diophantine question of finding, given a polynomial $f \in \mathbb{Q}[x, y]$, the integers n such that *all* roots of $f(n, y) = 0$ are rational. We have the following theorem, first proved however by Yuri Bilu in 1988 :

THEOREM 4 (Bilu). — *If Σ has positive genus then the set of $m \in \mathbb{Z}$ such that $F(m, y)$ has only rational roots is finite and effectively computable.*

In [Bi1] an equivalent result with a different statement is derived (over general number fields). In a more recent approach Bilu proves his statement as a special case of a very interesting criterion of him which implies effective versions of Siegel's theorem, provided certain supplementary conditions are satisfied (these concern finding covers of a curve, unramified outside a given set A, and such that the A-units have rank ≥ 2); see [Bi3]. Also, Bilu computed explicitly some bounds in [Bi2]. We sketch below our point of view about such question, outlining how a proof of Theorem 4 follows from the above ramification criteria (this also appears as an appendix to [DZ2]). We remark that the methods apply to general number fields.

A second application, given in [Za], is also concerned with the distribution of the fields arising from the solution of $F(m, y) = 0$; it allows to estimate the number of positive integers $m \leq T$ such that some root of such equation generates a field containing a given number field E (the question originally arose from certain possible applications to analytic number theory, due to M. Nair and A. Perelli). Without uniformity requirements such a problem may be easily solved by means of the known quantitative versions of H.I.T.; made exception for finitely many fields E depending on F, the estimate takes the form $O(\sqrt{T})$. However, use of the present methods allows an estimate of type $O(c^{\frac{\log T}{\log \log T}} \sqrt{T})$, $(c = c(F))$, uniformly in the field E, with much better bounds as soon as certain mild supplementary assumptions on F are verified. We note that such a result implies a version of H.I.T., uniform over number fields in the following sense :

THEOREM 5. — Let $f \in \mathbb{Q}[x, y]$ be irreducible over a number field L, let $i(L)$ be the number of subfields of L and let $T \geq 10$. Then the set of positive integers $m \leq T$ such that $F(m, y)$ is reducible over L contains at most $i(L)c^{\frac{\log T}{\log \log T}} \sqrt{T}$ elements, where c depends only on f.

In fact a classical procedure (see e.g. [Sch]) reduces the problem of the reducibility of $f(m, y)$ over L to the verification whether $F(m, y) = 0$ has a root in L, F being a certain polynomial associated to f. Now, if $F(m, y) = 0$ has a root in L, this root generates a subfield as in the statement. Made exception for finitely many fields depending on f, we can apply the above mentioned estimates, which are completely uniform with respect to the field. For each of the remaining finitely many possibilities, the ordinary version of H.I.T. gives an estimate $O(\sqrt{T})$ for the number of relevant integers up to T, and the conclusion follows.

Acknowledgement. It is a pleasure to thank D. Bertrand and P. Dèbes for helpful conversations, for the interest they have shown towards the above questions and for their kind encouragement.

2. — Proofs

We begin with a couple of lemmas, crucial for the proof of all the above theorems. Preliminary to the arguments we choose, once and for all, points $P_m \in \mathcal{C}$, $m \in \mathbb{N}$ such that $x(P_m) = m$. In the sequel we shall specialize at P_m algebraic functions $\psi \in \Sigma$, tacitly assuming that m is sufficiently large for the specialization to be defined, in which case $\psi(P_m) \in \Sigma(m)$. We also denote by Δ_m the absolute discriminant of $\Sigma(m)$.

LEMMA 1. — There exists p_0 such that if $p > p_0$ is a prime which ramifies in $\Sigma(m)$, then $p|G(m)$.

(This corresponds to Lemma 3.1 in [DZ2]. The result is connected with the Chevalley-Weil Theorem.) The proof runs as follows. Choose an integral basis $e_1, ..., e_h$ for the integers of Σ over $\mathbb{Q}[x]$. It may be easily checked that, since the e_i span Σ over $\mathbb{Q}(x)$, the algebraic numbers $e_i(P_m)$ span $\Sigma(m)$ over $\mathbb{Q}$. We may assume them to be algebraic integers, by multiplying if necessary by a nonzero rational integer independent of m. Let $s = [\Sigma(m) : \mathbb{Q}]$ and renumber indices to assume that $e_1(P_m), ..., e_s(P_m)$ are linearly independent over $\mathbb{Q}$. Let σ_j run over $Gal(\Sigma/\mathbb{Q}(x))$ and let τ_i run over $Gal(\Sigma(m)/\mathbb{Q})$. We may renumber indices so that $\tau_i(e_t(P_m)) = (\sigma_i e_t)(P_m)$ for $i \leq s, t \leq h$. Let E_m be the $s \times h$ matrix with such entries. Then every $s \times s$ minor is divisible by $\sqrt{\Delta_m}$, namely the quotient is an algebraic integer. We may multiply G by a rational number to assume that $G(x) = \det^2(\sigma_i e_t)$. Specialize now at P_m. The left side gives $G(m)$, while the first s rows of the matrix on the right side give E_m. Expanding the determinant along the upper $s \times s$ minors we obtain that $G(m)$ is divisible by Δ_m, i.e. the Lemma.

LEMMA 2. — Let L be a number field, let $g \in L[x]$ be a factor of G, irreducible over L and let e be the ramification index of g in $L\Sigma^{(4)}$. Let $\wp$ be a prime ideal of L, of sufficiently large norm, such that $\mathrm{ord}_\wp g(m) \not\equiv 0 (\mathrm{mod}\ e)$. Then $\wp$ ramifies in $L\Sigma(m)$.

(This corresponds to Remark 2 after Lemma 3.2 in [DZ2]. As remarked in the Introduction and in [DZ2], it may be derived from Weil's Decomposition Theorem, starting from the observation that the divisor of zeros of g, viewed as a function on $\mathcal{C}$ is divisible by e). To outline a direct proof, let B be the integral closure of $L[x]$ in $L\Sigma$ and write

$$(2) \qquad\qquad g(x)B = \Omega^e,$$

for some ideal $\Omega \in B$, generated by $\omega_1, ..., \omega_r$, say, normalized to take algebraic integer values at P_m. Let $B(m)$ be the ring of integers in $L\Sigma(m)$ and define $\Omega(m)$ as the ideal in $B(m)$ generated by the $\omega_i(P_m)$. From (2) we get :

$$g(x) = \sum_{(i)} R_{(i)} \omega_1^{i_1} ... \omega_r^{i_r},$$

where (i) runs over the r-tuples of natural numbers with sum e and where $R_{(i)} \in B$. We may assume that $a R_{(i)}(P_m)$ is an algebraic integer, for a suitable rational integer $a \neq 0$ independent of m. Specializing at P_m we

---

[4] namely the common ramification index of the prime ideals above $g(x)$ in the integral closure of $L[x]$ in $L\Sigma$

obtain that $ag(m) \subset (\Omega(m))^c$. To obtain an opposite inclusion observe that (2) implies, for each (i) as above

$$\omega_1^{i_1}...\omega_r^{i_r} = S_{(i)}g(x),$$

for some $S_{(i)} \in B$. As before we get, after specializing, $b(\Omega(m))^e \subset g(m)B(m)$, for suitable nonzero $b \in \mathbb{N}$. Let now Q be a prime ideal of $B(m)$ lying above $\wp$, and write $v(\Lambda)$ for the order at Q of an ideal $\Lambda \subset B(m)$. We have $ev(\Omega(m)) \leq v((ag(m)))$ and $v((g(m))) \leq v((b)) + ev(\Omega(m))$. We may suppose that $\wp$ does not lie above primes dividing ab, so $v((a)) = v((b)) = 0$ and $v((g(m))) = ev(\Omega(m))$ is a multiple of e. On the other hand $g(m) \in L$ so, if $\wp$ were unramified in $B(m)$, then $\operatorname{ord}_\wp g(m) = v((g(m)))$ would also be divisible by e. This completes the proof.

It will be noted that the normality of $L\Sigma$ over $L(x)$ is crucial to deduce that the ideal generated by $g(x)$ becomes a perfect power in B. In the general case we would have prime ideal factors with different exponents and a corresponding factorisation would hold after specialization, in accordance with Weil Decomposition. This provides information on the factorisation of $g(m)$, which is nontrivial provided we know the size of some factor. This is the case for instance when the g.c.d. of the exponents is > 1, in which case $g(m)$ will be nearly a perfect power. Otherwise one can use a rather delicate theorem of Bombieri [Bo] which supplements Weil's theorem with an estimate of the size of each factor. An equivalent result, formulated however differently, was obtained by Dèbes in [De1] and [De2]. Combining the principle of Lemma 2 with such tools, and using Belyi's theorem (see *e.g.* [Se]), one gets a version of Elkies' argument [El] deriving an effective Faltings's Theorem from the *abc*-conjecture (see [Schm1], p. 205 for the case of the rational field) : given a nonsingular curve $\mathcal{C}$ defined over a number field, one first uses Belyi's Theorem to obtain a cover $x : \mathcal{C} \to \mathbb{P}^1$ of degree n, say, unramified outside $0, 1, \infty$. We can write the divisor of x (resp. $1 - x$) in the form $e_1 P_1 + ... e_r P_r - g_1 R_1 - ... - g_t R_t$ (resp. $f_1 Q_1 + ... + f_s Q_s - g_1 R_1 - ... - g_t R_t$), where e_i, f_j, g_h are positive integers and the P_i, Q_j, R_h are distinct points on $\mathcal{C}$. The Hurwitz genus formula easily gives, for the genus γ of $\mathcal{C}$, $2\gamma - 2 + r + s + t = n$. Let $P \in \mathcal{C}$ be a rational point and consider the value $x(P)$. Weil's Decomposition Theorem implies a factorisation $x(P) = c\phi_1^{e_1}...\phi_r^{e_r}\psi_1^{-g_1}...\psi_t^{-g_t}$ (and similarly for $1 - x(P)$), where c is a factor of bounded height and the ϕ_i, ψ_j are algebraic integers (this step roughly corresponds to Lemma 2 above, applied to the irreducible factors x, $1 - x$ of the discriminant.) The above mentioned theorem of Bombieri allows to estimate the height of the factors in terms of the height of $x(P)$ (for instance, when $x(P)$ does not approach 0 or ∞ - in the archimedean sense - each factor has height $\approx h(x(P))/n$). Introducing such informations into the abc-type equation $1 = x(P) + (1 - x(P))$ and using the abc conjecture one obtains $h(x(P)) \leq (1 + \epsilon)(\frac{r+s+t}{n})h(x(P)) + O_\epsilon(1) \leq (1 + \epsilon)(\frac{n+2-2\gamma}{n})h(x(P)) + O(1)$ and a bound for $h(x(P))$ follows as soon as $\gamma \geq 2$.

Before giving the applications to Theorems 1, 2, 3, let us briefly show that $\log D^*(n) \gg \log D(n) + O(\sqrt{n})$, providing the comparison announced in the Introduction. The inequality $D^*(n) \leq D(n)$ is trivial. To go in the opposite direction, let m be such that $\mathbb{Q}(\theta_m) \subset K^*(m - 1)$. Then the normal closure (over $\mathbb{Q}$) of $\mathbb{Q}(\theta_m)$ is contained in the normal closure of $K^*(m - 1)$, which is the composite of the $\Sigma(j)$ for $j \leq m - 1$. Now, if $F(m, y)$ is irreducible, we then have $\Sigma(m) \subset \prod_{j \leq m-1} \Sigma(j)$. Plainly there are $\gg \log D(n)$ values of $m \leq n$ for which such inclusion does not hold. So, if $I(n)$ denotes the number of $m \leq n$ such that $F(m, y)$ is reducible, there are $\gg \log D(n) + O(I(n))$ values of $m \leq n$ such that the original inclusion does not hold. At each such m we have $D^*(m) \geq 2D^*(m - 1)$. Combining with the bound $I(n) \ll n^{1/2}$ given by the quantitative form of H.I.T. we get the claim (even weaker results would do for our purposes).

So it will suffice to estimate $D(n)$. We shall do that by finding many integers $j \leq n$ such that some prime p ramifies in $\Sigma(j)$ but not in $\Sigma(i)$ for $i < j$: this will prove that $D(j) \geq 2D(j - 1)$. Let T be very large and consider the set of primes $p \in [T/4, T/2]$ such that p divides some value $G(m)$. Certainly p will divide $S(m)$ and, using the fact that S is a squarefree polynomial and that p is large, it may be easily shown that p will occur with the first power in $S(m')$, say, where $m' = m$ or $m' = m + p$. By Lemma 2 (with $L = \mathbb{Q}$) p will ramify in $\Sigma(m')$. We then may consider the minimum positive integer r such that p ramifies in $\Sigma(r)$, denoting it by $j(p)$. Certainly $j(p) \leq 2p \leq T$. On the other hand if such a p ramifies in $\Sigma(j)$ then $p|G(j)$ by Lemma 1. If $j \leq T$ is given, (and $G(j) \neq 0$), this cannot hold for more than $O(1)$ such primes. So there are $O(1)$ primes such that $j(p)$ is a given integer $\leq T$, whence the number of the distinct integers among the $j(p)$'s will be $\gg$ than the number of the above primes. But it is known from analytic number theory that the number of primes in $[T/4, T/2]$ which divide some value on $\mathbb{N}$ of a given nonconstant polynomial is $\gg \frac{T}{\log T}$, and this concludes the proof of Theorem 1.

The proof of Theorem 2 is similar, but a little more delicate. Let g be the irreducible factor mentioned in the statement. An old result of Nagell (referred to in [DZ2]) implies that $\#\{p \geq T, p|g(m) \text{ for some } m \leq T\} \gg T$. Now either an easy argument, if $\deg g = 2$, or a deep theorem of Hooley on squarefree values of cubic polynomials[4], if $\deg g = 3$, implies that

---

[4] actually a slightly different version of Hooley's Theorem, as given in [Ho],

$\#\{p \geq T, \ p\|g(m) \text{ for some } m \leq T\} \gg T$. This inequality allows to derive Theorem 2 by the same arguments as above.

We shall now sketch the proof of Theorem 3 in the very special case when F is cubic in y and $\mathcal{G} = S_3$ (naturally the prime p in the statement is now 3); this simple case is sufficiently illustrative of the general method but rather less complicated in detail.

In such cubic case we may assume for our purposes that $F(x,y) = y^3 - 3b(x)y - 2c(x)$, where $b, c \in \mathbb{Q}(x)$. We now let $k = \mathbb{Q}(e^{\frac{2\pi i}{3}})$. Also, set $a(x) = c^2(x) - b^3(x)$; a is not a square in $\mathbb{C}(x)$ in view of our assumption on $\mathcal{G}$. Write $a = a_1 a_2^2$ with squarefree nonconstant $a_1 \in \mathbb{Q}[x]$. All the zeros of a_1 appear among the zeros of $S(x)$, assumed to be rational. Also, Cardano's formulae show that $k\Sigma = k(x, (c + \sqrt{a})^{\frac{1}{3}})$.

By combining a simple sieve method[5] with the quantitative version of H.I.T., already used in the proof of Thm.1, we may easily construct a sequence $M = \{m_j\}$ such that (i) the Galois group of $k\Sigma(m_j)$ over k is S_3, (ii) $m_j \ll j$ and (iii) $a_1(m_j) = c_j s_j$ where c_j has only prime factors bounded by p_0 and the s_j are pairwise distinct squarefree numbers all of whose prime factors are $> p_0$.

Put $\alpha := c + \sqrt{a}$, and choose once and for all, in an arbitrary manner, determinations for the values $\alpha(m_i) := \alpha_i$, say, and for the cube roots of these numbers. Also, define $F_h = k(\alpha_1, ..., \alpha_h)$. We have $k\Sigma(m_h) \subset F_h(\alpha_h^{1/3})$. Assume that h is such that

$$(3) \qquad k\Sigma(m_h) \subset \prod_{i=1}^{h-1} k\Sigma(m_i).$$

The field on the right is contained in $F_h(\alpha_1^{1/3}, ..., \alpha_{h-1}^{1/3})$ so by Kummer theory applied with F_h as base field we have

$$(4) \qquad \alpha_h = \prod_{i=1}^{h-1} \alpha_i^{e_i} \phi^3$$

for suitable integers e_i and $\phi \in F_h$. Let N denote the norm from F_h to $k(\alpha_h)$, and consider $N(\alpha_i)$ for $i < h$. Certainly $\alpha_i \notin k(\alpha_h)$ since, by condition (iii) in the construction of M, the numbers $a(m_i)$ and $a(m_h)$ have distinct squarefree parts, both > 1 by condition (i). So $N(\alpha_i)$ is a power of the

---

is needed

[5] it is here that condition (i) of Schinzel's conjecture proves useful

absolute norm of α_i, which equals $(c(m_i) + \sqrt{a(m_i)})(c(m_i) - \sqrt{a(m_i)}) = b^3(m_i)$. In particular $N(\alpha_i)$ is the cube of a rational number. So, taking the norm N of (4) we deduce that $\alpha_h^{[F_h:k(\alpha_h)]}$ would be the cube of an element of $k(\alpha_h)$, and the same would hold for α_h, since the absolute degree of F_h is plainly a power of 2. But then $k\Sigma(m_h)$ would be quadratic over k, violating condition (i) for the sequence M.

We have proved that (3) cannot hold, whence $D(m_h) \geq 2D(m_{h-1})$, and a bound $\log D(n) \gg n$ follows from condition (ii) for M. To obtain the more precise result with the absolute constant in the lower bound one has to refine the construction of M. Assume, to start with, $\deg a_1 > 1$. First of all, one can easily prove that, forgetting the requirement about the distinctness of the s_j, and making p_0 bigger, the density of M may be gotten as near to 1 as we want. Now an elementary argument involving diophantine approximation shows that for only a few pairs i, j we may have $s_i = s_j$. Throwing away the corresponding m_i's the remaining sequence will have the same density as before and will satify all required conditions. The case of linear a_1 is easier and may be reduced essentially to the distribution of squarefree numbers in an arithmetic progression $rm + s$ with coprime r, s. The value of the constant comes out from this case. For details see Remark 3 in [DZ3].

In the general case both the choices for α and for the determinations for the values at m_h of the algebraic functions involved are more delicate, and the last ones are no longer arbitrary. The analogue of the crucial fact that the norm of $c + \sqrt{a}$ is the cube of the rational function is played by Lemma 4 of [DZ3], mentioned in the Introduction. The role of Lemmas 1, 2, invisible in the cubic case, but otherwise fundamental, appears in the need to show, through comparison of ramification, that the specializations at m_i of a certain subfield of $k\Sigma$ (namely $k(x, \alpha)$ in the cubic case) are pairwise distinct.

Let us now discuss the proof of Theorem 4. We may suppose that k is so large to contain all finite ramification points of x. Denote them by $\xi_1, ..., \xi_r$. Let $e_j \geq 2$ be the common value of the ramification indices of points above ξ_j in C and apply Lemma 2 with $L = k$, $g(x) = x - \xi_j$, $e = e_j$. Assume that all roots of $F(m, y) = 0$ are rational. Then $k\Sigma(m) = k$, so no ideal $\wp$ as in the statement of that lemma may ramify. It follows that the fractional ideal generated by $m - \xi_j$ is an e_j-th power in k, made exception for a factor having finitely many possibilities. A classical argument involving the finiteness of the class group of k and Dirichlet's unit theorem allows to write, for each j,

$$m - \xi_j = \rho_j \mu_j^{e_j}$$

where the ρ_j have finitely many possible values, which can be effectively computed, and where the μ_j are algebraic integers in k. So we may suppose the ρ_j to be fixed and investigate the solutions of the above system in algebraic integers $m, \mu_1, ..., \mu_r$ in k.

If $r \geq 2$ and $e_1 \geq 3$, say, then eliminating m from the first two equations gives an equation

$$\rho_2\mu_2^{e_2} = \rho_1\mu_1^{e_1} + (\xi_1 - \xi_2)$$

whose solutions may be found by Baker's method.

If $r \geq 3$ and $e_j = 2$ for all j multiplication of the first three equations gives an elliptic equation

$$\rho\mu^2 = (m - \xi_1)(m - \xi_2)(m - \xi_3)$$

which again can be solved effectively.

In conclusion the solutions of each such system may be found except when either $r = 1$ or $r = e_1 = e_2 = 2$. We show that these cases cannot occur. In fact applying Hurwitz genus formula to the field extension $k\Sigma/k(x)$ of degree d, say, we get for the positive genus g of $k\Sigma$,

$$2g - 2 = -2d + \sum_{i=1}^{r}(e_i - 1)\frac{d}{e_i} + \sum_{P|\infty}(e(P) - 1),$$

the last sum running over the points of C above the infinite point of $k(x)$. This term is clearly bounded by $d-1$, whence $2 \leq 2g \leq 1+(r-1-\sum_{i=1}^{r}\frac{1}{e_i})d$. The right hand side is ≤ 1 in both the above cases, a contradiction.

We observe that, for the proof to work, the full force of the normality of $k\Sigma$ over k has not been used. In fact the method allows to find the integral points on algebraic curves of positive genus provided only certain conditions on ramification (which hold automatically in the present case in view of Hurwitz's theorem) are satisfied. We also remark that all the procedures involved before the application of Baker's theorems (namely the computation of bases for ideals in function fields, of generators for class groups and unit groups etc.) are known to be effective, thanks to results by Coates and Schmidt (see e.g. [Schm2]).

We conclude by sketching how to obtain the announced estimate for the number of $m \leq T$ such that some root of $F(m, y) = 0$ generates a field containing a given number field E. As observed in the Introduction, such bounds lead to a proof of Theorem 5.

It will be sufficient to replace E with its normal closure and to prove the following inequality, excluding the finitely many fields contained in k; namely, if $E \not\subset k$ we have

$$S(T) := \#\{m \leq T : E \subset k\Sigma(m)\} \ll c^{\frac{\log T}{\log \log T}}\sqrt{T}$$

uniformly in E, where $c = c(F)$.

First one lets m run through a sequence such that the lattice of fields between $k\Sigma(m)$ and $\mathbb{Q}$ is the same as the lattice between $k\Sigma$ and $\mathbb{Q}(x)$. This can be done by H.I.T. and moreover the excluded sequence will contain $\ll \sqrt{T}$ elements up to T and thus will not affect the final result. Now, if such an m is counted in $S(T)$, the field E will equal some field $\Gamma(m)$, where Γ is a normal extension of $\mathbb{Q}(x)$, contained in $k\Sigma$. Moreover Γ will not be a constant field extension of $\mathbb{Q}(x)$, since $E \not\subset k$. Γ has finitely many possibilities whence it may be supposed to be fixed. Replacing then $k\Sigma$ with Γ leads to a similar problem, so we may in fact bound the number of times $E = k\Sigma(m)$. Let g_i run through the irreducible (over $\mathbb{Q}$) factors of G, and let Δ be the discriminant of E. Also, let $g^* = \prod g_i$. By Lemma 2 we have, for each such m, i,

(5) $$g_i(m) = t_i z_i^{e_i}, \qquad e_i \geq 2,$$

where t_i are e_i-th power free integers, the z_i are integers and where the prime factors of t_i are either small or divide Δ. Let s be the greatest squarefree divisor of Δ. If s is small, say $s \leq T$, then each of the t_i has few possibilities. For each such possibility the number of solutions of each equation (5) with $m \leq T$, may be estimated as being $\ll \sqrt{T} \log T$ uniformly in t_i; this can be done either by appealing to certain deep theorems of Evertse and Silverman (see [Schm1], p.142) or, more simply, by considering the reduction of the equation modulo many primes congruent to 1 $\pmod{e_i}$ (and not dividing t_i); for each such prime one may apply Weil's estimates for the number of solutions to deduce that the reduction of m has few possibilities. Finally a large sieve collects all the informations and gives the required estimate (this approach was used by S.D.Cohen in connection with H.I.T., see [Coh] or [Se], §12). When, on the contrary, $s > T$, we use Lemma 1 to show that s divides, except possibly for a bounded factor, the value $g^*(m)$. If $p_1 < \ldots < p_h$ are (large) primes dividing s and $p_1 \ldots p_{h-1} \leq T < p_1 \ldots p_h$ then we consider the congruence $g^*(m) \equiv 0$ $\pmod{p_1 \ldots p_h}$. This has $\ll c^h$ solutions. But $h \ll \frac{\log T}{\log \log T}$. This completes the proof.

In fact in [Za] details are carried out in the more general case (needed for the applications of Nair and Perelli) where m runs through an arithmetic progression $qn + r$, $n \leq T$. The results remain uniform in r and in q too, provided q has, say, $\ll \log^{\frac{h}{4}} T$ prime factors (complete uniformity in q is perhaps true but leads to difficult problems already in basic cases). This case gives additional combinatorial complications. Also, a uniform version (with respect to r, q) of H.I.T. is now needed already at the beginning of the proof, and it may be accomplished by Cohen's method, which is based on reduction to various moduli, and is so well suited for investigating the behaviour in progressions.

As mentioned in the Introduction much better estimates may be obtained in some cases. Assuming for instance that every field $\Gamma \neq \mathbb{Q}(x) \subset \Gamma \subset k\Sigma$ has positive genus, and assuming there is at least one irrational ramification point (analogously as in Thm. 2), the factor $\sqrt{T}$ disappears. The proof of this result given in [Za] is similar to the above one, but now the use of the Evertse-Silverman estimates is crucial.

Manuscrit reçu le 21 décembre 1994

References

[Ba] A. BAKER. — *Transcendental Number Theory*, Cambridge Univ. Press, 1975.

[Bi1] Yu. BILU. — *Effective analysis of a new class of Diophantine equations* (Russian), Vestsi Akad. Navuk BSSR, Ser. Fiz. Math. Navuk, (1988), no. 6, 34-39, **125** (Math.Rew. 90j :11038).

[Bi2] Yu. BILU. — *Effective Analysis of Integral Points on Algebraic Curves*, thesis, Beer Sheva, 1993.

[Bi3] Yu. BILU. — *Effective Analysis of Integral Points on Algebraic Curves*, preprint Univ. Bordeaux 2, 1993.

[Bo] E. BOMBIERI. — *On Weil's "Théorème de décomposition"*, Amer. J. Math., **105** (1983), 295–308.

[Coh] S.D. COHEN. — *The distribution of Galois groups and Hilbert's Irreducibility Theorem*, Proc. London Math. Soc. **43** (1981), 227–250.

[De1] P. DÈBES. — *Spécializations de polynomes*, Math. rep. Acad. Sc., Royal Soc. Canada, Vol. V, no. 6, (Dec. 1983).

[De2] P. DÈBES. — *G-fonctions et Théorème d'irréductibilité de Hilbert*, Acta Arith., **47**, 4 (1986).

[De3] P. DÈBES. — *Hilbert's subsets and s-integral points*, preprint, 1993.

[De4] P. DÈBES. — *Resultats recents lies au théorème d'irréductibilité de Hilbert* , dans Séminaire de Théorie des Nombres, Paris 1985-86, Progress in Mathematics, vol. **71**, Birkhauser.

[De5] P. DÈBES. — *On a problem of Dvornicich and Zannier*, preprint, 1994.

[DZ1] R. DVORNICICH, U. ZANNIER. — *Fields containing values of algebraic functions*, preprint Univ.Pisa, $n^o.44$, novembre 1983.

[DZ2] R. DVORNICICH, U. ZANNIER. — *Fields containing values of algebraic functions*, Annali Sc.Normale Sup. Cl.Sci. Serie IV, Vol. **XXI**, Fasc. 3 (1994), 421–443.

[DZ3] R. DVORNICICH, U. ZANNIER. — *Fields containing values of algebraic functions II. (On a conjecture of Schinzel)*, preprint 1994 (to appear in Acta Arithmetica).

[El] N. ELKIES. — *ABC implies Mordell*, Duke Math. J. Int. Res. Not. **7** (1991), 95–109.

[Ho] C. HOOLEY. — *Applications of sieve methods to the Theory of Numbers*, Cambridge Tracts in Mathematics, Cambridge Univ. Press, 1976.

[La] S. LANG. — *Fundamentals of Diophantine Geometry* , Springer Verlag, 1983.

[Sch] A. SCHINZEL. — *Selected topics on polynomials* , Univ.of Michigan Press, Ann Arbor, 1982.

[Schm1] W.M. SCHMIDT. — *Diophantine Approximations and Diophantine Equations*, Springer Verlag LN 1467, 1991.

[Schm2] W.M. SCHMIDT. — *Construction and estimation of bases in function fields*, J. Number Theory, **39** (1991), 181-224.

[SZ] A. SCHINZEL, U. ZANNIER. — *The least admissible value of the parameter in Hilbert's Irreducibility Theorem*, Acta Arith. **LXIX**. 3 (1995), 293-302.

[Se] J.-P. SERRE. — *Lectures on the Mordell-Weil Theorem*, Vieweg, 1988.

[We1] A. WEIL. — *L'arithmétique sur une courbe algébrique*, C.R. **185** (1927), 1426-1428 = Coll.Papers, Vol. I, Second Printing, Springer Verlag, 1980, 9-10.

[We2] A. WEIL. — *L'arithmétique sur les courbes algébriques*, Acta Math. **52** (1928), 281-315 = Coll.Papers, Vol. I, Second Printing, Springer Verlag, 1980, 11-46.

[Za] U. ZANNIER. — *On the number of times a root of $f(n, x) = 0$ generates a field containing a given number field*, preprint 1993.

Umberto Zannier,
Ist.Univ.Arch. D.C.A.
S.Croce, 191
30135 Venezia (Italy)
zannier@dimi.uniud.it

Printed in the United States
By Bookmasters